Study Guide

James E. Brady
St. John's University

to accompany

CHEMISTRY
Matter and Its Changes

Fourth Edition

James E. Brady
St. John's University

Frederick Senese
Frostburg State University

W9-AKS-817

WILEY

JOHN WILEY & SONS, INC.

Cover image: Boron Nitride image courtesy of Zettl Research Group, University of
California at Berkeley, and Lawrence Berkeley National Laboratory.

To order books or for customer service call 1-800-CALL-WILEY (225-5945).

ISBN 0-471-21519-8

Printed in the United States of America

10 9 8 7 6 5 4 3 2 1

Printed and bound by Hamilton Printing, Inc.

PREFACE

Our goal in preparing this Study Guide was to provide the student with a structured review of important concepts and problem solving approaches. We begin with a preliminary chapter that introduces students to the text and to the Study Guide. Here we explain how to use the text and Study Guide together most effectively, and we explain the importance of regular class attendance and how to develop proper study habits. One of the principal features of the textbook is the organized approach to problem solving employing the chemical toolbox analogy. Because this approach is likely to be new to the student, we discuss in some detail how this analogy can help students expand their problem solving skills.

Each of the remaining chapters in the Study Guide begins with a brief overview of the chapter contents, followed by a list of Learning Objectives. Because students tend to study one section at a time, we divide each chapter in the Study Guide into sections that match one-for-one the sections in the text. Each section provides a review of the topics covered in the text. Here we call to students' attention key concepts and important facts. In many places, additional explanations of difficult topics are provided, and where students often find particular difficulty, additional worked examples are given.

In keeping with our aim of providing students with frequent opportunities to hone their skills and test their knowledge, almost all sections of the Study Guide include a brief Self-Test that consists of questions and problems that supplement those in the text. The answers to all the Self-Test exercises appear at the ends of the chapters. Many sections also contain a Thinking It Through question of the type found in the textbook, and for each there is a worked-out answer at the end of the chapter.

Following the Self-Test there is a list of new terms introduced in the section. As an exercise, the student is encouraged to write out the definitions of these terms in his or her notebook.

As an additional aid in problem solving, tables listing the Chemical Tools and their functions as well as summaries of important equations and other useful information are found on separate pages at the ends of chapters. The aim is to provide the student with another means to reinforce the key problem solving concepts.

James E. Brady

CONTENTS

Before You Begin...

Before you begin your general chemistry course, read the next several pages. They're designed to tell you how to use this study guide and to give you a few tips on improving your study habits.

How to Use the Study Guide

This book has been written to parallel the topics covered in your text, *Chemistry: Matter and Its Changes,* Fourth Edition. Each chapter of the Study Guide begins with a very brief overview of the chapter contents followed by a list of learning objectives. Read these before beginning a chapter, and then read them again after you've finished to be sure you have met the goals described. For each section in the textbook, you will find a corresponding section in the Study Guide. In the Study Guide, the sections are divided into **Review, Thinking It Through, Self-Test** and **New Terms**.

After you've read a section in the text, turn to the study guide and read the **Review**. This will point out specific ideas that you should be sure you have learned. Sometimes you will be referred back to the text to review topics there. Sometimes there will be additional worked-out sample problems. Work with the Review and the text together to be sure you have mastered the material before going on.

In some sections you will find questions titled **Thinking It Through**. The goal of these questions is to allow you to test your ability in figuring out *how* to solve problems. The emphasis is on the *method*, not the *answer*. (We will have more to say about this later.) In most sections you will also find a short **Self-Test** to enable you to test your knowledge and problem-solving ability. The answers to all of the Thinking It Through and Self-Test questions are located at the ends of the chapters in the Study Guide. However, you should try to answer the Thinking It Through and Self-Test questions without looking up the answers. A space is left after each Self-Test question so that you can write in your answers and then check them all after you've finished.

Chemical vocabulary

An important aspect of learning chemistry is becoming familiar with the language. There are many cases where lack of understanding can be traced to a lack of familiarity with some of the terms used in a discussion or a problem. A great deal of effort was made in your textbook to adequately define terms before using them in discussions. Once a term has been defined, however, it is normally used with the assumption that you've learned its meaning. It's important, therefore, to learn new terms as they appear, and for that reason, most of them are set in boldface type in the text. At the end of each section of the study guide there is a list of these **New Terms**. To test your knowledge of them, you are asked to write out their meanings. This will help you review them later when you prepare for quizzes or examinations. At the end of the textbook there is a Glossary which you can use to be sure you understand the meanings of the new terms. The Glossary also directs you to sections in the textbook where the terms are discussed.

Study Habits

You say you want to get an A in chemistry? That's not as impossible as you may have been led to believe, but it's going to take some work. Chemistry is not an easy subject—it involves a mix of memorizing facts, understanding theory, and solving problems. There is a lot of material to be covered, but it won't overwhelm you if you *stay up to date*. Don't fall behind, because if you do, you are likely to find that you can't catch up. Your key to success, then, is *efficient* study, so your precious study time isn't wasted.

Efficient study requires a regular routine, not hard study one night and nothing the next. At first, it's difficult to train yourself, but after a short time you will be surprised to find that your study routine has become a study habit, and your chances of success in chemistry, or any other subject, will be greatly improved.

To help you get more out of class, try to devote a few minutes the evening before to reading, in the text, the topics that you will cover the next day. Read the material quickly just to get a feel for what the topics are about. Don't worry if you don't understand everything; the idea at this stage is to be aware of what your teacher will be talking about.

Your lecture instructor and your textbook serve to complement one another; they provide you with two views of the same subject. Try to attend lecture regularly and take notes during class. These should include not only those things your teacher writes on the blackboard, but also the important points he or she makes verbally. If you pay attention carefully to what your teacher is saying in class, your notes will probably be somewhat sketchy. They should, however, give an indication of the major ideas. After class, when you have a few minutes, look over your notes and try to fill in the bare spots while the lecture is still fresh in your mind. This will save you a lot of time later when you finally get around to studying your notes in detail.

In the evening (or whatever part of the day you close yourself off from the rest of the world to really study intensely) review your class notes once again. Use the text and study guide as directed above and really try to learn the material presented to you that day. When you find yourself really struggling with a topic, get on your computer and go to our web site, www.wiley.com/college/brady, where you will find additional tutorial help on difficult concepts. If you have prepared before class and briefly reviewed the notes afterward, you'll be surprised at how quickly and how well your concentrated study time will progress. You may even find yourself enjoying chemistry!

As you study, continue to fill in the bare spots in your class notes. Write out the definitions of new terms in your notebook. In this way, when it comes time for an exam you should be able to review for it simply from your notes.

At this point you're probably thinking that there isn't enough time to do all the things described above. Actually, the preparation before class and brief review of the notes shortly after class take very little time and will probably save more time than they consume.

Well, you're on your way to an A. There are a few other things that can help you get there. If you possibly can, spend about 30 minutes to an hour at the end of a week to review the week's work. Psychologists have found that a few brief exposures to a subject are more effective at fixing them in the mind than a "cram" session before an exam. The brief time spent at the end of a week can save you hours just before an exam (efficiency!). Try it (you'll like it); it works.

There are some people (you may be one of them) who still have difficulty with chemistry even though they do follow good study habits. Often this is because of weaknesses in their earlier education. A lot of the material on our web site is designed to help you if you fall into this group. It has been carefully crafted by experienced teachers who know what your problems are, so we urge you to take advantage of the assets that are available to you there.

If, after following intensive study and working with the web-based material, you are still fuzzy about something, speak to your teacher about it. Try to clear up these problems before they get worse. Sometimes, by having study sessions with fellow classmates you can help each other over stumbling blocks. Group study is very effective, because if you find you can explain something to someone else, you really know the subject. But if you can't explain a topic, then it requires more study.

Problem Solving—Using Chemical Tools

Your course in chemistry provides a unique opportunity for you to develop and sharpen your problem solving skills. Just as in life outside the classroom, the problems you will encounter in chemistry are not only numerical ones. In chemistry, you will also find problems related to theory and the application of concepts. The techniques that we apply to these various kinds of problems do not differ much, and one of the goals of your textbook and this Study Guide is to provide a framework within which you can learn to solve all sorts of problems effectively.

If you've read the "To the Student" message at the beginning of the textbook, you learned that we view solving a chemistry problem as not much different than solving a problem in auto repair. Both involve the application of specific tools that accomplish specific tasks. A mechanic uses tools such as screwdrivers and wrenches; you will learn to use a different set of tools—ones that we might call *chemical tools*.

Chemical tools are the simple one-step tasks that you will learn how to do, such as changing units from feet to meters, or degrees Fahrenheit to degrees Celsius. Solving more complex problems just involves combining simple tools in various ways. The secret to solving complex problems, therefore, is learning how to choose the chemical tools that must be used.

Building a chemical toolbox

Our first goal is to clearly identify the tools you will have at your disposal. As you study the text, the concepts you will need to solve problems are marked by an icon in the margin when they are introduced. (To see what the icon looks like, refer to the "To the Student" message at the beginning of the textbook.) The chemical tools are summarized at the end of a textbook chapter in a section titled *Tools You Have Learned* and they are also collected in table form at the end of each of the chapters in this Study Guide.

Solving problems

In both the text and the Study Guide there are worked examples that illustrate a wide variety of problems and their solutions. You will notice that each begins with a section titled *Analysis*. The Analysis section describes the thinking that goes into solving the problem and identifies the tools needed to do the job. This is the most important step in problem solving, because once you've figured out *how* to solve the problem, the rest is easy. (Be sure to study the Examples thoroughly, and also be sure to work on the Practice Exercises that follow the Examples.)

The Analysis step is where you determine which of the chemical tools that you've learned will be applied to the specific problem at hand. Let's look at how you might go about this when working on a problem you haven't seen before—one for which the solution is not immediately obvious. To do this, we will look at a problem of the type you will encounter in Chapter 4. If you've had a previous course in chemistry, you will recognize many of the concepts presented. If they are unfamiliar, don't be concerned. The goal at this time is to illustrate *how* the chemical tools approach can be used to help find a solution to a problem.

Problem
Assemble all the information needed to determine the number of grams of Al that will react with 900 molecules of O_2 to form Al_2O_3, and then describe how the information can be used to find the answer.

The first step in solving the problem is determining what kind of problem it is. In this case, it is a problem dealing with a subject we call *stoichiometry*. (Don't worry, you will learn about all this later.)

Now that we have identified the *kind* of problem, we look over the tools that apply to stoichiometry problems. These are given in the table at the top of the next page.

Tools that apply to stoichiometry:

Tool	How it Works
Atomic mass	We use atomic mass to convert between grams and moles for an element.
Formula mass (molecular mass)	We use formula mass (or molecular mass) to convert between grams and moles for a compound.
Chemical formula	The formula gives the atom ratio in a compound and the mole ratio of elements in a compound.
Chemical equation	A balanced equation gives mole ratios between the various substances in a reaction.
Avogadro's number	We use this to convert between number of particles and moles.

This isn't a very large set of tools, and we can be fairly confident that these will be sufficient to solve the problem. Therefore, we next examine the problem to identify the quantities that relate to the tools we have at hand. Notice that we've drawn boxes around the quantities.

Assemble all the information needed to determine the number of $\boxed{\text{grams of Al}}$ that will react with $\boxed{\text{900 molecules of } O_2}$ to form $\boxed{Al_2O_3}$ and then describe how the information can be used to find the answer.

Now we begin to assign specific numbers to quantities as we assemble the final set of tools we will use to solve the problem. We've collected the information in a table just to make it easier for you to follow. Notice that we have not used all the tools related to stoichiometry. Instead, we have selected just the tools that apply to the quantities in the problem.

Quantity in question	Tool related to it	Relationship
$\boxed{\text{grams of Al}}$	atomic mass	27 g Al = 1 mol Al
$\boxed{\text{900 molecules of } O_2}$	Avogadro's number	6.02×10^{23} molecules O_2 = 1 mol O_2
$\boxed{Al_2O_3}$ $\boxed{O_2}$	chemical formula	2 mol Al = 3 mol O 1 mol O_2 = 2 mol O

The information in the column at the right is what we use to obtain the answer. As you will learn, we can use a method called the *factor label method* to make sure the units of the answer work out correctly. The proper setup of the solution is

$$900 \text{ molecules } O_2 \times \frac{1 \text{ mole } O_2}{6.02 \times 10^{23} \text{ molecules } O_2} \times \frac{2 \text{ mol O}}{1 \text{ mol } O_2} \times \frac{2 \text{ mol Al}}{3 \text{ mole O}} \times \frac{27 \text{ g Al}}{1 \text{ mol Al}} = \text{answer}$$

As you can see, we have not actually calculated the answer. Nevertheless, we really have *solved* the problem; we just haven't done the dirty work of doing the arithmetic.

At the end of most chapters in the textbook, and in some of the sections in the Study Guide, you will find questions titled Thinking It Through. These questions ask you to figure out what you need to know to solve various problems, but not what the answers are. The goal is to make you *think* about how to solve the problems without having to worry about the answers. They are worthwhile exercises and you should be sure to work on them. As you will see, some are pretty difficult. But as they say, "No pain, no gain!"

We realize, of course, that many problems have more than one path to the answer. We understand that after correctly analyzing a problem and after recognizing what tools must be used, intermediate calculations and thought processes can validly follow more than one *order*. Therefore, you might choose a path in which the order of the steps is different from ours. This is why we provide answers to the Thinking It Through exercises, so that you can have the reinforcement (and the reward) of comparing answers when your method differs from ours. For the Thinking It Through exercises in the study guide, the answers are at the ends of the chapters; for those in the textbook, the answers are available on the web site.

Time to Begin

As you begin your study of chemistry, we wish you well. Move on to the course now, and good luck on getting that A!

Chapter 1

Atoms and Elements:
The Building Blocks of Chemistry

This first chapter introduces you to some basic concepts that you will need in order to understand future discussions in class and in the textbook. You will also need them to function effectively in the laboratory part of your course.

If you've had a prior course in chemistry, much of what we discuss in this chapter will seem familiar. Nevertheless, be sure you really understand all of it fully and can do the assigned homework. In particular, be sure you've learned the meanings of the bold-faced terms in the text as well as equations that are highlighted by a yellow background and statements that are in boxes, such as the one on page 13.

We begin the chapter by explaining why chemistry is important to you. You will learn what chemistry is about, namely, chemicals and chemical reactions. You will also learn about the scientific method, which describes how scientists learn about nature.

Chemical substances are identified by their properties, and in this chapter you will learn how we categorize properties. We will introduce you to the concepts of elements and compounds and how the atomic theory devised by John Dalton explained observable chemical laws. We will delve into the internal structures of atoms to study subatomic particles and you will be introduced to the periodic table—a most useful device for correlating and remembering chemical facts.

Learning Objectives

As you study this chapter, keep in mind the following objectives:

1   To learn how chemistry is important in the study of all of the physical and biological sciences.

2   To learn how science develops through the application of the scientific method. In particular, you should learn the distinction between an observation and a conclusion and between a law and a theory.

3   To learn how matter is identified by its properties, and how properties are classified.

4   To learn how we use the atomic theory as a way to make the connection between the macroscopic world that we experience with our senses and what takes place at the submicroscopic level of atoms and molecules.

5   To learn the properties of the states of matter and how they are accounted for by examining them at the submicroscopic, atomic level.

6   To learn what an element is and how elements are identified by their chemical symbols.

7   To learn the definition of a compound.

8   To learn how elements, compounds, and mixtures differ.

9   To learn the laws of chemical combination—the law of conservation of mass and the law of definite proportions.

10   To begin to learn how we systematically approach problem solving through analysis, solution, and checking the result to see if it makes sense (i.e., is the answer reasonable?).

11   To learn the postulates of Dalton's atomic theory and how it was based on experimental observations.

12   To learn the law of multiple proportions.

13   To learn how relative atomic masses were determined and how they relate to the mass of carbon-12.

14   To learn how atomic masses can be calculated from isotopic abundances.

15   To learn about the basic features of atomic structure. This includes the concept of subatomic particles (electrons, protons, and neutrons) and where they are found within an atom.

16   To learn how to write the symbol of an isotope, showing its atomic number and mass number.

17   To learn the basis for Mendeleev's original periodic table of the elements.

18   To learn the basis for the modern periodic table of the elements. You should also be sure to learn the special terminology used to describe various aspects of the periodic table that are introduced in this chapter.

19   To learn how elements can be classified as metals, nonmetals, and metalloids according to their properties.

20   To learn how the periodic table can help to correlate properties with an element's position in the table.

1.1   Chemistry is important for anyone studying the sciences

Review

The central theme of this section is that chemistry is a science that all physical and biological scientists must know something about. All the sciences study the same natural world; only their perspectives differ. Chemistry is unique, however, in that it seeks to provide an understanding of the chemical substances that serve as the subjects of all the other sciences.

If you are not a chemistry major, you might ponder for a moment why it is that you are required to take chemistry as part of the requirements for your major.

New Terms

Write the definitions of the following terms, which were introduced in this section. If necessary, refer to the Glossary at the end of the text.

chemistry

matter

1.2   The scientific method helps us build models of nature

Review

The sequence of steps described by the scientific method is little more than a formal description of how people logically analyze any problem, scientific or otherwise. Observations are made in order to collect data (empirical facts) from which we often can draw conclusions. Generalizations come from the analysis of data and can lead to laws, which are concise statements about the behavior of chemical or physical systems. Laws, however, offer no explanations about *why* nature behaves the way it does. Tentative explanations are called hypotheses; tested explanations are called theories. The scientific method consists of collecting data in experiments, formulating theories, and testing the theories by more experimentation. Based on the results of new experiments, the theories are refined, tested further, refined again, and so on.

When scientists create theories, they build mental images that help them visualize how nature operates. The atomic theory is one of the most useful models of the intimate structure of matter. According to the theory, all substances are made up of very tiny atoms that combine in various combinations to form molecules. Often, chemists represent molecules by drawings of the kind illustrated in Figure 1.2.

Self-Test

1.   Identify each of the following as an observation or a conclusion.

   (a)   If you drop a stone, a force called gravity causes the stone to fall to the ground.

   _____

   (b)   Water boils if it is heated to a temperature of 212 °F. _____

2.   Identify each of the following statements as either a law or a theory.

   (a)   In general, what goes up must come down.   _____
   (b)   The ice ages resulted from the tilting of the earth's rotation axis, which was caused by the earth being hit by very large meteors.

   _____

3.   What does *empirical* mean? _____

   _____

4.   What is the difference between an atom and a molecule?

   _____

New Terms

Write the definitions of the following terms, which were introduced in this section. If necessary, refer to the Glossary at the end of the text.

observation	conclusion	empirical facts
data	scientific law	theoretical model
hypothesis	theory	scientific method
atom	molecule	

1.3   Properties of materials can be classified in different ways

Review

In the text, we see that the properties of substances can be classified in two ways. One is to divide them into either physical properties or chemical properties. The other is to divide them into intensive or extensive properties.

Physical and chemical properties

Physical properties are ones that can be observed without changing the chemical makeup of a substance. In general, physical properties can be specified without reference to another chemical substance. Examples are an object's color, or the temperature at which it melts, or its volume. Any change that occurs during the observation of a physical property is a physical change, as when a substance changes from solid to liquid when we observe its melting point.

When we describe a chemical property of a substance, we describe how the substance reacts chemically with something else. Such a chemical reaction (chemical change) produces new chemical substances, so after we've observed a chemical property, the substance is no longer the same; it has changed into a different substance. For example, a chemical property of iron is that it rusts when in contact with air and moisture. When we observe this property, the iron changes to rust and no longer has the same appearance.

Intensive and extensive properties

Extensive properties, such as mass or volume, depend on the size of the sample of matter being examined. Although extensive properties are important for a given sample, they are not especially useful for identifying substances. More useful are intensive properties, because all samples of a given substance have identical values for its intensive properties. For example, if we were asked whether a sample of a liquid was water, we would examine its properties and compare them to those of a known sample of water. If we were to find that the mass of the liquid is 12.0 g, we still would not be any closer to knowing whether or not the sample is water. By itself, the mass is of no value, because different samples of water, or any other liquid, have different masses. However, if we further note that the liquid is clear, has no color, has no odor, and freezes at 0 °C, we would strongly suspect the sample to be water. This is because *all* samples of pure water, regardless of size, are clear, colorless, odorless, and freeze at 0 °C.

Self-Test

5. Identify the following as chemical properties or physical properties.

    (a) Nitroglycerine explodes if it is heated.   _____chemical_____

    (b) Gold is a yellow metal.   _____physical_____

6. Sodium is a soft, silvery metal that melts at 97.8 °C. It burns with a yellow light in the presence of chlorine gas to give the compound sodium chloride (table salt).

    (a) What are some physical properties of sodium? _____

    _____

    (b) Give a chemical property of sodium. _____

    _____

7.   (a)   Give two examples of intensive properties. _____

_____

(b)   Give two examples of extensive properties. _____

_____

New Terms

Write the definitions of the following terms, which were introduced in this section. If necessary, refer to the Glossary at the end of the text.

physical property   physical change

chemical reaction   chemical property

chemical change   extensive property

intensive property

1.4   Materials are described by their properties

Review

Substances are classified according to their properties. One such classification is physical state—solid, liquid, or gas. Study Figure 1.6 to see how they differ in the way their atomic sized particles are organized.

Chemically, substances are classified as elements, compounds, or mixtures of elements and/or compounds. Elements are the basic building blocks of all substances and cannot be decomposed into simpler substances by chemical reactions. Each element is identified by its chemical symbol. Your teacher will probably expect you to learn the symbols of a number of the more common elements.

Compounds are formed from elements. They contain elements in fixed (constant) proportions by mass. Stated differently, the proportions of the elements in a compound cannot be changed. Mixtures are formed from elements and/or compounds and can be of variable composition. A mixture composed of two or more phases is said to be heterogeneous; a mixture composed of just a single phase is a solution and is said to be homogeneous.

Separating a compound into its components (elements) involves a chemical change, just as in the combination of the elements to form the compound. Separation of a mixture usually can be accomplished by a physical change—a change in which the chemical identities of the components aren't altered. Be sure to study Figure 1.10 on page 12.

Self-Test

8.   A student placed some sand and salt in a beaker and stirred them with a glass rod.

(a)   After stirring, does the beaker contain a compound or a mixture?

_____

(b)   How many phases are present? _____

(c) Can you suggest a method of separating the components in the beaker? Does this method involve a chemical or a physical change?

_____

_____

_____

_____

9. What is the chemical symbol for:

(a) carbon _____   (d) nickel _____

(b) chlorine _____   (e) chromium _____

(c) copper _____   (f) phosphorus _____

10. Write the name of the element with the symbol

(a) S _____

(b) Na _____

(c) Ca _____

(d) Br _____

New Terms

Write the definitions of the following terms, which were introduced in this section. If necessary, refer to the Glossary at the end of the text.

decomposition	mixture	solution
element	homogeneous	phase
compound	heterogeneous	physical change
pure substance	chemical symbol	

1.5   Atoms of an element have properties in common

Review

Although the concept of atoms was not new when Dalton presented his atomic theory, the theory nevertheless revolutionized chemistry. This is because it was based on two *experimentally observed* laws relating to the compositions of substances and chemical reactions. We call them laws of chemical combination.

Be sure you study the definitions of the law of conservation of mass and the law of definite proportions given on page 13. Notice that the laws refer to experimentally measurable quantities—the masses of substances in compounds and in chemical reactions.

The example below illustrates how we can use the law of definite proportions and some proportional reasoning to work a chemical calculation.

Example 1.1   Proportional Reasoning and the Law of Definite Proportions

Rust is a compound of iron and oxygen. A particular sample of rust was found to contain 1.25 g of iron and 0.536 g of oxygen. If a 3.56 g sample of iron were to form rust, how many grams of rust would be formed?

Analysis:
According to the law of definite proportions, all samples of rust will have the same mass ratio of iron to oxygen. We see that in one sample, the ratio is 1.25 g iron to 0.536 g oxygen. The second sample of rust will have iron and oxygen in the same ratio, so we will do some proportional reasoning to find how much oxygen must combine with 3.56 g of iron to have the same iron to oxygen ratio. To find the total mass of rust, we just add the mass of oxygen to the mass of iron.

Solution:
If we let the unknown mass of oxygen in the rust be represented by x, we can set up two ratios of oxygen to iron.

Known rust sample   New rust sample

$$\frac{0.536 \text{ g oxygen}}{1.25 \text{ g iron}} \qquad\qquad \frac{x}{3.56 \text{ g iron}}$$

According to the law of definite proportions, these two ratios must be the same, so we can equate them.

$$\frac{0.536 \text{ g oxygen}}{1.25 \text{ g iron}} = \frac{x}{3.56 \text{ g iron}}$$

Next, we solve for x.

$$x = 3.56 \text{ g iron} \times \frac{0.536 \text{ g oxygen}}{1.25 \text{ g iron}} = 1.53 \text{ g oxygen}$$

Thus, the new rust sample will contain 3.56 g iron and 1.53 g of oxygen. The total mass of the rust will be the sum of these two masses.

$$\text{Total mass of rust} = 3.56 \text{ g} + 1.53 \text{ g} = 5.09 \text{ g}$$

Is the Answer Reasonable?
Let's compare the masses of iron in the two samples. The second mass (3.56 g) is almost three times larger than the first (1.25 g), so we would expect that the amount of oxygen in the second sample would be almost three times that in the first. Three times 0.536 g would be a little larger than 1.5 g, so the mass of oxygen we calculated seems to be about right. For the total mass, it's easy to check the arithmetic to see that it's okay.

Daltons atomic theory

Study the postulates of Dalton's theory. You should be able to describe how they explain the laws of chemical combination mentioned above. The law of multiple proportions (page 16), which was discovered as a result of Dalton's theory, is éasily explained on the basis of the atomic theory. Study Figure 1.11 to obtain a clear understanding of this law. Study the discussion of the scanning tunneling microscope, a device that provides direct experimental evidence for the existence of atoms.

Isotopes

Contrary to Dalton's theory, not all the atoms of a given element have exactly the same mass. Atoms of an element with slightly different masses are said to be isotopes of the element. The existence of isotopes did not affect the validity of Dalton's theory because of two reasons.

1   Any sample of an element large enough to see has an enormous number of atoms.

2   The proportions of the different isotopes in samples of an element are uniform from sample to sample.

As a result, an element behaves as if its atoms have masses corresponding to the *average mass* of the isotopes.

Carbon-12 atomic mass scale

A major feature of Dalton's theory was the notion that atoms of each element have a characteristic atomic mass (sometimes called *atomic weight*). The atomic masses we use today are relative masses, with carbon-12 (an isotope of carbon) being the foundation of the mass scale: one atom of carbon-12 has a mass of *exactly* 12 u (atomic mass units).

$$1 \text{ atom } {}^{12}C = 12 \text{ u (exactly)}$$

For example, scientists have found that magnesium atoms are, on average, a little more than twice as heavy as atoms of carbon-12. Therefore, if we assign one atom of ${}^{12}C$ a mass of 12 u, the mass of a magnesium atom must be slightly larger than 24 u. (More precisely, the average atomic mass of magnesium is 24.305 u). This means that if we have a sample of carbon with a mass of 12 g, and we want a sample of magnesium *with as many atoms*, we have to take a sample of 24.305 g of magnesium.

Average atomic masses are calculated from isotopic abundances and accurate mass of each of the isotopes. Be sure to study Example 1.2 and work Practice Exercises 3 through 5.

Self-Test

11. State the law of conservation of mass.   _____

_____

12. State the law of definite proportions.   _____

_____

13. In a 4.00 g sample of table salt, NaCl, there is 1.57 g of Na. Use the law of definite proportions to calculate the mass of Cl in a 6.00 g of NaCl.

_____

14. If the ratio by mass of the elements in the carbon dioxide in a sample obtained in New York City is 2.66 g O to 1.00 g C, what ratio would be present in a sample of this compound taken in Los Angeles?

_____

15. A vinegar sample purchased in one store was found to have a mass ratio of oxygen to carbon of 173 g O to 1.00 g C. A sample bought in a different store had a ratio of 185 g O to 1.00 g C. Do these data suggest that vinegar is a compound or a mixture? Why?

_____

_____

16. In the compound sodium chloride, there are 1.45 g of chlorine for every 1.00 g of sodium. Suppose that 2.00 g of chlorine were mixed with 1.00 g of sodium in some appropriate reaction vessel. After the reaction between them is as complete as possible:

    (a)   What is the total mass of chemicals in the reaction vessel?

    (b)   How many grams of chlorine have reacted? _____

    (c)   How many grams of chlorine remain unreacted?

17.   According to Dalton, all atoms of the same element

    (a) are indestructible   (c) have identical masses   (e) *a, b* and *c*
    (b) are very small   (d) *a* and *b* only

18.   Covellite is a soft, indigo-blue mineral in which copper is combined with sulfur in a ratio of 1.000 g S to 0.506 g Cu. Chalococite is a soft, blackish mineral composed of copper and sulfur in a ratio of 1.000 g S to 0.253 g Cu. Do these two minerals illustrate the law of multiple proportions? Explain.

19.   How do isotopes of the same element differ chemically?

20.   How do isotopes of the same element differ physically?

21.   The atomic mass of gold (rounded) is 197. How much heavier are gold atoms than carbon-12 atoms?

22.   The element iridium is composed of 37.3% of ^{191}Ir, which has a mass of 190.960 u, and 62.7% ^{193}Ir, which has a mass of 192.963 u. What is the average atomic mass of iridium?

New Terms

law of conservation of mass	law of definite proportions	atom
law of multiple proportions	gram (g)	molecule
scanning tunneling microscope	atomic mass	atomic weight
isotope	atomic mass unit (amu, u)	

1.6   Atoms are composed of subatomic particles

Review

At the center of an atom is a tiny particle called a nucleus, which is composed of subatomic particles called protons and neutrons. Protons are positively charged and neutrons are electrically neutral. Surrounding the nucleus are electrons, which are negatively charged. The amount of charge carried by a proton and an electron is the same; however, the *kind* of charge (positive and negative) is different. Each proton has one unit of positive charge, and each electron has one unit of negative charge. Because of this, in a neutral atom the number of electrons surrounding the nucleus equals the number of protons in the nucleus.

Protons and neutrons (collectively referred to as nucleons) have nearly the same mass (approximately 1 atomic mass unit) and are much heavier than electrons. As a result, nearly all the mass of an atom is contained within its nucleus.

All of the atoms in a given element have the same number of protons, no matter if the element consists of several isotopes. Atoms of different elements have different numbers of protons. This makes the number of protons—called the atomic number (Z)—something unique for each element, and it corresponds to the size of the positive charge on the nuclei of the element.

The isotopes of an element differ only in their numbers of neutrons, so they differ only in mass. The sum of the protons and neutrons—called the mass number (A)—is one way to specify which isotope of an element is being discussed. Thus, the name carbon-12 identifies the one isotope of carbon that has a mass number of 12. (Because carbon's atomic number is 6, meaning six protons, the nucleus of a carbon-12 atom must have 6 neutrons so that the total number of nucleons is 12.)

Sometimes it is desirable to specify the mass number and the atomic number when writing the symbol of an isotope. This is illustrated on page 25. Some other examples are shown below.

$$^{35}_{17}\text{Cl} \qquad\qquad ^{190}_{77}\text{Ir} \qquad\qquad ^{87}_{37}\text{Rb}$$

17 protons	77 protons	37 protons
18 neutrons	113 neutrons	50 neutrons
17 electrons	77 electrons	37 protons

Self-Test

23. Name the three subatomic particles. _____

24. Name the nucleons.   _____

25. Name the particles that are in the nucleus. _____

26. An atom with a mass number of 25 has an atomic number of 12. How many of the following particles does it have?

    (a) electrons _____   (c) neutrons   _____

    (b) protons   _____   (d) nuclei _____

27. An atom with 14 electrons has 15 neutrons. What is its approximate atomic mass?

    _____

28. An isotope of oxygen has 10 neutrons. Write its symbol.   _____

29. Which is the only atom that has an atomic mass that is exactly a whole number?_____

New Terms

Write the definitions of the following terms, which were introduced in this section. If necessary, refer to the Glossary at the end of the text.

nucleus	nucleon	proton
neutron	atomic number	mass number

1.7   The periodic table is used to organize and correlate facts

Review

The number of facts in chemistry is enormous. Besides the many similarities among various elements, many differences also exist. Progress in understanding and explaining these similarities and differences required a search for order and organization. The product of this search—the modern periodic table—has become our primary tool for organizing chemical facts.

Mendeleev discovered that similar properties are repeated at regular intervals when the elements are arranged in order of increasing atomic mass. By breaking this sequence at the right places, Mendeleev arranged elements with similar properties in vertical columns (groups) within his periodic table. Mendeleev's genius was his insistence on having elements with similar properties in the same column, even though it sometimes meant leaving empty spaces for (presumably) undiscovered elements.

The modern periodic table

Atomic number forms the basis for the sequence of the elements in the modern periodic table shown on page 28 and on the inside front cover of the textbook. The rows are called periods and the columns are called groups, just as in Mendeleev's original table. Sometimes the term *family* is used when speaking of a group of elements. The periods are numbered with Arabic numerals (1, 2, etc.). In the North American form of the table, which we shall use throughout the remainder of the text, the groups are given Roman numerals and a letter (e.g., Group IIA). However, the IUPAC has recommended that the groups be numbered sequentially from left to right with Arabic numerals. You should be aware of this because you may encounter this numbering elsewhere.

In the table that we will use, the representative elements are the A-group elements. The B-group elements are the transition elements. The two long rows of elements below the main body of the table are the inner transition elements (the lanthanides and actinides). Examine Figure 1.15 on page 28 to be sure you understand where they properly fit into the periodic table. Also, remember the names of the following families:

Group IA	Alkali metals	Group VIA	Halogens
Group IIA	Alkaline earth metals	Group VIII	Noble gases

Metals, nonmetals, and metalloids

Metals are *shiny*, are *good conductors of heat and electricity*, and many are *malleable* and *ductile*. They are found in the lower left portion of the periodic table and make up most of the known elements. (Study Figure 1.17 on page 30.)

Nonmetals lack the luster of metals, are nonconductors of electricity, and are poor conductors of heat. Many nonmetals are gases (H_2, O_2, N_2, F_2, Cl_2 and the elements of Group VIII) and those that are solids tend to be brittle. Nonmetals are found in the upper right portion of the periodic table.

Metalloids are semiconductors, but resemble nonmetals in many of their properties. They are located on opposite sides of the step-like line running from boron (B) to astatine (At) in the periodic table.

We can use the periodic table as a tool to help us compare the metallic and nonmetallic properties of elements. Within the periodic table, elements become less metallic (more nonmetallic) going from left to right in a period and more metallic going from top to bottom in a group.

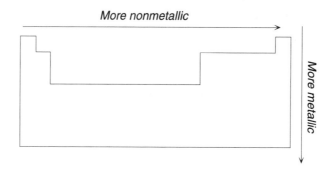

Self-Test

30. What problem did Mendeleev face with the elements iodine and tellurium? How did he solve it?

   _____

31. Argon and potassium do not fit in atomic mass-order in the periodic table. What does this suggest about Mendeleev's basis for constructing the periodic table?

   _____

32. According to the IUPAC system, in which groups are the following elements found?

   (a)   lithium   _____

   (b)   iron   _____

   (c)   silicon   _____

   (d)   sulfur   _____

33. Among the elements Mg, Al, Cr, U, Kr, K, Br, and Ce,

   (a)   Which are representative elements?   _____

   (b)   Which is a transition element?   _____

   (c)   Which is a halogen?   _____

   (d)   Which is a noble gas?   _____

   (e)   Which is an alkaline earth metal?   _____

   (f)   Which is an alkali metal?   _____

(g)   Which is an actinide element?   _____

(h)   Which is a lanthanide element?   _____

34.   Fill in the blanks.

(a)   The ability of copper to be drawn into wire depends on its …

_____.

(b)   Gold can be hammered into very thin sheets because of its …

_____.

(c)   The property that makes mercury useful in thermometers is …

_____.

(d)   The reason that tungsten is used as filaments in electric light bulbs is …

_____.

(e)   A nonmetal having a yellow color is _____.

(f)   Two elements that are liquids at room temperature are …

_____ and _____

(g)   The color of bromine is _____.

(h)   The reason that helium is used to inflate the Goodyear blimp rather than hydrogen is …

_____

(i)   Sodium is rarely seen as a free metal because _____

_____.

(j)   Two common semiconductors used in electronic devices are…

_____ and _____.

35.   Which element is more metallic, Ga or Ge? _____

36.   Which element is more nonmetallic, As or P? _____

New Terms

Write the definitions of the following terms, which were introduced in this section. If necessary, refer to the Glossary at the end of the text.

group	inner transition elements
period	lanthanide elements
periodic table	actinide elements
alkali metals (alkalis)	family of elements
alkaline earth metals	representative elements

New Terms (continued)

noble gases (inert gases)   transition elements

halogens   metal

nonmetal   malleability

metalloid   ductility

semiconductor

Answers to Self-Test Questions

1. (a) conclusion (b) observation
2. (a) law (b) theory
3. Based on observation (experiment) or experience.
4. Atoms are basic building blocks of chemical substances and combine to form more complex particles called molecules.
5. (a) chemical property (b) physical property
6. (a) soft, silvery, melts at 97.8 °C (b) Burns in presence of chlorine to give sodium chloride.
7. (a) color, melting point, boiling point, odor (b) mass, volume
8. (a) a mixture, (b) two, (c) Add water, which dissolves the salt but not the sand. This involves a physical change.
9. (a) C (b) Cl (c) Cu (d) Ni (e) Cr (f) P
10. (a) sulfur (b) sodium (c) calcium (d) bromine
11. and 12. Consult the definitions on page 13 of the textbook.
13. 3.65 g of Cl
14. A ratio of 2.66 g O to 1.00 g C.
15. A mixture. The oxygen-to-carbon ratio is not constant from one sample to another.
16. (a) 3.00 g total,
    (b) 1.45 g Cl react,
    (c) 0.55 g Cl unreacted
17. (e)
18. Yes, the amounts of copper that combine with 1.00 g S in the two compounds are in the ratio of 0.506 to 0.253, which is a ratio of 2 to 1, simple whole numbers.
19. They don't differ chemically.
20. They differ slightly in their masses.
21. 16.4 times as heavy
22. 192
23. electron, proton, neutron
24. proton and neutron
25. proton and neutron
26. (a) 12, (b) 12, (c) 13, (d) 1
27. 29 u
28. $^{18}_{8}O$
29. ^{12}C

30. If placed in the table in order of atomic weight, their properties do not match those of other elements in the same columns. He put them in the table according to their properties rather than according to their atomic weights.

31. Atomic weight is not really the basis for the periodic law.

32. (a) Group 1, (b) Group 8, (c) Group 14, (d) Group 16

33. (a) Mg, Al, Kr, K, Br   (b) Cr (c) Br
     (d) Kr (e) Mg (f) K (g) U (h) Ce

34. (a) ductility, (b) malleability, (c) low melting point and fairly high boiling point, (d) it conducts electricity and has a very high melting point.
     (e) sulfur, (f) bromine, mercury, (g) red-brown, (h) helium is very unreactive, (i) it is very reactive toward oxygen, (j) silicon and germanium

35. Ga

36. P

Tools you have learned

Consider removing this chart from the Study Guide so you can have it handy when tackling homework problems.

Tool	How it Works
Law of definite proportions	If we know the mass ratio of the elements in one sample of a compound, we know the ratio will be the same in a different sample of the same compound.
Law of conservation of mass	The total mass of chemicals present before a reaction starts equals the total mass after the reaction is finished. We can use this law to check whether we have accounted for all the substances formed in a reaction.
Periodic Table	From an element's position in the periodic table, we can tell whether it's a metal, nonmetal, or metalloid.

Chapter 2
Compounds and Chemical Reactions

In Chapter 1, you learned about elements and atoms. We now turn our attention to compounds, which are substances formed when two or more elements combine. Compounds can be roughly divided into two general types, molecular and ionic. We will discuss how the two types differ, the kinds of elements that form each type, and the characteristic properties the help us identify each. We also describe the systematic method used to name molecular and ionic substances.

Learning Objectives

As you study this chapter, keep in mind the following objectives:

1　To learn how chemical formulas are used to describe chemical substances.

2　To learn the formulas of the common diatomic elements.

3　To learn how to count atoms in a formula.

4　To learn how to interpret the formula of a hydrate.

5　To learn the meaning of the terms *reactant* and *product*.

6　To learn how to use coefficients and formulas to count atoms in chemical equations.

7　To learn how to use the symbols (*s*), (*l*), and (*g*) to express the physical states of reactants and products, and (*aq*) to specify a substance dissolved in aqueous solution.

8　To learn the meaning of kinetic and potential energy, and the law of conservation of energy.

9　To learn the difference between heat and temperature.

10　To understand how temperature changes in a chemical reaction are related to changes in the potential energies of the atoms and molecules involved.

11　To learn how molecular compounds are formed when nonmetallic elements combine.

12　To learn how to use the periodic table to remember the formulas of the hydrogen compounds of nonmetals.

13　To learn how to write the formulas for alkane hydrocarbons.

14　To learn how to name molecular compounds.

15　To understand how ionic and molecular compounds differ at the submicroscopic, atomic level.

16　To learn the kinds of elements that combine to form ionic compounds.

17　To learn how to use the periodic table to predict formulas for ionic compounds.

18　To learn the rules for writing formulas for ionic compounds.

19　To learn the symbols and charges of ions of some important transition and post-transition metals.

✳ 20   To learn the formulas and charges of important polyatomic ions.

21   To learn the system used to name ionic compounds.

22   To learn the characteristic properties of ionic and molecular compounds.

2.1   Elements combine to form compounds

Review

There are two important topics to learn in this section. First, it's important to understand that when a chemical reaction takes place, the observable properties of the chemicals usually change significantly. As a reaction progresses, new substances are formed (with their unique properties) as the original substances (with their own unique properties) disappear.

The second topic is a skill that's critical to much of the rest of the course. You must be able to interpret a chemical formula in terms of the number of atoms of each element present. An example is given below.

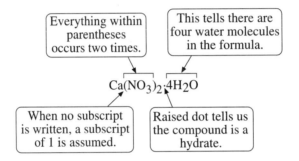

This formula shows 1 calcium atom, 2 nitrogen atoms, 8 hydrogen atoms, and 10 oxygen atoms (4 from the four H_2O molecules plus 6 from the two NO_3 units within parentheses).

A number of the nonmetals occur as diatomic molecules in the *free*, uncombined state (i.e., not combined with any other element). Be sure to study the margin table on page 42.

Self-Test

1. A simple experiment you can perform in your kitchen at home or in an apartment is to add a small amount of milk of magnesia to some vinegar in a glass. Stir the mixture and observe what happens. Then add some milk of magnesia to the same amount of water and stir. What evidence did *you* observe that suggests that there is a chemical reaction between the milk of magnesia and the vinegar?

    _____

    _____

2. Drop an Alka Seltzer tablet into a glass of water. Observe what happens. What evidence is there that a chemical reaction is taking place?

    _____

3.   How many atoms of each element are represented in the formula of...

(a) Al_2Cl_6 _____

(b) $CaSiO_3$ _____

(c) $K_2Cr_2O_7$ _____

(d) $C_{12}H_{22}O_{11}$ _____

(e) $Na_2CO_3 \cdot 10H_2O$ _____

(f) $(NH_4)_2SO_4$ _____

(g) Al_2O_3 _____

(h) $CaSO_4 \cdot 2H_2O$ _____

(i) $Ca_3(PO_4)_2$ _____

4.   Write the formulas of the diatomic nonmetals.

$H_2, O_2, N_2, \; I_2, Br_2, F_2, Cl_2,$

New Terms

Write the definitions of the following terms, which were introduced in this section. If necessary, refer to the Glossary at the end of the text.

chemical formula	free element	diatomic molecule
hydrate	dehydrate	anhydrous

2.2   Chemical equations describe what happens in chemical reactions

Review

The *reactants* in a chemical equation appear on the left side of the arrow and are the substances present before the reaction begins. The *products* appear on the right side of the arrow and are the substances formed by the reaction. An equation is balanced by placing numbers called coefficients (often called *stoichiometric coefficients*) in front of the chemical formulas so that for each element there are the same number of atoms on both sides of the arrow. You will learn more about balancing equations later; for now you just need to recognize when an equation is balanced. This means you have to be able to count atoms in the formulas. Remember that in a chemical equation, a coefficient multiplies everything in the formula that follows. Consider, for example, the following:

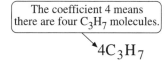

The coefficient 4 means there are four C_3H_7 molecules.

$$4C_3H_7$$

Therefore, in the expression $4C_3H_7$ we have $4 \times 3 = 12$ atom of carbon and $4 \times 7 = 28$ atoms of hydrogen.

Sometimes in a chemical equation we specify the physical states of the chemicals involved or whether they are dissolved in a solution. The following symbols are used and are placed after the chemical formula:

solid	(*s*)	liquid	(*l*)
gas	(*g*)	aqueous solution	(*aq*)

Self-Test

5. How many atoms of each kind are represented by the expression ...

    (a) $3KMnO_4$ _____ (b) $5Al_2(SO_4)_3$ _____

6. How many atoms of each element are found on each side of the following equations? Are the equations balanced?

    (a) $Al_2O_3 + 6HCl \rightarrow 2AlCl_3 + 3H_2O$

    _____

    (b) $2(NH_4)_3PO_4 + 3CaO \rightarrow 3NH_3 + Ca_3PO_4 + 3H_2O$

    _____

7. Rewrite the equation in part (a) of the preceding question to show that the Al_2O_3 is a solid, the HCl and $AlCl_3$ are dissolved in water, and that H_2O is a liquid.

    _____

New Terms

Write the definitions of the following terms, which were introduced in this section. If necessary, refer to the Glossary at the end of the text.

chemical equation	coefficients	reactants
products	balanced equation	

2.3   Energy is an important part of chemical change

Review

Energy is something an object has if it has the capability of performing work. There are two kinds of energy an object can have. Kinetic energy is energy of motion and can be calculated from the object's mass (*m*) and velocity (*v*) by the equation $KE = 1/2\ mv^2$. Potential energy is stored energy. The potential energy stored in chemicals, which can be released in chemical reactions, is sometimes called chemical energy. The law of conservation of energy states that energy cannot be destroyed, but only changed from one form to another.

In an object, the tiny particles are constantly jiggling about and have kinetic energy because of their motions. The temperature of an object is proportional to the average kinetic energy of its particles. A hot object has particles with a larger average KE than a cold object. This means that the particles that make up the hot object are moving faster, on average, than those in the cold object.

Heat is energy (thermal energy) that's transferred from a warmer object to a cooler one. What is actually transferred is kinetic energy. This occurs as the fast moving molecules of the hot object collide with and lose KE to the slower molecules in the cool object. Over time, the average KE of the molecules in the hot object decreases and the temperature of the hot object decreases. Similarly, the average KE of the molecules of the cool object increases, so the temperature of the cool object increases. Eventually, both objects come to the same temperature when the average KE of the molecules becomes the same in both.

In chemical reactions there are changes in the potential energies of the atoms in the chemicals undergoing reaction. If the potential energy decreases, the kinetic energy of the atoms must increase, because energy cannot just disappear. The rise in kinetic energy translates into a rise in temperature. Thus, when we see a chemical reaction that becomes hot (for example, a fire), we know that the potential energy of the chemicals is decreasing and the kinetic energy of the atoms is increasing.

Self-Test

8. How does the kinetic energy of a 1000 lb car traveling at 60 mph compare with the kinetic energy of a 4000 lb car traveling at the same speed?

$\frac{1}{2} \cdot 1000 \cdot 60^2$ $\frac{1}{2} \cdot 4000 \cdot 60^2$

9. How does the kinetic energy of a 2000 lb car moving at 20 mph compare with the kinetic energy of the same car traveling at 60 mph?

$\frac{1}{2} \cdot 2000 \cdot 20^2$ $\frac{1}{2} \cdot 2000 \cdot 60^2$

10. How does the kinetic energy of a 4000 lb truck moving at 60 mph compare to the kinetic energy of a 2000 lb car moving at 30 mph?

11. What is the difference between potential energy and chemical energy?

P.E. in *Samties* is *stored* in *chemicles* and *released* in rxns

12. When the substance urea dissolves in water, the mixture becomes cool.

(a) What is happening to the kinetic energy of the substances as the solution forms?

↓

(b) What is happening to the potential energy of the substances as the solution forms?

↑

New Terms

Write the definitions of the following terms, which were introduced in this section. If necessary, refer to the Glossary at the end of the text.

energy	kinetic energy	potential energy
chemical energy	temperature	heat
thermal energy	law of conservation of energy	

↳ energy transferred as heat

2.4   Molecular compounds contain neutral particles called molecules

Review

Molecules are electrically neutral particles composed of two or more atoms. Evidence for the existence of molecules is Brownian motion, which is believed to be caused by the battering about of tiny particles by collision with molecules of a liquid. Molecules are held together by chemical bonds between their atoms. These bonds arise from the sharing of electrons between atoms, which we will discuss more fully in later chapters. The chemical formulas we write for molecules are called **molecular formulas**; they describe the number of atoms of each kind in one molecule of a substance.

It is useful to remember that nonmetals combine with nonmetals to form molecular compounds. Examples presented in this section include the simple hydrogen compounds of the nonmetals. Also notice that most of the nonmetals exist in their elemental forms as molecules. Be sure you learn the ones that are diatomic: H_2, N_2, O_2, F_2, Cl_2, Br_2, and I_2. (Only the noble gases occur in nature as single atoms.)

Compounds of nonmetals with hydrogen

The formulas of the simple hydrogen compounds of the nonmetals are given in Table 2.1 on page 51. Notice that we can use the periodic table to figure out their formulas: For a nonmetal in a particular group in the periodic table, the number of hydrogens in the formula equals the number of steps to the right we have to go to get to Group VIII (the noble gases). Thus, for any element in Group VIIA, we have to move *one* step to the right to get to Group VIII; the formula of the hydrogen compound contains *one* hydrogen (e.g., HF). Similarly, for an element in Group IVA, we have to move *four* elements to the right to get to Group VIII, and each element in Group IVA forms a hydrogen compound with *four* hydrogens (e.g., CH_4).

Compounds of carbon

Carbon and hydrogen form compounds called hydrocarbons in which atoms of carbon are attached to each other, often in chains of varying lengths. Hydrocarbons make up the chief constituents of petroleum, and they serve as the foundation of a class of substances called organic compounds. Organic chemistry is the study of hydrocarbons and compounds formed from them by substituting various groups of atoms in place of hydrogen atoms.

Remember the general formula for an *alkane* hydrocarbon, C_nH_{2n+2}, where n is the number of carbon atoms. Butane (the fuel in cigarette lighters) is a four-carbon alkane. Its formula is $C_4H_{2(4)+2} = C_4H_{10}$.

An important class of organic compounds is the alcohols, obtained by substituting OH in place of a hydrogen atom. Be sure you know the formulas for methanol (methyl alcohol), CH_3OH, and ethanol (ethyl alcohol) C_2H_5OH.

Self-Test

13.   Without referring to anything but the periodic table, write the formula for the simplest compound formed from hydrogen and

   (a)   sulfur H_2S

   (b)   bromine HBr

   (c)   phosphorus PH_3

(d) carbon ___CH_4___

14. Write the formula for the alkane hydrocarbon that has

(a) five carbon atoms ___C_5H_{12}___

(b) seven carbon atoms ___C_7H_{16}___

New Terms

molecule   Brownian motion   molecular formula

organic chemistry   hydrocarbon   alkane

2.5 Naming molecular compounds follows a system

Review

This is the first section in the book that discusses naming compounds (nomenclature). We begin with simple inorganic molecular compounds. To name them, you must learn the Greek prefixes given on page 54.

For binary compounds of two elements, the first is specified by giving its English name, preceded by a prefix if necessary. The second element is specified by appending the suffix -ide to the stem of the elements name. This is illustrated on page 55. Study Example 2.2.

For binary nonmetal hydrides, it is not necessary to use Greek prefixes to specify the number of hydrogens in the formula. Once we know the other element, we can figure out how many hydrogens go with it.

Common names

Some well-known compounds, such as H_2O (water) and NH_3 (ammonia) were given names before a systematic system was developed. In most instances, we do not attempt to name them according to the IUPAC rules and simply use their common names.

Self-Test

15. Name the following compounds.

(a) N_2O_5 _____

(b) H_2Se _____

(c) P_2S_3 _____

(d) $SbCl_5$ _____

(e) IF_7 _____

16. Write formulas for the following.

(a) dinitrogen trioxide _____

(b) disulfurdichloride _____

(c)   selenium tetrachloride   _____

(d)   nitrogen monoxide   _____

(e)   oxygen difluoride   _____

New Terms

Write the definitions of the following terms, which were introduced in this section. If necessary, refer to the Glossary at the end of the text.

nomenclature   inorganic compound   binary compound

2.6   Ionic compounds are composed of charged particles called ions

Review

Metals combine with nonmetals to form ionic compounds by the transfer of electrons from one atom to another. This produces electrically charged particles that we call ions. Study the diagram that illustrates the transfer of an electron from a sodium atom to a chlorine atom.

In an ionic compound we cannot identify individual groups of ions as belonging to each other. We simply specify in the formula the smallest whole-number ratio of the ions. The group of ions specified in such a formula is called a formula unit.

Self-Test

17.   In terms of their composition, what is the major difference between a molecular compound and an ionic compound?

_____

_____

18.   (a)   What name is used for the tiny individual particles that

occur in molecular compounds?   _____

(b)   What is the name for the individual particles that make up

an ionic compound?   _____

19.   What kind of compound (ionic or molecular) would be expected if the following pairs of atoms combine?

(a)   magnesium and bromine   _____

(b)   sulfur and nitrogen   _____

New Terms

ion   ionic compound   formula unit

2.7 The formulas of many ionic compounds can be predicted

Review

Cations (positive ions) are formed by metals when their atoms lose electrons. Anions (negative ions) are formed by nonmetals when their atoms acquire electrons.

Table 2.3 on page 59 illustrates how the periodic table can help you remember the ions formed by the representative metals and nonmetals. The number of positive charges on a cation equals the group number; the number of negative charges on an anion equals the number of spaces to the right we have to go to get to a noble gas in the periodic table. Be sure you understand how the formulas of the ions in Table 2.3 correlate with the locations of the elements in the periodic table.

Ionic compounds always contain positive ions and negative ions in ratios that give electrically neutral substances. In an ionic compound, the ions are arranged around one another in a way that maximizes the attractions between oppositely charged ions and minimizes the repulsions between like-charged ions. Since we can't identify unique molecules in an ionic compound, we always write their formulas with the smallest set of whole-number subscripts. This identifies one formula unit of the substance.

Writing the formulas for ionic compounds correctly is a skill you must master, which is why the Rules for Writing Formulas of Ionic Compounds on page 60 are identified by the icon in the margin. The rules are simple to apply, so be sure to study Example 2.3 on page 60 and work Practice Exercise 10.

The transition and post-transition metals generally are able to form two or more different ions, and there are no simple rules that enable us to figure them out. Therefore, you must simply memorize the symbols (including charges) for these ions which are given in Table 2.4.

Table 2.5 contains the formulas and names of frequently encountered polyatomic ions—ions composed of more than one atom. These should also be memorized. Practice writing both their names and their formulas. Be sure to learn their charges, too. (It might help to prepare a set of flash cards.) When you feel you have learned the contents of Tables 2.4 and 2.5 and can use the periodic table to write the formulas of the ions in Table 2.3, try the Self-Test below.

Self-Test

20. Without looking at Table 2.3, but using the periodic table on the inside front cover of your textbook, write the formula for the ion formed by each of the following elements:

    (a) K   _____   (e) Ba   _____

    (b) Al   _____   (f) Na   _____

    (c) N   _____   (g) Br   _____

    (d) Mg   _____   (h) Se   _____

21. What is the general formula for an ion formed by a metal from

    (a) Group IA _____   (b) Group IIA _____

22. What is the general formula for an ion formed by a nonmetal from

    (a) Group VA _____   (b) Group VIIA _____

23. Write the formula for the ionic compound formed by

   (a)   Na and S Na_2S

   (b)   Sr and F SrF_2

   (c)   Al and Se Al_2Se_3

   (d)   Mg and Cl $MgCl_2$

24. Write the formula for the ionic compound formed from

   (a)   sodium ion and perchlorate ion   _____

   (b)   barium ion and hydroxide ion   _____

   (c)   ammonium ion and dichromate ion   _____

   (d)   magnesium ion and sulfite ion   _____

   (e)   nickel ion and phosphate ion   _____

   (f)   silver ion and sulfate ion   _____

25. Write the formulas for *two* different ionic compounds of

   (a)   iron and oxygen   _____   _____

   (b)   copper and chlorine   _____   _____

   (c)   mercury and chlorine   _____   _____

   (d)   tin and sulfur   _____   _____

   (e)   lead and oxygen   _____   _____

New Terms

Write the definitions of the following terms, which were introduced in this section. If necessary, refer to the Glossary at the end of the text.

   cation   anion   post-transition metal   polyatomic ion

2.8   Naming ionic compounds also follows a system

Review

Once again, our goal is to have a system that allows us to derive a unique name for each compound and that enables us to derive the formula given the name. In writing the name of an ionic compound, the cation is always named first followed by the anion. The rules are designed to permit us to derive the names of cations and anions. Once we have them, we write them in the order just mentioned.

Binary compounds containing a metal and a nonmetal

If the cation is of metal that forms only one positive ion, it simply is given the English name of the metal. For example, the cation Mg^{2+} is specified as *magnesium*. When the metal forms two or more cations, the preferred method for naming the cation is called the Stock system. In this case, the number of positive charges on the cation is specified by placing a Roman numeral in parentheses following the English name for the metal. Thus, Fe^{2+} is the called the iron(II) ion and Fe^{3+} is the iron(III) ion. Notice that there is no space between the name of the metal and the parentheses containing the Roman numeral.

Monatomic (one-atom) anions of nonmetals are named by adding the suffix *-ide* to the stem of the nonmetal name. Note that hydroxide, OH^-, and cyanide, CN^- are the only two polyatomic ions ending in *-ide*. If the name of the anion ends in *-ide* and it isn't hydroxide or cyanide, then you can be sure it's a monatomic anion.

Ionic compounds that contain polyatomic ions

Here we simply use the name of the polyatomic cation or anion instead of the name of a monatomic ion. It's essential that you learn the names, formulas, and charges of the polyatomic ions in Table 2.5. This is especially so when you have to write the formula given the name of a compound.

The old system for naming cations

The old system for naming cations of metals that form more than one ion is described in Facets of Chemistry 2.1. It uses the ending *-ous* for the cation with the lower charge and the ending *-ic* for the one with the higher charge. Except for mercury, if the symbol for the metal is derived from the element's Latin name, then the Latin stem is used in specifying the metal (ferrous ion and ferric ion, for example).

Self-Test

26. Name these compounds.

    (a) CaI_2 _____

    (b) $FeBr_3$ _____

    (c) $Fe(NO_3)_2$ _____

    (d) KNO_2 _____

    (e) $Cu(BrO_3)_2$ _____

    (f) CaH_2PO_4 _____

    (g) $KHSO_3$ _____

    (h) $Ca(C_2H_3O_2)_2$ _____

27. Write the formulas for these compounds.

    (a) barium arsenide   _____

    (b) sodium perbromate   _____

    (c) manganese(IV) oxide   _____

    (d) sodium permanganate   _____

(e)   aluminum nitrate   _____

(f)   copper(II) iodate   _____

(g)   cobalt(II) acetate   _____

(h)   lead(II) chloride   _____

28   The following compounds are named using the older system of nomenclature. Write their formulas.

(a)   cobaltous nitrate   _____

(b)   ferric chloride   _____

2.9   Molecular and ionic compounds have characteristic properties

Review

Properties of molecular compounds

Many molecular compounds are relatively soft and easily deformed when struck, and they tend to have relatively low melting points. When melted, they do not conduct electricity because the liquid does not contain charge particles.

Properties of ionic compounds

The physical properties of ionic compounds are the result of the strong attractive forces that exist between ions of opposite charge as well as the strong repulsive forces between ions of the same charge. Ionic solids tend to have high melting points (i.e., they melt at high temperatures), they tend to be brittle and easily shattered when struck. In the solid state, ionic compounds do not conduct electricity, but they do conduct when melted. This is because melting frees the ions and allows them to move—a requirement for electrical conduction.

Self-Test

29.   The compound $SnCl_2$ is ionic, but $SnCl_4$ is molecular.

(a)   Which compound has the higher melting point?   _____

(b)   Which compound is more brittle?   _____

(c)   Which compound forms the softer solid?   _____

(d)   Which compound conducts electricity when melted? _____

(e)   Does either compound conduct electricity as a solid?_____

New Terms

None

Answers to Self-Test Questions

1. You should have observed that the milk of magnesia dissolves in the vinegar, but not in plain water.
2. The fizzing is evidence that a reaction is taking place.
3. (a) 2Al, 6Cl (b) 1Ca, 1Si, 3O (c) 2K, 2Cr, 7O (d) 12C, 22H, 11O (e) 2Na, 1C, 13O, 20H
    (f) 2N, 8H, 1S, 4O   (g) 2Al, 3O (h) 1Ca, 1S, 6O, 4H (i) 3Ca, 2P, 8O
4. H_2, N_2, O_2, F_2, Cl_2, Br_2, I_2
5. (a) 3K, 3Mn, 12O   (b) 10Al, 15S, 60O
6. (a) On the left: 2Al, 3O, 6H, 6Cl; on the right: 2Al, 3O, 6H, 6Cl (Balanced) (b) On the left: 6N, 24H, 2P, 11O, 3Ca; On the right: 3N, 15H, 1P, 7O, 3Ca (Not balanced)
7. $2Al_2O_3(s) + 6HCl(aq) \rightarrow 2AlCl_3(aq) + 3H_2O(l)$
8. The KE of the 1000-lb car is one-fourth of the KE of the 4000-lb car.
9. At 60 mph the KE is 9 times larger than at 20 mph.
10. The heavier truck has eight times the KE of the lighter truck.
11. They are the same kind of energy.
12. (a) KE decreases, (b) PE increases
13. (a) H_2S, (b) HBr, (c) PH_3 (or H_3P), (d) CH_4 (or H_4C)
14. (a) C_5H_{12}, (b) C_7H_{16}
15. (a) dinitrogen pentaoxide, (b) hydrogen selenide, (c) diphosphorus trisulfide,
    (d) antimony pentachloride, (e) iodine heptafluoride
16. (a) N_2O_3, (b) S_2Cl_2, (c) $SeCl_4$, (d) NO, (e) OF_2
17. A molecular compound is made up of neutral particles (molecules); an ionic compound is composed of charged particles (ions).
18. (a) molecules,   (b) ions
19. (a) ionic,   (b) molecular
20. (a) K^+, (b) Al^{3+}, (c) N^{3-}, (d) Mg^{2+}, (e) Ba^{2+}, (f) Na^+, (g) Br^-, (h) Se^{2-}
21. (a) M^+, (b) M^{2+}
22. (a) X^{3-}, (b) X^-
23. (a) Na_2S, (b) SrF_2, (c) Al_2Se_3
24. (a) $NaClO_4$, (b) $Ba(OH)_2$, (c) $(NH_4)_2Cr_2O_7$,
    (d) $MgSO_3$, (e) $Ni_3(PO_4)_2$, (f) Ag_2SO_4
25. (a) FeO and Fe_2O_3, (b) CuCl and $CuCl_2$,
    (c) Hg_2Cl_2 and $HgCl_2$, (d) SnS and SnS_2, (e) PbO and PbO_2
26. (a) calcium iodide, (b) iron(III) bromide,   (c) iron(II) nitrate,   (d) potassium nitrite,
    (e) copper(II) bromate,   (f) calcium dihydrogen phosphate, (g) potassium hydrogen sulfite,
    (h) calcium acetate
27. (a) Ba_3As_2, (b) $NaBrO_4$, (c) MnO_2, (d) $NaMnO_4$, (e) $Al(NO_3)_3$, (f) $Cu(IO_3)_2$,
    (g) $Co(C_2H_3O_2)_2$, (h) $PbCl_2$
28. (a) $Co(NO_3)_2$, (b) $FeCl_3$
29. (a) $SnCl_2$, (b) $SnCl_2$, (c) $SnCl_4$, (d) $SnCl_2$, (e) No.

Tools you have learned

Consider removing this chart from the Study Guide so you can have it handy when tackling homework problems.

Tool	*How it Works*
Chemical formula	Subscripts in a formula specify the number of atoms of each element in one formula unit of the substance. This gives us *atom ratios* that we will find useful when we deal with the compositions of compounds in Chapter 4.
Rules for naming molecular compounds	They allow us to write the name of a molecular compound given its chemical formula, and to write the formula given the name.
Rules for writing formulas of ionic compounds	The rules permit us to write correct chemical formulas for ionic compounds. You will need to learn to use the periodic table (see below) to remember the charges on the cations and anions of the representative metals and nonmetals. You also should learn the ions formed by the transition and post-transition metals in Table 2.4, and it is essential that you learn the names and formulas (including charges) of the polyatomic ions in Table 2.5.
Rules for naming ionic compounds	These rules allow us to write the name of an ionic compound given its chemical formula, and to write the formula given the name.
Periodic Table	From a nonmetal's position in the periodic table we can write the formula of its simple hydride. For the nonmetals and the metals in Groups IA and IIA, we can use the elements' positions in the periodic table to obtain the charges on their ions.

Chapter **3**

Measurement

An essential part of the scientific method is the acquisition of data. This is done by making observations, and for them to be truly useful, they usually must be numerical. We call numerical observations *measurements*. In this chapter you will learn what distinguishes a numerical measurement from a simple mathematical number. You will also learn the system of units we use in science to express measured quantities, and you will learn a method that will aid you in performing calculations that involve the conversion of units.

Learning Objectives

As you study this chapter, keep in mind the following objectives:

1 To learn the difference between a number and a measurement.

2 To learn the difference between qualitative and quantitative measurements.

3 To learn the base SI units.

4 To learn how derived units are obtained from base units.

5 To learn how the sizes of units are modified by decimal multipliers.

6 To learn the commonly-used laboratory units for length, volume, mass, and temperature.

7 To learn how to convert among Celsius, Fahrenheit, and Kelvin temperatures.

8 To learn the meaning of significant figures and how to count the number of significant figures in a reported measurement.

9 To learn the sources of error in measurements.

10 To understand the difference between *accuracy* and *precision*.

11 To learn how to determine the number of significant figures in the results of calculations involving multiplication, division, addition, and subtraction.

12 To learn how to use a relationship between units to construct a conversion factor used in unit conversions.

13 To learn how to apply the factor-label method in unit conversion problems.

14 To learn the definition of density and to use density in calculations.

15 To learn the definition of specific gravity and to use specific gravity in calculations.

3.1 Measurements are quantitative observations

Review

This section introduces you to the concept of measurement and differentiates between numbers in a mathematical sense and numbers that arise from measurement. In mathematics, a number such as 2.6 is considered to be exact, meaning it contains no error. Mathematics makes no distinction between 2.6 and 2.600000. In science, where numbers arise from measurement, these two representations are not equivalent because measurements are inexact; they are subject to *error* (also called *uncertainty*). As you learn later in this chapter, the quantities 2.6 and 2.600000 imply different amounts of uncertainty.

We do find exact relationships in science, but they generally arise from definitions (e.g., the number of seconds in a minute) or from a direct count (e.g., the number of meatballs in your bowl of spaghetti).

Measurements also include units that describe the nature of the measured quantity and its size. Units are an absolutely essential part of any measurement, so whenever you report a measured value, you **must** include its units as well.

New Terms

Write the definitions of the following terms, which were introduced in this section. If necessary, refer to the Glossary at the end of the text.

measurement	qualitative	quantitative
unit	exact	inexact

3.2   Measurements always include units

Review

A measurement always has two parts, a number and a unit. When we write the value of a measurement, we always leave a space between the number and the unit.

3 meters (correct)   3meters (incorrect)

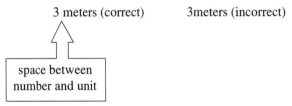

The units used in science, such as the meter above, are based on the precisely defined International System of Units, or SI. Learn the SI base units and their symbols for length, mass, time, and temperature (Table 3.1). The mole, another essential unit, will be defined in Chapter 4.

Quantities other than those given in Table 3.1 are obtained from the base quantities by mathematical operations, and their units (called derived units) are obtained from the base units by the same operations. Study Example 3.1 to see how derived units are made by combining base units. Notice in particular that units undergo the same kinds of mathematical operations that numbers do. For example, volume is a product of three length units

$$\text{length} \times \text{width} \times \text{height} = \text{volume}$$

The unit for volume is the product of the units for length, width, and height.

$$\text{meter} \times \text{meter} \times \text{meter} = \text{meter}^3$$

$$\text{m} \times \text{m} \times \text{m} = \text{m}^3$$

Similarly, speed is expressed as a ratio of distance divided by time. The SI base unit for distance is the meter and the base unit for time is the second. Therefore

$$\text{speed} = \frac{\text{distance}}{\text{time}} = \frac{\text{meter}}{\text{second}} = \frac{\text{m}}{\text{s}}$$

Often, the base units (or the derived units that come from them) are too large or too small to be used conveniently. For example, if we were to use cubic meters to express the volumes of liquids that we measure in the laboratory, we would find ourselves using very small numbers such as 0.000025 m^3 or 0.000050 m^3. Because they have so many zeros, these values are difficult to comprehend. To make life easier for us, the SI has a simple way of making larger or smaller units out of the basic ones. This is done with the decimal multipliers and SI prefixes given in Table 3.3 on page 84. Be sure you learn the ones in the table below. (These are the ones in bold colored type in Table 3.3.)

SI Prefixes and Decimal Multipliers

Prefix	Symbol	Multiplication Factor
mega	M	10^6
kilo	k	10^3
deci	d	10^{-1}
centi	c	10^{-2}
milli	m	10^{-3}
micro	μ	10^{-6}
nano	n	10^{-9}
pico	p	10^{-12}

The SI prefixes and decimal multipliers are tools we use to scale units to convenient sizes and to translate between differently sized units. You will see how this is done in Section 3.5. Notice that each prefix stands for a particular decimal multiplier. Thus *kilo* means "× 1000" or "× 10^3." This lets us translate a quantity into the value that it has when expressed in terms of the base units. For example, suppose we wanted to know how many meters are in 25 kilometers (25 km). Since kilo (k) means "× 1000," we substitute "× 1000" for "k."

$$25 \text{ km} = 25 \times 1000 \text{ m}$$

$$\boxed{\text{k means} \times 1000}$$

$$25 \text{ km} = 25,000 \text{ m}$$

Similarly, a length of 25 millimeters (25 mm) would be

$$25 \text{ mm} = 25 \times 0.001 \text{ m}$$

$$= 0.025 \text{ m}$$

Common laboratory units of measurement

In the lab, the most common measurements are those of length, volume, mass, and temperature.

Length is usually measured in units of centimeters (cm) or millimeters (mm). Remember the following:

$$1 \text{ m} = 100 \text{ cm} = 1000 \text{ mm}$$

$$1 \text{ cm} = 10 \text{ mm}$$

Volume is conveniently measured in cubic centimeters (cm^3). Frequently we use the nonSI unit called the liter (L), which is slightly larger than a quart. Remember that 1 L = 1000 cm^3. Often, glassware is graduated in milliliters (mL): 1000 mL = 1 L. It is also important to remember that 1cm^3 = 1 mL.

Although the SI base unit of mass is the kilogram, in the lab we usually report mass measurements in units of gram (g). Common laboratory balances are graduated in grams.

Temperature is measured with a thermometer, and in the sciences it is measured in units of degrees Celsius (°C). The Celsius and Fahrenheit degree units are of different sizes; five degree units on the Celsius scale correspond to nine degree units on the Fahrenheit scale. Equation 3.1 on page 88 enables you to make conversions between °C and °F.

The SI unit of temperature is the kelvin (K). Zero on the Kelvin scale corresponds to −273 °C (rounded to the nearest degree) and is called absolute zero, because it is the coldest temperature. (Notice that the name of the SI *temperature scale* is capitalized; the name of the *unit* measured on that scale, the kelvin, is not capitalized.) The kelvin and Celsius degree are the same size, so a *temperature change* of 10 K, for example, is the same as a temperature change of 10 °C. Be sure you can convert from °C to K (Equation 3.3 usually, but Equation 3.2 when more precision is required), because when the temperature is needed in a calculation, it nearly always must be expressed in kelvins. In mathematical equations, the capital letter T is used to stand for the Kelvin temperature.

Self-Test

1. Give the SI base unit and its abbreviation for

    (a)   mass   _____

    (b)   length   _____

    (c)   time   _____

    (d)   amount of substance   _____

    (e)   temperature   _____

2. Torque (pronounced "tork") is a quantity that describes a twisting force, such as that applied to a nut or a bolt by a wrench. It is a product of distance × force and in English units is normally given in foot pounds. The SI derived unit for force is the Newton (symbol, N). What is the SI derived unit for torque?

    Answer _____

3. Fill in the blanks with the correct prefixes.

    (a)   1 _____ gram = 0.01 gram   (e)   1 pm = _____ m

    (b)   1 _____ meter = 10^{-9} meter   (f)   1 μg = _____ g

    (c)   1 _____ g = 0.001 g   (g)   1 dm = _____ m

    (d)   1 _____ m = 1000 m

4. Fill in the blanks with the correct numbers.

    (a)   63 dm = _____ m   (c)   2450 nm = _____ m

    (b)   0.023 Mg = _____ g   (d)   2487 cm = _____ m

5. Fill in the blanks with the correct numbers.

    (a) ___135___ cm = 1.35 m   (g) _____ mL = 0.022 L

    (b) ___2240___ mm = 22.4 cm   (h) _____ L = 346 mL

    (c) _____ cm = 32.6 mm   (i) _____ L = 2.41 mL

    (d) _____ mL = 1.250 L   (j) _____ K = 25 °C

    (e) _____ cm^3 = 246 mL   (k) _____ °C = 265 K

    (f) _____ L = 525 cm^3 (l) _____ °C = 300 K

6. (a) What Celsius temperature corresponds to 23 °F?   _____

    (b) What Fahrenheit temperature equals 10 °C?   _____

New Terms

Write the definitions of the following terms, which were introduced in this section. If necessary, refer to the Glossary at the end of the text.

cubic meter (m^3)   milliliter (mL)

International System of Units (SI)   base unit

derived unit   decimal multiplier

kilogram (kg)   gram (g)

meter (m)   centimeter (cm)

millimeter (mm)   liter (L)

Fahrenheit Scale   Celsius scale

Kelvin temperature scale   kelvin (K)

absolute zero

3.3   Measurements always contain some uncertainty

Review

Measurements are inexact and contain errors that arise from various sources. When we write the result of a measurement, we use the concept of significant figures to provide an estimate of how certain we are of the measured value. For a given measurement, its significant figures include all the digits known for sure *plus* the first digit that contains some uncertainty.

Expressing a measurement to the correct number of significant figures allows us to convey to someone else who sees the measured value how precise the measurement is. For example, a measured length of 32.47 cm has four significant figures. It tells us that the 3, 2, and 4 are known with certainty and the measuring instrument allowed the hundredths place to be *estimated* to be a 7. Since no digit is reported in the thousandths place, it is assumed that no estimate of that place could be obtained. In general, the larger the number of significant figures in a measurement, the greater is its precision.

Accuracy refers to how closely a measured quantity is to the actual correct value. *Precision* refers to how closely repeated measurements of the same quantity are to each other.

Self-Test

7.   Which term, accuracy or precision, is related to the number of significant figures in a measured quantity?

_____

New Terms

Write the definitions of the following terms, which were introduced in this section. If necessary, refer to the Glossary at the end of the text.

error in measurement	significant figures
significant digits	average (or mean)
accuracy	precision

3.4   Measurements are written using the significant figure convention

Review

One of the skills you need to learn is counting the number of significant figures in a measured quantity. Usually, this is a simple matter. However, zeros sometimes cause a problem. Note the statements on page 94, that describe when zeros are counted as significant figures and when they are not. Notice that scientific notation eliminates confusion if zeros appear at the end of a number without a decimal point.

The rules given on page 95 are tools we use to correctly express the number of significant figures in a computed quantity.

Multiplication and Division

The answer cannot contain more significant figures than the factor that has the fewest number of significant figures. Study the example on page 95.

Addition and Subtraction

The answer is rounded to the same number of decimal places as the quantity with the fewest decimal places. See the example, also on page 96.

When using exact numbers in calculations, they can be considered to have as many significant figures as desired. They contain no uncertainty.

Example 3.1   Calculations Using Significant Figures

Assume that all of the numbers in the following expression come from measurement. Compute the answer to the correct number of significant figures.

$$(3.25 \times 10.46) + 2.44 = ?$$

Analysis:
We have to apply the rules for both multiplication and addition. To perform the calculation, the multiplication step is done first, so we will have to use the rule for significant figures in multiplication to determine the number of significant figures in the product (3.25 × 10.46). Then we perform the addition, applying the rule for significant figures in a sum.

Solution:
When we perform the multiplication, we obtain 33.995. This should be rounded to three significant figures because that's how many there are in 3.25. The result is 34.0, which is then added to 2.44

34.0	one decimal place
+ 2.44	two decimal places
36.4	rounded to one decimal place

The correct answer, therefore, is 36.4

Is the Answer Reasonable?
There's no simple check. Keep in mind, however, that the order in which the rules are applied is the same as the order in which the different parts of the calculation are performed.

Self-Test

8. Perform the following arithmetic and express the answers to the proper number of significant figures (assume all numbers come from measurements).

(a) 4.87×3.1 _____

(b) $8.4 \div 21.02$ _____

(c) $14.35 + 0.022$ _____

(d) $145.3 - 4.68$ _____

New Terms

Write the definition of the following term, which was introduced in this section. If necessary, refer to the Glossary at the end of the text.

exact number

3.5 Units are converted using the factor-label method

Review

The factor label method is a procedure we use to set up the arithmetic correctly when solving a problem. It is based on the ability of units to cancel from numerator and denominator of a fraction, as described earlier. To apply this method, we view a problem as a conversion from some initial set of units to a final desired set of units. For example, if we need to know how many inches are in 1.5 ft, we begin by restating the problem as

$$1.5 \text{ ft} = ? \text{ in.}$$

To change the units from ft to in., we multiply the given quantity, 1.5 ft, by a conversion factor that will change the units from ft to in.

$$1.5 \text{ ft} \times \left(\frac{\text{conversion}}{\text{factor}} \right) = \text{answer in inches}$$

A **conversion factor** is a fraction that we form from a relationship between units. In this case, the relationship is between feet and inches, which can be expressed by the equation

$$1 \text{ ft} = 12 \text{ in.}$$

We can make a conversion factor by dividing both sides of the equation by 1 ft

$$\frac{1 \text{ ft}}{1 \text{ ft}} = \frac{12 \text{ in.}}{1 \text{ ft}}$$

Notice that the quantities in the numerator and denominator on the left cancel, so we can write

$$1 = \frac{1 \text{ ft}}{1 \text{ ft}} = \frac{12 \text{ in.}}{1 \text{ ft}}$$

Because the fraction 12 in./1 ft is numerically equivalent to 1, we can multiply something by it and not change its magnitude. (Multiplying something by 1 doesn't change its size.) Using this as our conversion factor, then, gives

$$1.5 \text{ ft} \times \frac{12 \text{ in.}}{1 \text{ ft}} = \text{answer in inches}$$

Notice that the units ft cancel[1], leaving us with units of inches, which are the units we want for the answer. Performing the arithmetic gives 18 in., which is the correct answer.

$$1.5 \text{ ft} \times \frac{12 \text{ in.}}{1 \text{ ft}} = 18 \text{ in.}$$

The relationship between feet and inches that we employed above can actually be used to form *two* different conversion factors. One is formed by dividing both sides by 1 ft, as we've done. The other is formed by dividing both sides by 12 in.

$$\frac{1 \text{ ft}}{12 \text{ in.}} = \frac{12 \text{ in.}}{12 \text{ in.}} = 1$$

Notice that if we were to use this second conversion factor in performing the conversion, the units "ft" do not cancel. Instead, the units work out to be ft²/in., which make no sense. The answer, of course, is also incorrect.

$$1.5 \text{ ft} \times \frac{1 \text{ ft}}{12 \text{ in.}} = 0.12 \text{ ft}^2/\text{in.}$$

This illustrates one of the greatest benefits in using the factor label method — if the problem is set up incorrectly, the units will not cancel properly to give the desired units of the answer. If the units are wrong, we can

---

[1] A quantity such as 1.5 ft can be viewed as a fraction in which the denominator is 1.

$$1.5 \text{ ft} = \frac{1.5 \text{ ft}}{1}$$

This places the unit ft in the numerator where it can be canceled by the unit ft in the denominator of the conversion factor.

also be sure that the answer is wrong. Thus, the units have informed us that we've made an error and that we must rework the setup of the problem.

Forming and applying conversion factors

In general, when we can express a relationship between units in the form of an equation, two conversion factors can be derived from it. For example, if we wished to convert between minutes and seconds, we would use the following.

60 s = 1 min

Two fractions
(conversion factors)
can be formed.

$$\frac{60 \text{ s}}{1 \text{ min}} \qquad \frac{1 \text{ min}}{60 \text{ s}}$$

The connection between feet and inches used earlier, and between minutes and seconds are both derived from definitions. Often we will use relationships that come to use in this way. We can also form conversion factors from relationships established by measurements. For instance, in an earlier example we described a situation in which we've determined that a car is able to travel 274 miles using 12.6 gallons of fuel. In making these measurements, we've established a relationship between distance traveled and fuel consumed that we can express in an equation format.

274 mile ⟺ 12.6 gallon

Notice, however, that we've used the symbol ⟺ rather than an equals sign. This is because miles can't actually "equal" gallons; they're two different kinds of units. The symbol ⟺ is read as "is equivalent to." In other words, for this car, "traveling 274 miles *is equivalent to* using 12.6 gallons of fuel." Although the symbol is different, it behaves like an equal sign when we wish to form conversion factors. Thus,

274 mile ⟺ 12.6 gallon

Two fractions
(conversion factors)
can be formed.

$$\frac{274 \text{ mile}}{12.6 \text{ gallon}} \qquad \frac{12.6 \text{ gallon}}{274 \text{ mile}}$$

Once you've learned how to form conversion factors, the next step is learning how to select the correct one for use in a calculation. Here we let the units be our guide. For instance, suppose we wished to know how many gallons of fuel we would need to travel 850 miles in the car described above. Let's begin by stating the problem in the form of an equation (or an equivalence, to be more exact)[2].

850 mile ⟺ ? gallon

To find the answer by the factor label method, we will multiply the 850 mile by a conversion factor to convert to gallons.

---

[2] Don't be overly concerned about whether you use an equals sign of the symbol ⟺ . The setup of the problem and finding the answer follows the same route.

$$850 \text{ mile} \times \left(\begin{array}{c} \text{conversion} \\ \text{factor} \end{array} \right) \Leftrightarrow ? \text{ gallon}$$

The conversion factor we select has to eliminate the unit mile by cancellation. Only one of the two conversion factors does this, so that's the one we select.

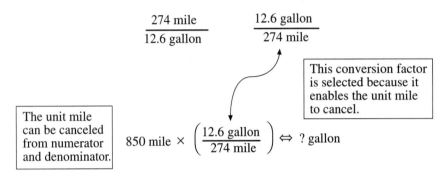

Once we've chosen the conversion factor, we cancel the units and perform the arithmetic.

$$850 \text{ mile} \times \left(\frac{12.6 \text{ gallon}}{274 \text{ mile}} \right) \Leftrightarrow 39.1 \text{ gallon}$$

Notice that if we had selected the other conversion factor, the units would not work out correctly.

The unit mile does
not cancel.

$$850 \text{ mile} \times \frac{274 \text{ mile}}{12.6 \text{ gallon}} \Leftrightarrow ? \frac{\text{mile}^2}{\text{gallon}}$$

As you see in this example, the cancellation of units serves as a guide in selecting conversion factors as we set up the solution to the problem. In fact, the need to cancel particular units in arriving at the answer to a problem often serves as a clue to the kinds of information and relationships we have to find. For example, when we have arrived at the stage

$$850 \text{ mile} \times \left(\begin{array}{c} \text{conversion} \\ \text{factor} \end{array} \right) \Leftrightarrow ? \text{ gallon}$$

we would recognize that to obtain an answer we must have a relationship between gallons of fuel used and miles traveled. If this information were not already available, we would have to obtain it by some means before we could solve the problem.

The importance of correct relationships between units

In setting up problems using the factor label method, it's essential that we use proper and correct relationships between the units. In converting between feet and inches, we need to use the relationship

$$1 \text{ ft} = 12 \text{ in.}$$

If we didn't know this relationship, we couldn't just make one up, such as

$$5 \text{ ft} = 24 \text{ in.}$$

Although we could use it to get the units to cancel, the answer is bound to be wrong because the relationship we've used in incorrect. Establishing correct relationships between units is one of the difficulties students often have in applying this approach. Before you form and apply a conversion factor, ask yourself "is this relationship between the units correct?"

The factor label method and the "Toolbox" approach

In our earlier discussion of how the toolbox concept can be applied to problem solving, we emphasized the importance of analyzing a problem so that critical links and the tools that are needed to obtain a solution can be identified. If the problem involves calculations, the tools are usually relationships between units which can be stated in equation form. Once these statements have been assembled, the factor label method provides guidance in assembling the arithmetic. By assuring the units cancel correctly, the factor label method gives us confidence that we are doing the correct math. In a sense, then, the factor label method is itself a tool that we use to assemble the arithmetic correctly. With this in mind now, let's take a look at an example that illustrates how we apply the factor-label method.

Example 3.2 Calculations Using The Factor-Label Method

During a cross country trip on Interstate I-80, a driver became concerned that her speedometer was inaccurate. To avoid a speeding ticket she decided to check the speedometer. She adjusted the speed of her car until the speedometer read 65 mph and set the car on speed control so the speed would remain constant. She then used the stopwatch feature of her wristwatch to time how long it took to travel exactly five miles, using the mile markers along the side of the road to measure her progress. She found that it took 4 minutes and 24 seconds to travel the five miles. She then pulled into a rest stop to calculate her actual speed over the five-mile stretch. What was her actual speed in miles per hour (mph)?

Analysis:

The first, and most important step in solving a problem is the **analysis**. Here we study the data given, search for any critical links that might exist, and decide how we will proceed with the solution. We look for unit relationships (our tools), which we will use to solve the problem. After doing all this, we come to the easy part — setting up the **solution** and doing necessary calculations. We encourage you to think of problem solving as involving these two distinct stages: analysis first, followed by solution.

    This problem involves a lot of words, so let's begin by extracting the critical information.

> It took **4 minutes and 24 seconds**
> to travel **5.00 miles**
> How fast was she going, in **miles per hour**?

As we've noted, the word "per" can be interpreted to mean "divided by," so our answer must have the units: mile "divided by" hour

$$\text{units of answer} = \frac{\text{mile}}{\text{hour}}$$

Understanding that we need to assemble the data in a way to give us this desired ratio of units is really our critical link in solving this problem. We see that we need to have distance in the numerator and time in the denominator. Let's use the data to set this up.

$$\text{speed} = \frac{5.00 \text{ mile}}{4 \text{ minutes and 24 seconds}}$$

Now we can see that we're not too far from having the answer. We need to change the denominator into units of hour. Let's change 24 seconds to minutes, add the result to the 4 minutes, and then change the total minutes to hours. To do this, we need the relationships between seconds and minutes and between minutes and hours. (These are our tools.) We'll state them in equation form:

$$1 \text{ minute} = 60 \text{ second}$$

$$1 \text{ hour} = 60 \text{ minute}$$

Now that we know how we're going to proceed, we can work out the solution.

Solution:

First, we convert 24 seconds to minutes. Although this is really a simple exercise, let's illustrate how we perform the conversion using the factor label method. We use the relationship 1 minute = 60 second to form a conversion factor that will cancel second and give the desired unit, minute.

$$24 \text{ second} \times \frac{1 \text{ minute}}{60 \text{ second}} = 0.40 \text{ minute}$$

The total time required to go five miles was therefore 4.40 minute. Now we convert this to hour using 1 hour = 60 minute. Once again, we form a conversion factor so units cancel correctly.

$$4.40 \text{ minute} \times \frac{1 \text{ hour}}{60 \text{ minute}} = 0.0733 \text{ hour}$$

Now we're able to calculate her speed by dividing the distance, 5.00 mile, by the time, 0.0733 hour.

$$\text{speed} = \frac{5.00 \text{ mile}}{0.0733 \text{ hour}} = 68.2 \frac{\text{mile}}{\text{hour}} = 68.2 \text{ mph}$$

The answer, therefore, is that the driver's speed was 68.2 mph.

Is the Answer Reasonable?

Aside from checking the arithmetic, can we feel confident that we've solved the problem correctly? We can if the answer seems reasonable, and here it does. We don't expect that the speedometer will be grossly in error and therefore we expect that the answer should be fairly close to 65 mph, which it is.

Stringing conversion factors together

The factor label method is especially useful in multi-step calculations where two or more conversion factors are needed to go from the starting units to those of the answer. For example, suppose you wanted to know how may minutes there are in 2.00 weeks.

$$2.00 \text{ week} = ? \text{ minutes}$$

We could do this in a one step calculation if we knew how many minutes are in one week, but most people don't keep such numbers in their heads. However, we do know how many days there are in a week, and we know how many hours there are in a day and how many minutes in an hour.

$$1 \text{ week} = 7 \text{ day}$$

$$1 \text{ day} = 24 \text{ hour}$$

$$1 \text{ hour} = 60 \text{ minute}$$

We can apply these relationships one after another, being sure the units cancel, until we have the final desired unit, minute

$$\boxed{1 \text{ week } = 7 \text{ day}} \qquad \boxed{1 \text{ day } = 24 \text{ hour}} \qquad \boxed{1 \text{ hour } = 60 \text{ minute}}$$

$$2.00 \text{ week } \times \frac{7 \text{ day}}{1 \text{ week}} \times \frac{24 \text{ hour}}{1 \text{ day}} \times \frac{60 \text{ minute}}{1 \text{ hour}} = 2.02 \times 10^4 \text{ minute}$$

When we do this, we say that we are "stringing together " the conversion factors.

Following the path through the units

In a multi-step calculation, a way to be sure you have all the information necessary to set up the string of conversion factors is to follow the "path" through the units, starting with the units given and going through until you reach the units of the answer. For instance, in the list of relationships we used in the preceding example,

$$1 \text{ week } = 7 \text{ day}$$
$$1 \text{ day } = 24 \text{ hour}$$
$$1 \text{ hour } = 60 \text{ minute}$$

we can follow the path: week → day → hour → minute. Thus, the first equation takes us from *week* to *day*, the second from *day* to *hour*, and the third from *hour* to *minute*. If one of these relationships were missing (for example, the middle one), the path would be interrupted.

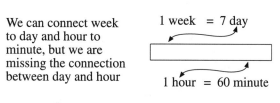

We can connect week to day and hour to minute, but we are missing the connection between day and hour

$$1 \text{ week } = 7 \text{ day}$$
$$1 \text{ hour } = 60 \text{ minute}$$

$$\text{week } \to \text{ day } \overset{?}{-} \text{ hour } \to \text{ minute}$$

We see that to solve the problem, we need an additional relationship that connects the units day and hour.

Thinking It Through

For the following, identify the information needed to solve the problem and show what must be done with it.

1   The distance between the Earth and the moon is approximately 239,000 miles. Radio waves travel through space at the speed of light, 3.00×10^8 m/s. How many seconds does it take for a radio wave to go to the moon?

Self-Test

9.   The distance between two houses was measured to be 0.27 miles. What is the distance expressed in yards (1 mile = 5280 ft)?

_____

10.   What conversion factor could you use to convert

(a)   62 lb to kg _____   (b)   45 in.^{2} to cm^2 _____

New Terms

Write the definitions of the following terms, which were introduced in this section. If necessary, refer to the Glossary at the end of the text.

factor-label method   dimensional analysis

conversion factor

3.6   Density is a useful intensive property

Review

Density is the ratio of an object's mass to its volume, as expressed in Equation 3.7 on page 100. Determining the density therefore involves two measurements and a brief calculation. In performing the calculation, be careful about significant figures and the significant-figure rules that we follow in multiplication and division.

Density is an intensive property, meaning that its value for a given substance is the same regardless of the size of the sample. Increasing the sample size increases both the mass *and* the volume, but their ratio remains the same. Density also serves as a tool for converting between the mass and volume for a sample of a substance. For example, methanol (a substance used in "dry gas" and as the fuel in "canned heat") has a density of 0.791 g/mL, which tells us that each milliliter of the liquid has a mass of 0.791 g. This information provides a relationship between mass and volume.

$$0.791 \text{ g methanol} \Leftrightarrow 1.00 \text{ mL methanol}$$

We can use this equivalence to construct two conversion factors.

$$\frac{0.791 \text{ g methanol}}{1.00 \text{ mL methanol}} \quad \text{and} \quad \frac{1.00 \text{ mL methanol}}{0.791 \text{ g methanol}}$$

Suppose we wanted the mass of 12.0 mL of methanol. We would use the first factor so the units mL will cancel.

$$12.0 \text{ mL methanol} \times \frac{0.791 \text{ g methanol}}{1.00 \text{ mL methanol}} = 9.49 \text{ g methanol}$$

If we wanted the volume occupied by, say 15.0 g of methanol, we would use the second factor.

$$15.0 \text{ g methanol} \times \frac{1.00 \text{ mL methanol}}{0.791 \text{ g methanol}} = 19.0 \text{ mL methanol}$$

The factor we choose simply depends on the units that we want to cancel.

Specific gravity

A quantity related to density is specific gravity—the ratio of the density of a substance to the density of water. If we have the specific gravity of a substance, we can get its density in any units we wish simply by multiplying the specific gravity of the substance by the density of water in the desired units.

$$(\text{sp. gr.})_{\text{substance}} \times d_{\text{H}_2\text{O}} = d_{\text{substance}}$$

Thinking It Through

For the following, identify the information needed to solve the problem and show what must be done with it.

2   A rectangular piece of steel 12.0 in. long and 18.2 in. wide has a mass of 12.5 lb. Steel has a specific gravity of 7.88. Water has a density of 62.4 lb/ft^3. What is the thickness of the piece of steel expressed in inches?

Self-Test

11. Why is density more useful for identifying a substance than either the mass or the volume alone?_

_____

_____

12. A sample of copper was found to have a mass of 34.2 g and a volume of 3.82 cm^3.

    (a)   What is the density of copper in g/cm^3? _____

    (b)   What is the specific gravity of copper? _____

    (c)   What is the density of copper in units of lb/ft^3? (Refer to Example 3.9 on page 103 of the text for the necessary data.)

                                                 _____

13. Aluminum has a density of 2.70 g/cm^3.

    (a)   What is the mass of 8.30 cm^3 of aluminum?   _____

    (b)   What is the volume in mL of 25.0 g of aluminum?   _____

New Terms

Write the definitions of the following terms, which were introduced in this section. If necessary, refer to the Glossary at the end of the text.

property	intensive property
physical property	chemical property
extensive property	density
specific gravity	

Solutions to Thinking It Through

1   We must relate the distance, 239,000 miles, to time in seconds. The speed of light can be used to convert between meters and seconds, so we need a relationship that will take us from miles to meters. There are several ways to obtain this, but the simplest is to use the relationship between miles and kilometers given in the table of conversions on the inside rear cover of the textbook.

$$1 \text{ mi} = 1.609 \text{ km} = 1.609 \times 10^3 \text{ m}$$

The setup of the problem according to the factor label method is

$$293,000 \text{ mi} \times \frac{1.609 \times 10^3 \text{ m}}{1 \text{ mi}} \times \frac{1 \text{ s}}{3.00 \times 10^8 \text{ m}} \Rightarrow \text{answer}$$

Notice that all the units will cancel correctly to give the desired units for the answer. (If you work through to the answer, you should get 1.57 s.)

2   For a rectangular object, volume = length × width × thickness. We are given the length and width, so to calculate thickness we need the volume. We've been given the specific gravity of the steel and the density of water. To see how we might use these quantities, we go back to the definitions.

$$d = \frac{m}{V} \quad \text{and} \quad \text{sp. gr.} = \frac{d_{\text{substance}}}{d_{\text{water}}}$$

From the density of the water and the specific gravity of water, we can find the density of the steel in units of lb/ft^3. The density of the steel equals its specific gravity multiplied by the density of water. The density so calculated is then used along with the mass of the steel to calculate the volume of the steel object, and then the thickness is obtained by dividing the volume by the product of the length × width, both expressed in ft. The thickness in units of ft is then changed to in. (The answer to the problem is 0.20 in.)

Answers to Self-Test Questions

1.   (a) kilogram, kg, (b) meter, m, (c) second, s, (d) ampere, A, (e) kelvin, K
2.   newton × meter or N m
3.   (a) centi (b) nano, (c) milli, (d) kilo, (e) 10^{-12}, (f) 10^{-6}, (g) 10^{-1},
4.   (a) 6.3, (b) 23,000, (c) 0.00000245, (d) 24.87
5.   (a) 135, (b) 224, (c) 3.26, (d) 1250, (e) 246, (f) 0.525, (g) 22, (h) 0.346,
       (i) 0.00241, (j) 298, (k) –8, (l) 27
6.   (a) –5 °C, (b) 50 °F
7.   Precision
8.   (a) 15, (b) 0.40, (c) 14.37, (d) 140.6
9.   475.2 yd rounds to 480 yd (2 sig. fig.)
10.   (a) $\dfrac{1 \text{ kg}}{2.205 \text{ lb}}$, (b) $\left(\dfrac{2.54 \text{ cm}}{1 \text{ in.}}\right)^2 = \dfrac{6.45 \text{ cm}^2}{1 \text{ in.}^2}$
11.   Density doesn't depend on sample size; mass and volume do depend on the size of the sample.
12.   (a) 8.95 g/cm^3, (b) 8.95, (c) 558 lb/ft^3
13.   (a) 22.4 g, (b) 9.26 mL

Tools you have learned

Consider removing this chart from the Study Guide so you can have it handy when tackling homework problems.

Tool	*How it Works*
SI Prefixes mega (10^6), kilo (10^3), deci (10^{-1}), centi (10^{-2}), milli (10^{-3}), micro (10^{-6}), nano (10^{-9}), pico (10^{-12})	We use these prefixes to create larger and smaller units. The prefixes enable us to form conversion factors for converting between differently sized units.
Units in laboratory measurements 1 cm = 10 mm, 1 L = 1000 mL	To convert between commonly used laboratory units of measurement.
Temperature conversions Equations 3.1, 3.2, and 3.3	To convert among temperature units. Be especially sure you can convert between Celsius and Kelvin temperatures, because that will be needed most in this course.
Rules for counting significant figures	To determine the number of significant figures in a number.
Rules for arithmetic and significant figures	To round answers to the correct number of significant figures.
Factor-label method	We use this method when a problem can be viewed as a conversion between units. It helps us set up the arithmetic in numerical problems by assembling necessary relationships into conversion factors and applying them to obtain the desired units of the answer.
Density $d = \dfrac{m}{V}$	To calculate the density of a substance from measurements of mass and volume. We use density when it is necessary to convert between mass and volume for a substance.
Specific gravity sp. gr.. $= \dfrac{d_{\text{substance}}}{d_{\text{water}}}$	To convert density from one set of units to a different set of units.

Chapter 4

The Mole: Connecting the Macroscopic and Molecular Worlds

Atoms and molecules react and combine in whole-number ratios. Therefore, to deal with formulas and reactions quantitatively, we must have a way to measure and count numbers of atoms and molecules. However, these particles are so tiny that we can't count them in the same way that we count knives and forks. Instead, we count them using a balance, taking advantage of the fact that each element has atoms with a characteristic atomic mass. The mole concept is the critical link that allows us to relate numbers of atoms and molecules to mass measurements we can perform in the laboratory.

In this chapter you will learn how to measure moles of atoms and molecules, how to determine chemical formulas from experimental data, and how to use balanced chemical equations to relate amounts of substances involved in chemical reactions. All of these are aspects of *stoichiometry*. The most important skill to develop is the ability to use the mole concept to perform quantitative reasoning.

Learning Objectives

In this chapter, you should keep in mind the following goals:

1   To understand how we can count atoms or molecules by measuring mass, provided we know the masses of individual particles.

2   To learn how to calculate formula mass and molecular mass from atomic masses and chemical formulas.

3   To learn how the mole concept links mass to numbers of atoms or molecules.

4   To learn the value of Avogadro's number and to learn how to use it in calculations.

5   To become adept at performing calculations converting grams to moles, and vice versa, both for atoms and molecules.

6   To learn how to use a chemical formula to relate numbers of moles of the elements in the substance.

7   To learn how to use a formula to calculate the mass of an element in a sample of a compound.

8   To be able to calculate the percentage composition of a compound from its chemical formula or from the results of an elemental analysis.

9   To be able to calculate an empirical formula from mass data or percentage composition obtained in a chemical analysis.

10   To be able to obtain a molecular formula from an empirical formula and an experimentally determined molecular mass.

11   To learn how to interpret the coefficients of a balanced chemical equation in terms of moles of reactants and products.

12   To learn how to balance chemical equations by "inspection."

13   To understand the concept of a limiting reactant and to learn how to determine which is the limiting reactant when amounts of two or more reactants are specified in a problem.

14   To predict the theoretical yield in a reaction and to calculate the percentage yield if the actual yield of products is known.

4.1   Use large-scale measurements to count tiny objects

Review

The main point of this section is that we can use the known mass of an object to count the number of such objects in a large collection of them. We simply measure the total mass of all the objects and then divide by the mass of one of them.

In counting atoms of an element in a lab-sized sample, we need the atomic mass of the element expressed in atomic mass units, u, as well as the mass in grams of the atomic mass unit. This is illustrated in Example 4.1. Similar counts for molecules can be made by calculating the molecular mass. Study Example 4.2. Notice that once we have the molecular mass, the calculation becomes very similar to that in Example 4.1.

$50\, \text{atom Au} \times$

Self-Test

1.   What is the mass, expressed in atomic mass units, of 50 atoms of gold?

     _____

2.   How many atoms of silver are in a bracelet that weighs 34.0 g? $34.0\,g\,Ag \times \dfrac{6.02 \times 10^{23}\,\text{Atoms}}{107.86}$

     $1.90\,E23$

3.   How many molecules of CO_2 are present in a sample that has a mass of 264 u?

     $6\,CO_2$ $264\,u \times \dfrac{1\,\text{molecule } CO_2}{76.01\,u}$

4.   What is the molecular mass of acetone, $(CH_3)_2CO$, a solvent in nail polish?

     $58.1\,u$

5.   Acetone has a density of 0.788 g cm^{-3}. How many molecules of acetone are in 15.0 cm^3 of acetone?

     1.22×10^{23}

     $$\dfrac{0.788\,g}{cm^3} \times \dfrac{mol}{58.1\,g} \times \dfrac{6.02 \times 10^{23}\,v}{mol} \times \dfrac{15\,cm^3}{}$$

New Terms

Write the definitions of the following terms, which were introduced in the introduction and this section. If necessary, refer to the Glossary at the end of the text.

     stoichiometry   formula mass   molecular mass

4.2   The mole links mass to number of atoms or molecules

Review

In the preceding section, you learned how to relate numbers of individual atoms or molecules to the lab-sized mass unit gram. When we work with realistic samples of substances, they contain enormous numbers of atoms or molecules, and we are usually not concerned with the actual number of particles. Instead, we're interested in the *ratios* of their numbers. For example, in a molecule of CO_2, the ratio of C to O atoms is 1 to 2. This same atom ratio will exist regardless of the sample size. If the sample had 100 molecules of CO_2, it would be composed of 100 atoms of C and 200 atoms of O; we still have a 1 to 2 atom ratio of C to O.

The mole is a quantity that allows us to conveniently measure, using a balance, large amounts of atoms (or molecules) in any desired ratio. A mole contains a fixed number of particles and has a mass, in grams, that is numerically the same as the atomic mass (for atoms) or molecular mass (for molecules). Thus, one mole of carbon atoms (1 mol C) has a mass of 12.01 g, and one mole of magnesium atoms (1 mol Mg) has a mass of 24.30 g. Similarly, the molecular mass of CO_2 is 44.02, so one mole of CO_2 molecules (1 mol CO_2) has a mass of 44.02 g.

The point of the preceding discussion is that a table of atomic masses (or a periodic table that incorporates atomic masses) provides us with a tool for relating moles of elements to masses measured in grams. All we need to do is look up the atomic mass of the element and we can write down how much one mole of that element weighs. For example, the periodic table gives the atomic mass of sodium (Na) as 22.98977, so we can write

$$1 \text{ mol Na} = 22.98977 \text{ g Na}$$

Usually, these are more significant figures than we require, so we round it as needed. If four significant figures are sufficient, then we write

$$1 \text{ mol Na} = 22.99 \text{ g Na}$$

Here are two more examples, also to four significant figures.

$$1 \text{ mol Cl} = 35.45 \text{ g Cl}$$

$$1 \text{ mol Fe} = 55.85 \text{ g Fe}$$

A similar situation applies when we're working with compounds. For example, if we wished to know the molar mass of hydrogen sulfide, H_2S, we would add the atomic masses of the elements to obtain the molecular mass. For H_2S, we obtain 34.082. Therefore, we can write

$$1 \text{ mol } H_2S = 34.082 \text{ g } H_2S$$

Let's look at a sample problem that requires us to relate grams to moles.

Example 4.1 Converting between grams and moles for an element.

Suppose we needed to measure 2.50 mol of sulfur for a particular experiment. How many grams of sulfur would we need?

Analysis:
We'll begin by stating the problem in equation form.

$$2.50 \text{ mol S} = ? \text{ g S}$$

To solve the problem, we need to relate moles of sulfur to grams of sulfur. The tool to accomplish this is the atomic mass of sulfur. In the periodic table, this is given as 32.066. To be sure to have enough precision in the calculation, we'll round this to one more significant figure than in the given data and write

$$1 \text{ mol S} = 32.07 \text{ g S}$$

Now we can use this to form a conversion factor to solve the problem.

Solution:
We form the conversion factor so mol S will cancel.

$$2.50 \text{ mol S} \times \frac{32.07 \text{ g S}}{1 \text{ mol S}} = 80.2 \text{ g S}$$

Is the Answer Reasonable?
Some simple arithmetic will answer the question. One mole of sulfur is about 32 g. We have between two and three moles of S, so the answer should lie between $2 \times 32 \text{ g} = 64 \text{ g}$ and $3 \times 32 \text{ g} = 96 \text{ g}$. Our answer, 80.2 g, is between these two values, so the answer does seem reasonable.

The mole and Avogadro's number

As mentioned above, the mole is a unit that stands for a fixed number of things. This number is called Avogadro's number and is equal to 6.02×10^{23}.

$$1 \text{ mol} = 6.02 \times 10^{23} \text{ things}$$

For most purposes, it isn't necessary to know the size of this number. Usually, it is sufficient to understand that if we have equal numbers of moles of two elements, they contain equal numbers of atoms. However, there are some instances when we do need to know the number of things in a mole, and we will study those now.

To understand when we need to use Avogadro's number, we need to examine how we view chemical substances from two perspectives—the large scale *macroscopic* world of the laboratory (where we deal with physical samples of substances and describe amounts in grams and moles) and the very tiny *submicroscopic* world of atoms and molecules (where we count individual atoms and molecules). As long as we stay entirely within one of these two views of matter, we don't need Avogadro's number. Thus, to calculate the number of *moles* of carbon in 5 *moles* of glucose, $C_6H_{12}O_6$, Avogadro's number isn't necessary. As we discuss in the next section, we simply use the chemical formula to tell us how many moles of carbon are in one mole of $C_6H_{12}O_6$. Similarly, if we wished to know how many *atoms* of carbon are in 5 *molecules* of glucose, $C_6H_{12}O_6$, we use the chemical formula to tell us how many carbon atoms are in one molecule of $C_6H_{12}O_6$, and then multiply by five. Once again, we don't need Avogadro's number.

Avogadro's number becomes necessary when we wish to relate an amount of something in the macroscopic world to an amount in the submicroscopic world. An example would be finding the mass in *grams* (a unit in the macroscopic world) of some number of carbon *atoms* (a unit in the submicroscopic world).

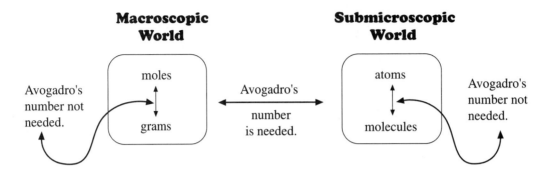

Besides understanding *when* you need to use Avogadro's number, you also need to know *how* to use it. As a tool in calculations, Avogadro's number provides a conversion between moles and individual units of a substance. For example, if we are dealing with an element such as sodium, we can write

$$1 \text{ mol Na} = 6.02 \times 10^{23} \text{ atom Na}$$

For the compound C_3H_8, we can write

$$1 \text{ mol } C_3H_8 = 6.02 \times 10^{23} \text{ molecule } C_3H_8$$

For an ionic compound such as NaCl, which doesn't contain discrete molecules, we write

$$1 \text{ mol NaCl} = 6.02 \times 10^{23} \text{ formula units NaCl}$$

There are two important things to observe in these equations. First, notice that we are relating *moles* in the macroscopic world to atoms, molecules, or formula units in the submicroscopic world. Second, notice that when we write these equations, we are very careful to specify, using chemical formulas, the exact nature of the substances involved. Now let's look at a sample problem.

Problem 4.2   Working with Avogadro's number

What is the mass in grams of 25 atoms of nickel?

Analysis:
We'll begin by expressing the problem in the form of an equation:

$$25 \text{ atom Ni} = ? \text{ g Ni}$$

The critical link in solving this problem is realizing that we are attempting to convert a unit in the submicroscopic world (atoms) to a unit in the macroscopic world (grams). Because we are connecting these two realms, we will need to use Avogadro's number as the tool. We are dealing with an element, so we write

$$1 \text{ mol Ni} = 6.02 \times 10^{23} \text{ atom Ni}$$

We can now go from the unit "mol Ni" to "atom Ni." Next, we need a tool to take us from "mol Ni" to "g Ni." This is provided by the atomic mass of nickel.

$$1 \text{ mol Ni} = 58.7 \text{ g Ni}$$

We now have a path from "atom Ni" to "g Ni," so we can proceed to the solution step.

Solution:
As usual, we assemble conversion factors from the relationships we've established in a way that will allow us to cancel units correctly. (Add the cancel marks yourself.)

$$25 \text{ atom Ni} \times \frac{1 \text{ mol Ni}}{6.02 \times 10^{23} \text{ atom Ni}} \times \frac{58.7 \text{ g Ni}}{1 \text{ mol Ni}} = 2.44 \times 10^{-21} \text{ g Ni}$$

Thus, 25 atoms of nickel have a mass of 2.44×10^{-21} g.

Is the Answer Reasonable?

In working these kinds of problems, the most common mistake is to use Avogadro's number incorrectly, or to not use it at all. Let's see how you can avoid these pitfalls.

We know atoms are very tiny and that individual atoms have very small masses. The answer we obtained is a very small number, so in that sense, it seems that our answer is reasonable.

If we had used Avogadro's number incorrectly, the answer would have been a huge number. Atoms do not have huge masses, so we should recognize our mistake. Similarly, if we had not used Avogadro's number at all, our calculation would give an answer that is not an extremely tiny number. That should sound an alarm and suggest that we'd better examine our method more closely.

Self-Test

6. Calculate the formula mass of $(NH_4)_3PO_4 \cdot 3H_2O$, and write the two conversion factors relating grams to moles for this compound.

7. For ammonia, $17.0 \text{ g NH}_3 \Leftrightarrow 6.02 \times 10^{23}$ molecules NH_3. Write the two conversion factors that this equivalency makes possible.

8. How many atoms of iron, Fe, are in 0.400 mol of iron?

    2.41 E 23

9. If an experiment calls for 0.500 mol of H_2SO_4 (sulfuric acid), how many grams of H_2SO_4 do we have to weigh out?

10. If 110 g of CO_2 are formed in an experiment, how many moles of CO_2 are formed?

11. How many grams of calcium are in 1.00 mol Ca? _____

12. How many moles of phosphorus atoms are in 1.00 g of phosphorus ?_____

13. What is the mass, in grams, of 0.250 mol of phenobarbital ($C_{12}H_{12}N_2O_3$)? _____

14. How many moles of pentachlorophenol (C_6HCl_5O) are in 1.00 g of C_6HCl_5O?

15. How many atoms are there in 0.520 moles of iron? _____

16. 25.0 moles of phosphoric acid (H_3PO_4) contain how many hydrogen atoms?

17.   What is the mass in grams of 245 atoms of sodium? _____

18.   How many atoms of uranium are there in 1.00 g of uranium metal?

_____

19.   How many H_2O molecules are there in a raindrop weighing 0.050 g?

_____

New Terms

Write the definitions of the following terms, which were introduced in this section. If necessary, refer to the Glossary at the end of the text.

mole   molar mass

Avogadro's number   Avogadro's constant

4.3   Chemical formulas relate amounts of substances in a compound

Review

The subscripts in a formula give us the *atom ratio* in which the elements are combined. Those same subscripts also give us the ratio by moles in which the elements are combined. For example, consider the substance propane, the gas used for cooking in rural areas and in nearly all gas barbecue grills. Propane consists of molecules that have the formula C_3H_8.

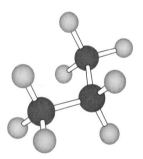

A molecule of propane, C_3H_8. It consists of three carbon atoms and eight hydrogen atoms joined together in a single particle that we call a molecule.

If we had one molecule of propane, it would contain 3 atoms of C and 8 atoms of H. If we had 10 propane molecules, all together they would contain 30 atoms of C and 80 atoms of H. Notice, however, that even though we have more atoms of C and H in the larger sample, they are in the same numerical C-to-H ratio, namely 3-to-8. In fact, in any sample of this compound, the atom *ratio* of C to H will be 3 to 8.

Now suppose we had one dozen propane molecules. In this sample we would find 3 dozen carbon atoms and 8 dozen hydrogen atoms[1].

---

[1] When we write "1 dozen C_3H_8" we mean "1 dozen C_3H_8 *molecules*," and when we write "3 dozen C" and "8 dozen H", we mean "3 dozen C *atoms*" and "8 dozen H *atoms*." We generally omit the word molecule or atom because the formula C_3H_8 stands for a molecule of C_3H_8, and the symbols C and H stand for atoms of C and H.

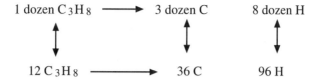

Notice that in a dozen molecules, the ratio of atoms *by the dozen* (3 to 8) is numerically the same as the atom ratio in one molecule of the compound.

As we discussed earlier, the mole is a quantity similar to the dozen, because it stands for a fixed number of things. The reasoning with moles is therefore the same as the reasoning with dozens. If we had one mole of C_3H_8, the ratio of atoms *by the mole* would be numerically the same as the atom ratio in one molecule.

$$1 \text{ mol } C_3H_8 \longrightarrow 3 \text{ mol C} \qquad 8 \text{ mol H}$$

$$1 \text{ dozen } C_3H_8 \longrightarrow 3 \text{ dozen C} \qquad 3 \text{ dozen H}$$

$$1 \text{ molecule } C_3H_8 \longrightarrow 3 \text{ atom C} \qquad 8 \text{ atom H}$$

Thus, whether we're dealing with individual particles (molecules and atoms), particles by the dozen, or particles by the mole, the ratios are the same.

Interpreting a chemical formula for problems in stoichiometry

In chemical problem solving, the chemical formula serves as a tool that establishes the ratios among atoms in a single molecule of the compound. It is also a tool that establishes the ratios among moles of atoms in a mole of the compound. We can use these relationships to construct equivalencies that we can use in chemical calculations. For propane, we can write the following, which apply when we are dealing with numbers of individual molecules and atoms.

$$1 \text{ molecule } C_3H_8 \Leftrightarrow 3 \text{ atom C}$$

$$1 \text{ molecule } C_3H_8 \Leftrightarrow 8 \text{ atom H}$$

$$3 \text{ atom C} \Leftrightarrow 8 \text{ atom H}$$

For example, the first equivalence would be used when we are interested in relating the number of carbon atoms to a certain number of C_3H_8 molecules. It tells us that for every 1 molecule of C_3H_8, we will find 3 atoms of C. If we were interested in the relation between molecules of C_3H_8 and atoms of H, then we would use the second equivalence, and if we wanted to relate atoms of C and atoms of H in a sample of propane, we would use the third equivalence.

Individual atoms and molecules are too small to work with in the laboratory, so we scale everything up to mole-sized quantities and write the following equivalencies.

$$1 \text{ mol } C_3H_8 \Leftrightarrow 3 \text{ mol C}$$

$$1 \text{ mol } C_3H_8 \Leftrightarrow 8 \text{ mol H}$$

$$3 \text{ mol C} \Leftrightarrow 8 \text{ mol H}$$

Notice that the numbers involved in these equivalencies are identical to those in the relationships given for the individual molecules and atoms; in both cases they are derived from the subscripts in the formula for propane.

$$1 \text{ mol } C_{3}H_{8}$$

3 mol C   8 mol H

Let's look at an example that illustrates how we use the concepts described above to solve problems.

Problem 4.3   Using the mole concept in stoichiometry

When butane burns in air, the carbon in the compound becomes incorporated entirely in molecules of carbon dioxide, CO_2. How many moles of CO_2 would be formed if 0.200 mol of C_4H_{10} is burned?

Analysis:
Let's begin by expressing the problem in an equation format:

$$0.200 \text{ mol } C_4H_{10} \Leftrightarrow ? \text{ mol } CO_2$$

The critical link in solving this problem is finding the amount of carbon (in moles) in the sample of C_4H_{10}. Because all the carbon in the C_4H_{10} goes into the CO_2, we can then use this amount of carbon to figure out how much CO_2 is formed. The tools we will use are the subscripts in the formulas of C_4H_{10} and CO_2. Here's an outline of the path we will use to solve the problem.

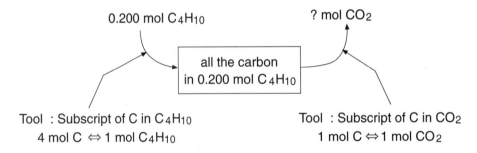

Tool : Subscript of C in C_4H_{10}
4 mol C $\Leftrightarrow$ 1 mol C_4H_{10}

Tool : Subscript of C in CO_2
1 mol C $\Leftrightarrow$ 1 mol CO_2

Solution:
We use the tools to form conversion factors. As usual, we are careful to select the factors that allow the units to cancel.

$$0.200 \text{ mol } C_4H_{10} \times \frac{4 \text{ mol } C}{1 \text{ mol } C_4H_{10}} \times \frac{1 \text{ mol } CO_2}{1 \text{ mol } C} = 0.800 \text{ mol } CO_2$$

The calculation tells us that 0.800 mol of CO_2 would be formed.

Is the Answer Reasonable?
Let's do some whole-number reasoning to see how the relative amounts of C_4H_{10} and CO_2 compare. If we had 1 mol C_4H_{10}, it would contain 4 mol C. But it takes only 1 mol of C to make 1 mol of CO_2, so 4 mol of C is enough to make 4 mol of CO_2. The conclusion, then, is that 1 mol of C_4H_{10} has enough carbon in it to make 4 mol of CO_2. Another way of looking at this is that the number of moles of CO_2 formed is four times the number of moles of C_4H_{10} that react. If we start with 0.200 mol C_4H_{10}, then the amount of CO_2 formed would be 4 × 0.200 = 0.800 mol. That's the answer we obtained.

In the preceding example, we worked entirely in moles. Translating between grams and moles is a task you learned how to do in Section 4.2. Study Example 4.9 in the textbook to see how we are able to combine these two skills to solve more complex problems.

Relationships between masses of elements in a compound

There are times when we need to be able to calculate the mass of an element in a sample of a compound. For example, suppose we wished to know how many grams of sodium are in 3.58 g of Na_2CO_3. One way to do the calculation is the following:

1   Use the molar mass of Na_2CO_3 to calculate the moles of Na_2CO_3 in the sample.

2   Next, use the subscripts in Na_2CO_3 to calculate the moles of Na in the sample.

3   Finally, use the atomic mass of Na to calculate the grams of Na in the sample.

There is a faster (and simpler) way to solve the problem, however. In the process of calculating the molar mass of Na_2CO_3 we obtain all the data we need to do the calculation in one step instead of three. This is because in calculating the formula mass we obtain the data required to establish a mass relationship between Na and Na_2CO_3.

Element	Moles	Mass
Sodium	2 mol	2×22.99 g $= 45.98$ g
Carbon	1 mol	1×12.01 g $= 12.01$ g
Oxygen	3 mol	3×16.00 g $= 48.00$ g
	Total mass of 1 mol Na_2CO_3	105.99 g

$$105.99 \text{ g } Na_2CO_3 \Leftrightarrow 45.98 \text{ g Na}$$

We can use this relationship to solve the problem.

$$3.58 \text{ g } Na_2CO_3 \times \frac{45.98 \text{ g Na}}{105.99 \text{ g } Na_2CO_3} = 1.55 \text{ g Na}$$

Thinking It Through

1   Compound X was known to be either ammonia (NH_3), hydrazine, (N_2H_4), or hydrazoic acid (HN_3). In a sample of 0.012 mol of compound X there was found to be 0.024 mol of N atoms and 0.048 mol of H atoms. What was compound X?

Self-Test

20.   How many mol of O and how many mol of H are present in 0.800 mol of H_2O?

_____

21.   How many mol of Ca, O, and H are present in 0.600 mol of $Ca(OH)_2$?

_____

22. How many g of O are present in 0.800 mol of H_2O?

_____

23. How many g of H are present in 0.600 mol of $Ca(OH)_2$?

_____

24. How many grams of phosphorus are needed to make 24.00 g of tetraphosphorus hexaoxide, P_4O_6? How many grams of O are needed?

_____

25. If a sample of $(NH_4)_3PO_4 \cdot 3H_2O$ contains 0.60 mol N, how many mol O does it contain?

_____

New Terms

None

4.4   Chemical formulas can be determined from experimental mass measurements

Review

Qualitative analysis is used in the lab to find out which elements are present in a substance. Then *quantitative analysis* is used to obtain their relative amounts, often expressed as the percentage by mass of each element. If a compound consists, say, of 60.26% carbon, then to make 100.00 g of the compound from its elements requires 60.26 *grams* of carbon. Another way of looking at a percentage like 60.26% carbon is that it is the number of grams of carbon that could be obtained by changing 100.00 g of the compound back into its elements.

We can find the *percentage by mass* of an element in a compound in two ways that involve two different kinds of data. One way is experimental; it uses the actual masses of the individual elements found in a weighed sample of the compound. Equation 4.1 in the text is used to perform the calculation.

The other method is theoretical; it uses the formula to calculate the percentages. The procedure is straightforward. Consider, for example, the compound Na_2CO_3. One mole of this substance contains two moles of sodium, one mole of carbon, and three moles of oxygen. We can use these amounts and the atomic masses of the elements to calculate the masses of each in one mole of the compound.

Element	Moles	Mass
Sodium	2 mol	2 × 22.99 g = 45.98 g
Carbon	1 mol	1 × 12.01 g = 12.01 g
Oxygen	3 mol	3 × 16.00 g = 48.00 g
Total mass		105.99 g

Once we have the masses of each element and the formula mass, we can calculate the percentages by mass of each. Let's start with the percent sodium.

$$\text{percent sodium} = \underbrace{\frac{45.88 \text{ g}}{105.99 \text{ g}}}_{} \times 100\% = 43.38\% \text{ Na}$$

mass of sodium in 1 mol Na_2CO_3

mass of 1 mol Na_2CO_3

Similarly, for carbon and oxygen,

$$\text{percent carbon} = \frac{12.01 \text{ g}}{105.99 \text{ g}} \times 100\% = 11.33\% \text{ C}$$

$$\text{percent oxygen} = \frac{48.00 \text{ g}}{105.99 \text{ g}} \times 100\% = 45.29\% \text{ O}$$

$$\text{Sum of percentages} = 100.00\%$$

Calculating a percentage composition doesn't involve a great deal of reasoning. It's just a matter of learning how to do the calculation. It is helpful to remember that the sum of the percentages should add up to 100 % (although sometimes the calculated sum will be slightly more or slightly less than 100 % because of rounding during the calculations).

A list of the percentages by mass of a compound's elements makes up the compound's *percentage composition*. As Example 4.12 in the text demonstrates, a calculated percentage composition can be compared with one obtained from experimental analysis to check the identity of an unknown substance.

Empirical and molecular formulas

When the smallest whole-number subscripts are used to describe the ratio of the atoms of different elements in a compound the resulting formula is an *empirical formula.* The empirical formula of caffeine, for example, is $C_4H_5N_2O$, but its *molecular formula,* the actual composition of one molecule, is $C_8H_{10}N_4O_2$.

To calculate an empirical formula, *the critical data needed are the masses of the elements in a sample of the compound.* These masses can be converted to moles, from which the ratios by moles can then be obtained. The ratios by moles have to be identical to the ratios by atoms, so the mole ratios give us the subscripts in the empirical formula.

Data to calculate an empirical formula can be obtained in several ways. A chemical analysis might give the masses directly, as illustrated in Example 4.13 in the text. Study Example 4.14 to see how we handle situations in which whole number subscripts are not obtained initially.

Sometimes the results of an analysis are provided as the percentages by mass of each of the elements. The percentages by mass of the elements in a compound are numerically equal to the *grams* of each element in 100 grams of the compound. We can therefore take the grams of each element in a 100-g sample (as given by the percentages), multiply each mass by a conversion factor obtained from the atomic masses to find the corresponding moles of each element in the sample. The moles are converted to mole ratios, from which we obtain the subscripts, as illustrated in Example 4.15.

In an indirect analysis, chemical reactions are carried out that separate the elements in a sample and capture them quantitatively (i.e., with out loss) in compounds of known composition. An example is a combustion analysis, in which a compound containing carbon and hydrogen is burned to give CO_2 and H_2O. The experimental data obtained in this way are masses not of elements but of compounds, so the masses of C and H in the original sample have to be calculated from the masses of CO_2 and H_2O formed in the combustion reaction. This is easy because we know that 1 mol of CO_2 contains 1 mol C; and 1 mol H_2O contains 2 mol of H. In other words, the calculations are nothing more than grams-to-moles-to-grams calculations.

Finding molecular formulas from empirical formulas

Often the experimentally measured molecular mass is not the same as the one we calculate from an empirical formula. It is some simple multiple of it, such as 2× or 3× or 4× and so forth. To find out which multiple, we divide the experimental molecular mass by the calculated empirical formula mass. That will give the *whole* number multiple we need (or one so close to a whole number we can round to it). Finally, multiply each subscript in the empirical formula by the whole number and the result is the molecular formula. For example, the true molecular mass of benzene is 78 and its empirical formula is CH (with a formula mass of 13). We simply divide 78 by 13 to get 6. This tells us that there are really 6 units of CH in each benzene molecule, and so the molecular formula is $C_{(1 \times 6)} H_{(1 \times 6)} = C_6H_6$.

Thinking It Through

2   A compound was known to contain only potassium and oxygen. When a sample with a mass of 0.2564 g was analyzed, there was found to be 0.2128 g of potassium. How many grams of oxygen were also present? What is the empirical formula of the compound?

Self-Test

26.   A 0.9278-g sample of a bright orange compound consists of 0.3683 g of chromium, 0.1628 g of sodium and 0.3967 g of oxygen. Calculate the percentage composition.

_____

27.   An analysis of a sample believed to be photographers hypo, $Na_2S_2O_3$, gave the following percentage composition: Na, 29.10%; S, 40.50%; and O, 30.38%. Do these data correspond to the formula given? Use a separate sheet of paper to solve the problem.

_____

28.   Carotene, the pigment responsible for the color of carrots, has a percentage composition of 89.49% C and 10.51% H. Its molecular mass was found to be 546.9 Calculate its empirical formula and its molecular formula.

_____

29.   When 0.8788 g of a liquid isolated from oil of balsam was burned completely, 2.839 g of CO_2 and 0.9272 g of H_2O were obtained. The molecular mass of the compound was found to be 136.2. Calculate the empirical and molecular formulas of this compound.

_____

New Terms

Write the definitions of the following terms, which were introduced in this section. If necessary, refer to the Glossary at the end of the text.

percentage composition   percentage by mass

empirical formula   molecular formula

4.5   Chemical equations link amounts of substances in a reaction

Review

You've already learned that mole relationships among the elements in a compound are given by the subscripts in the formula. In this section we see that quantitative relationships among substances that react and form during chemical reactions are provided by the coefficients in balanced chemical equations.

For stoichiometry problems dealing with chemical reactions, *balanced* chemical equations are essential. This is because the coefficients in an equation provide the *only way* to relate the relative amounts of reactants and products. This relationship is on both a molecular and mole basis. Consider, for example, the balanced equation for the combustion of methanol (CH_3OH), the fuel in "canned heat" products such as Sterno[®].

$$2CH_3OH + 3O_2 \rightarrow 2CO_2 + 4H_2O$$

On a molecular basis, we see that 2 molecules of CH_3OH combine with 3 molecules of O_2 to form 2 molecules of CO_2 and 4 molecules of H_2O. Scaling up to laboratory amounts, we find that when 2 moles of CH_3OH burn, they consume 3 moles of O_2 and form 2 moles of CO_2 and 4 moles of H_2O. Notice that the *mole ratios* among reactants and products are numerically equal to the ratios expressed in terms of the individual molecules of the reactants and products.

When we face a stoichiometry problem involving substances in a chemical reaction, the critical links are the coefficients of the balanced equation. These coefficients *alone* provide the relationships among moles of reactants and moles of products. For example, if we wished to relate the amounts of CO_2 and H_2O formed in the combustion of CH_3OH, we would use the coefficients of CO_2 and H_2O to establish the mole relationship

$$2CH_3OH + 3O_2 \rightarrow 2CO_2 + 4H_2O$$

$$2 \text{ mol } CO_2 \Leftrightarrow 4 \text{ mol } H_2O$$

Similarly, if we wished to relate the amount of O_2 consumed when a certain amount of CO_2 is formed in the reaction, we would use the coefficients of O_2 and CO_2 in the equation to write

$$3 \text{ mol } O_2 \Leftrightarrow 2 \text{ mol } CO_2$$

Examples 4.18 through 4.21 in the text illustrate the principles above. Let's look closely at two more examples.

Example 4.4   Calculations using balanced chemical equations

In a gas barbecue, the combustion of propane, C_3H_8, follows the equation

$$C_3H_8(g) + 5O_2(g) \rightarrow 3CO_2(g) + 4H_2O(g)$$

How many moles of water vapor are formed if 0.250 mol of C_3H_8 are burned?

Analysis:
Let's begin by stating the problem in equation form.

$$0.250 \text{ mol } C_3H_8 \Leftrightarrow ? \text{ mol } H_2O$$

To solve the problem we need to find a relationship between the amounts of these substances. Because the C_3H_8 and H_2O are involved in the combustion reaction, the tools we need are their coefficients in the balanced equation. The coefficients of C_3H_8 and H_2O give us the following

$$1 \text{ mol } C_3H_8 \iff 4 \text{ mol } H_2O$$

This provides the connection we need to convert moles of C_3H_8 to moles of H_2O.

Solution:
To solve the problem we apply the mole relationship between C_3H_8 and H_2O as a conversion factor.

$$0.250 \text{ mol } C_3H_8 \times \frac{4 \text{ mol } H_2O}{1 \text{ mol } C_3H_8} = 1.00 \text{ mol } H_2O$$

We conclude that combustion of 0.250 mol C_3H_8 will produce 1.00 mol H_2O.

Is the Answer Reasonable?
This is a pretty simple problem, and you could probably see the answer without having to work it out in such great detail. The coefficients tell us that the number of moles of water formed is four times the number of moles of propane consumed. Therefore, when 0.250 mol C_3H_8 is burned, we will get 4×0.250 mol = 1.00 mol of water as one of the products.

Example 4.5   Calculations using balanced chemical equations

Magnesium hydroxide, $Mg(OH)_2$ (the creamy white substance in milk of magnesia), reacts with hydrochloric acid, HCl (found in stomach acid), to give magnesium chloride and water. The reaction is

$$Mg(OH)_2(s) + 2HCl(aq) \rightarrow MgCl_2(aq) + 2H_2O$$

How many grams of $Mg(OH)_2$ are required to react completely with 12.8 g of HCl?

Analysis:
We'll begin by expressing the problem in the form of an equation.

$$12.8 \text{ g } HCl \iff ? \text{ g } Mg(OH)_2$$

We see that we are attempting to relate amounts of substances involved in a chemical reaction. This immediately suggests that we will need to use their coefficients in the balanced equation to relate them on a mole basis. Let's write that relationship.

$$1 \text{ mol } Mg(OH)_2 \iff 2 \text{ mol } HCl$$

We have part of the solution, but we are still missing some connections.

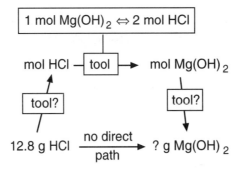

There is no direct path between grams of HCl and grams of $Mg(OH)_2$, but we do have the path between moles. To complete the connections between units, we need to relate grams to moles for both substances. By now you should realize that the tool to accomplish this is the formula mass. For HCl, the formula mass is 36.46 and for $Mg(OH)_2$ it is 58.32. Therefore, we can write

$$1 \text{ mol HCl} = 36.46 \text{ g HCl}$$

$$1 \text{ mol Mg(OH)}_2 = 58.32 \text{ g Mg(OH)}_2$$

Now we have a complete path from the units given (g HCl) to the units desired (g $Mg(OH)_2$), so we can proceed to the solution step.

Solution:

We have all the relationships we need, so we apply them as conversion factors, being sure the units cancel.

$$12.8 \text{ g HCl} \times \frac{1 \text{ mol HCl}}{36.46 \text{ g HCl}} \times \frac{1 \text{ mol Mg(OH)}_2}{2 \text{ mol HCl}} \times \frac{58.32 \text{ g Mg(OH)}_2}{1 \text{ mol Mg(OH)}_2} = 10.2 \text{ g Mg(OH)}_2$$

The calculations tell us that 12.8 g of HCl will require 10.2 g of $Mg(OH)_2$ for complete reaction.

Is the Answer Reasonable?

We can do some proportional reasoning and approximate arithmetic to get an idea whether our answer is "in the right ballpark." The equation tells us that 2 mol HCl are needed to react with 1 mol $Mg(OH)_2$. The mass of 2 mol HCl is approximately $2 \times 36 = 72$ g, and the mass of 1 mol of $Mg(OH)_2$ is approximately 60 g. Therefore, we expect that 72 g of HCl will react with approximately 60 g of $Mg(OH)_2$. In other words, the mass of $Mg(OH)_2$ consumed will be a little less than the mass of HCl that reacts. Our answer, 10.2 g $Mg(OH)_2$, is a little less than 12.8 g HCl, so the answer seems to be reasonable.

The preceding example demonstrates an important lesson about problems that deal with chemical reactions. The critical connections among the substances involved in a reaction are mole relationships provided by the coefficients of the balanced equation. Notice that this was the first relationship we established. Then we proceeded to find connections that take us from: (1) the given data to moles of the first substance, and (2) from moles of the second substance to the units of the desired answer. The following "flow diagram" shows the tools we used in the various steps in solving the problem.

With practice, you will find that all stoichiometry problems involving chemical reactions are pretty much the same. The first step is to establish the balanced chemical equation. Then, using the coefficients, write the mole relationship between the substances of interest. After this, look for any additional connections you may need between moles and the units of the given data and/or moles and the units of the answer. Once you've reached this point, the hard work is done. All that's left is to assemble the relationships into conversion factors and calculate the answer.

Thinking It Through

3   Sulfur dioxide will react with oxygen (O_2) under special conditions to form sulfur trioxide. How many moles of SO_2 are needed to react with 4.5 mol of O_2, and how many moles of SO_3 will form?

Self-Test

30.   Phosphorus burns according to the following equation.

$$4P + 5O_2 \rightarrow P_4O_{10}$$

(a)   How many moles of O_2 react with 10.0 mol of P?

_____

(b)   How many moles of P_4O_{10} can be made from 2.00 mol of P?

_____

(c)   To make 4.00 mol of P_4O_{10}, how many moles of O_2 are needed?

_____

(d)   To make 3.60 mol of P_4O_{10}, how many moles of P are needed?

_____

How many moles of O_2 are needed?

_____

31.   Arsenic combines with oxygen to form As_2O_3 according to the following equation.

$$4As + 3O_2 \rightarrow 2As_2O_3$$

(a)   To make 9.68 g As_2O_3, how many grams of As are needed?

_____

(b)   How many grams of O_2 are needed to make 8.92 g As_2O_3?

_____

(c)   If 5.85 g As are used, how many grams of As_2O_3 form?

_____

New Terms

None

4.6   Chemical equations cannot create or destroy atoms

Review

In Section 2.2, we introduced you to the concept of a balanced equation and coefficients preceding chemical formulas. We're now ready to begin to write equations and balance them. As you proceed, keep in mind that writing and balancing equations should be viewed as two separate steps; first write the formulas for all the reactants and products, then adjust the coefficients to achieve balance. To obtain a *balanced equation:*

1   Be sure all the chemical formulas are correct—both the atomic symbols and the *subscripts*.

2   Remember that all atoms among the reactants must end up somewhere among the products and that a *coefficient* is a multiplier for all of the atoms in a formula.

3   Balance elements other than H and O first.

4   Balance as a group those polyatomic ions that appear unchanged on both sides of the arrow.

5   Be sure you have the smallest set of whole number coefficients.

Self-Test

32.   Balance each of these equations.

(a) $2\,Li + H_2 \rightarrow 2\,LiH$ _____

(b) $4\,Na + O_2 \rightarrow 2\,Na_2O$ _____

(c) $2\,Sr + O_2 \rightarrow 2\,SrO$ _____

(d) $2\,HCl + SrO \rightarrow SrCl_2 + H_2O$ _____

(e) $2\,HBr + Na_2O \rightarrow 2\,NaBr + H_2O$ _____

(f) $2\,Al + 3\,S \rightarrow Al_2S_3$ _____

(g) $CH_4 + 4\,F_2 \rightarrow CF_4 + 4\,HF$ _____

(h) $CO_2 + 4\,H_2 \rightarrow CH_4 + 2\,H_2O$ _____

New Terms

None

4.7   The reactant in shortest supply limits the amount of product

Review

When substances are combined for a chemical reaction, they are not always mixed together in just the right proportions to get complete reaction. As a result, sometimes one of the reactants will be completely used up before the rest, and once this happens, no further products can form. The reaction will simply come to a halt. At this point, the reaction mixture will contain the products of the reaction plus some of the reactant that had not been able to react. The following diagram illustrates this for the reaction of carbon with oxygen to form carbon dioxide.

$$C + O_2 \rightarrow CO_2$$

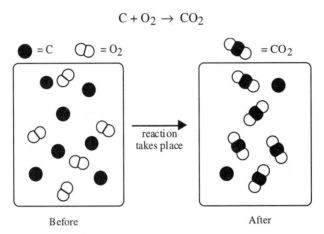

Before   After

Notice that in both the before and after views, there are the same numbers of atoms of C and O. However, there are not enough oxygen atoms to combine with all the carbon, so after the oxygen is all used up, no more CO_2 can form. At the end of the reaction, there is still some unreacted carbon along with the CO_2 that has formed.

In this reaction mixture, we call the oxygen the **limiting reactant** because we don't have enough of it to consume all the carbon; its amount is what *limits* the amount of product (CO_2) that forms. We might call the carbon the **excess reactant**, because it is present in *excess* amount — that is, in an amount greater than is needed to react with all the oxygen.

If we wish to predict the amount of product in a reaction, we need to know which of the reactants is the limiting reactant. For example, consider the following mixture of C and O_2. How many molecules of CO_2 can form from this mixture?

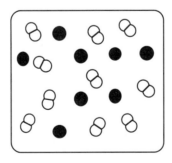

In this mixture, we have 10 molecules of O_2 and 8 atoms of C. It's pretty easy to see which is the limiting reactant. We can figure it out two ways (although we only need to do it one way if all we need to know is which is limiting). The reasoning we do here will be used again later, so be sure you understand it.

One way to find the limiting reactant is to examine the O_2. Each O_2 requires one C to form CO_2, so 10 O_2 molecules would require 10 carbon atoms. We have only 8 C atoms, so C must be the limiting reactant. It will be the one used up first and its amount limits the amount of CO_2 that can form.

The second way is to look at the carbon. Each carbon requires one O_2 molecule. Therefore, 8 C atoms require 8 O_2 molecules. We have more than 8 O_2 molecules, so there is an excess of O_2. If O_2 is present in excess, then carbon must be the limiting reactant.

Once we've identified the limiting reactant, we can figure out how much product will form and how much of which reactant is left over. *We always use the limiting reactant to determine how much product is formed.* When all 8 atoms of C react, they will yield 8 molecules of CO_2, so this is the amount of product that forms. There will be O_2 left over, and the amount left over is the difference between the amount of O_2 originally available (10 molecules) and the amount that reacts with the carbon (8 molecules). This difference, 2 molecules, is the amount of O_2 left over.

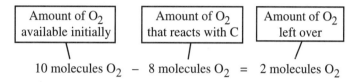

How to recognize a limiting reactant problem

There is a particular strategy we will employ in solving limiting reactant problems, but before we can use it, we must be able to recognize when a problem falls into this class. The key is in the way the problem is worded. The following are two stoichiometry problems that deal with the same chemical reaction, the reaction of $Mg(OH)_2$ with HCl.

$$Mg(OH)_2 + 2HCl \rightarrow MgCl_2 + 2H_2O$$

Problem 1

How many grams of $MgCl_2$ can be formed when 2.48 g of $Mg(OH)_2$ reacts with HCl?

Problem 2

How many grams of $MgCl_2$ can be formed when 2.48 g of $Mg(OH)_2$ are combined with 2.88 g of HCl?

In Problem 1, we are not told about the amount of HCl available, so the only way we can solve the problem is to assume that there is more than enough HCl available. We would therefore base the calculation on the amount of $Mg(OH)_2$ given in the problem. Problem 1 is *not* a limiting reactant problem.

Problem 2 *is* a limiting reactant problem. *Notice that we are given amounts of both reactants;* this is the clue that identifies it as a limiting reactant problem. We can't tell, reading the problem, which one of the reactants will be completely used up, and therefore we don't know at this point which reactant to use to calculate the amount of product. To solve the problem, we must first identify the limiting reactant and then use the amount of that reactant to calculate the amount of product that can form.

Strategy for Solving Limiting Reactant Problems

The approach we will take in solving limiting reactant problems is as follows:

1   We will first determine the number of moles of both reactants.

2   We will select one reactant and calculate the amount of the other that is required for complete reaction. This will require the coefficients of the balanced equation.

3   We will compare the amount needed for complete reaction with the amount available and determine which of the reactants is limiting. We described this reasoning above in our example of the reaction of C with O_2 to form CO_2. Review it if necessary.

4   We will use the amount of limiting reactant to calculate the amount of product formed. We can also calculate the amount of the excess reactant that is actually consumed and, by difference, determine the amount of excess reactant left over.

With this as background, let's work on Problem 2, which we'll restate.

Example 4.5 Solving limiting reactant problems

How many grams of $MgCl_2$ can be formed when 2.48 g of $Mg(OH)_2$ are combined with 2.88 g of HCl?

$$Mg(OH)_2 + 2HCl \rightarrow MgCl_2 + 2H_2O$$

Analysis and Solution:
You already know this is a limiting reactant problem, but in other situations, you should look for the clue that identifies it as a problem of this type — namely, that we are given amounts of both reactants. Then we proceed as described in the strategy. We'll combine the Analysis and Solution steps here because we can't simply assemble all the data and perform the calculations all at once.

The first step is to determine the number of moles of the reactants that are available. We are given grams, so we convert to moles using the formula masses. The relationships are

$$1 \text{ mol HCl} = 36.46 \text{ g HCl}$$

$$1 \text{ mol Mg(OH)}_2 = 58.32 \text{ g Mg(OH)}_2$$

Applying them, gives

$$2.48 \text{ g Mg(OH)}_2 \times \frac{1 \text{ mol Mg(OH)}_2}{58.32 \text{ g Mg(OH)}_2} = 0.0425 \text{ mol Mg(OH)}_2$$

$$2.88 \text{ g HCl} \times \frac{1 \text{ mol HCl}}{36.46 \text{ g HCl}} = 0.0790 \text{ mol HCl}$$

Next, we select one reactant and calculate the number of moles of the other required for complete reaction. For this step we need the coefficients of the balanced equation, from which we obtain

$$1 \text{ mol Mg(OH)}_2 \Leftrightarrow 2 \text{ mol HCl}$$

It doesn't matter which reactant we choose, so let's select the HCl and calculate how many moles of $Mg(OH)_2$ are needed to use it up.

| Amount of HCl available | | Amount of $Mg(OH)_2$ needed to use up all the HCl |

$$0.0790 \text{ mol HCl} \times \frac{1 \text{ mol Mg(OH)}_2}{2 \text{ mol HCl}} = 0.0395 \text{ mol Mg(OH)}_2$$

Now we compare the amount of $Mg(OH)_2$ that's available in the reaction mixture (0.0425 mol) with the amount of $Mg(OH)_2$ that's required to react with all the HCl.

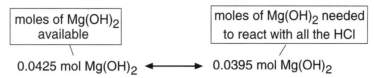

0.0425 mol Mg(OH)$_2$ ⟷ 0.0395 mol Mg(OH)$_2$

Notice that there is more Mg(OH)$_2$ available than is needed to use up all the HCl. This means that Mg(OH)$_2$ is the reactant in excess and that some Mg(OH)$_2$ will be left over. It also means that *HCl must be the limiting reactant.*

Now that we've identified the limiting reactant, we can calculate the amount of MgCl$_2$ that can form. We use the moles of Mg(OH)$_2$ to calculate the moles of MgCl$_2$ that can form, and then convert that to grams.

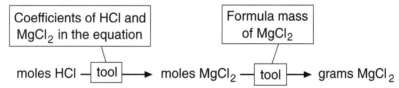

The chemical equation gives us

$$2 \text{ mol HCl} \Leftrightarrow 1 \text{ mol MgCl}_2$$

and the formula mass of MgCl$_2$ gives us

$$1 \text{ mol MgCl}_2 = 95.2 \text{ g MgCl}_2$$

We use these and the amount of the limiting reactant to calculate the mass of MgCl$_2$ that could be formed in the reaction.

$$0.0790 \text{ mol HCl} \times \frac{1 \text{ mol MgCl}_2}{2 \text{ mol HCl}} \times \frac{95.2 \text{ g MgCl}_2}{1 \text{ mol MgCl}_2} = 3.76 \text{ g MgCl}_2$$

The results tell us that the maximum amount of MgCl$_2$ that could form in the reaction using these amounts of reactants is 3.76 g.

Is the Answer Reasonable?

There are a few things we can check without too much trouble. Let's start with the number of moles of Mg(OH)$_2$ and HCl in the reaction mixture. For Mg(OH)$_2$, 0.1 mol would weigh 5.8 g, so 2.5 g is about half of 0.1 mol, or 0.05 mol; the value we obtained (0.0425 mol) seems about right. Similarly, 0.1 mol HCl weighs about 3.6 g, so 2.88 g HCl is a little less than 0.1 mol. The value we obtained (0.0790 mol) seems okay.

We can check the limiting reactant conclusion by working with the Mg(OH)$_2$. If 0.0425 mol Mg(OH)$_2$ were to react, it would require twice as many moles of HCl, or 0.0850 mol HCl. We have only 0.0790 mol HCl, so there isn't enough HCl to react with all the Mg(OH)$_2$. The reactant we "don't have enough of" is the limiting reactant, so once again, we conclude that HCl is the limiting reactant.

According to the coefficients in the equation, the number of moles of MgCl$_2$ that could form should equal half the number of moles of HCl that reacts. Half of 0.079 is about 0.04, so we expect about 0.04 moles of MgCl$_2$ could form. The formula mass of MgCl$_2$ is 95.2, so a mole of MgCl$_2$ weighs about 100 g. Therefore, 0.04 mol of MgCl$_2$ would weigh about 4 g. Our answer of 3.76 is not far from this, so the answer does seem to be reasonable.

In the preceding problem, we could also have been asked to calculate the amount of the excess reactant that would remain after the reaction is complete. This is a fairly simple calculation. We determined that the

HCl would consume 0.0395 mol of $Mg(OH)_2$. We also know that we begin with 0.0425 mol $Mg(OH)_2$. The difference between these two values is the amount of $Mg(OH)_2$ that will be left over.

Amount available in the reaction mixture	Amount that would react with the HCl	Amount left over

$$0.0425 \text{ mol } Mg(OH)_2 - 0.0395 \text{ mol } Mg(OH)_2 = 0.0030 \text{ mol } Mg(OH)_2$$

The amount of $Mg(OH)_2$ left over would be 0.0030 mol. If we wished to know its mass, we would multiply by the mass of a mole of $Mg(OH)_2$.

$$0.0030 \text{ mol } Mg(OH)_2 \times \frac{58.32 \text{ g } Mg(OH)_2}{1 \text{ mol } Mg(OH)_2} = 0.17 \text{ g } Mg(OH)_2$$

Thinking It Through

4   Aluminum and sulfur, when strongly heated, react to give aluminum sulfide, Al_2S_3. How many grams of Al_2S_3 can form from a mixture of 5.65 g of Al and 12.4 g of S?

Self-Test

33.   To prepare some $AlBr_3$, a chemist mixed 3.13 g of Al with 28.8 g of Br_2. The equation is

$$2Al + 3Br_2 \rightarrow 2AlBr_3$$

(a)   Identify the limiting reactant.   _Alumla_

(b)   How many grams of $AlBr_3$ can be made?   _30.93_

(c)   Of the reactant that was taken in excess, how many grams of it are left over?

New Terms

Write a definition of the following terms, which were introduced in this section. If necessary, refer to the Glossary at the end of the text.

limiting reactant   excess reactant

4.8   The predicted amount of product is just an estimate.

Review

Experiments designed to prepare a substance do not always go "perfectly." The *main reaction* often has one or more *competing reactions,* sometimes called *side reactions,* that channel some of the reactants into the formation of one or more *by products.* By products do not appear in the balanced equation for the main reaction. To see how well the experiment has gone, the chemist uses the balanced equation for the main reaction and the mass of the limiting reactant to calculate the "perfect" result, the *theoretical yield*, and then assess the success of forming the desired product in terms of a calculated *percentage yield*.

We can calculate a *percentage yield* of a product by the following equation:

$$\frac{\text{(amount of product actually obtained)}}{\begin{pmatrix} \text{amount of product predicted from the} \\ \text{stoichiometry calculation based on} \\ \text{the limiting reactant} \end{pmatrix}} \times 100\% = \text{percentage yield}$$

The quantity in the numerator is called the *actual yield* and the quantity in the denominator is the theoretical yield. "Amount" may be in moles or in grams, but it must be in the same unit in both the numerator and denominator. The calculation of the theoretical yield *must* be based on the limiting reactant (Section 4.7).

Thinking It Through

5   Aluminum and bromine combine to give aluminum bromide. When 10.00 g of Al was mixed with 80.00 g of Br_2 there was obtained 80.00 g of $AlBr_3$. Calculate the percentage yield of $AlBr_3$.

Self-Test

34.   Ethyl alcohol can be converted to diethyl ether (an anesthetic) by the following reaction.

$$2C_2H_5OH \rightarrow C_4H_{10}O + H_2O$$

ethyl alcohol   diethyl ether

Some of the diethyl ether is lost during its synthesis because it evaporates very easily. If 40.0 g of ethyl alcohol gives 28.2 g of diethyl ether, what is the percentage yield?

35.   Solid sodium carbonate, Na_2CO_3, was sprinkled onto a spill of hydrochloric acid, HCl, to destroy the acid by the following reaction.

$$Na_2CO_3 + 2HCl \rightarrow 2NaCl + H_2O + CO_2$$

If the spill contained 50.0 g of HCl (dissolved in water), was the HCl entirely destroyed by 50.0 g of Na_2CO_3? If not, how many grams of HCl were destroyed? If more Na_2CO_3 was used than necessary, then how many grams of it were in excess?

New Terms

Write the definitions of the following terms, which were introduced in this section. If necessary, refer to the Glossary at the end of the text.

by-product   actual yield

competing reaction   percentage yield

main reaction   theoretical yield

side reaction

Solutions to Thinking It Through

1   The formulas give the ratios by atoms (or by moles) of N and H in compound X. The experimental data tell us that the observed ratio is 0.024 mol N to 0.048 mol N, which is the same as 1:2. This ratio fits only N_2H_4, because 2:4 is the same as 1:2. Moreover, N_2H_4 is the only candidate for which the ratio of formula units (N_2H_4) to the atoms of N is 1:2 (0.012 : 0.024).

2   To find the mass of O in the sample, simply do a subtraction of the mass of K from the total mass.

$$0.2564 \text{ g} - 0.2128 \text{ g} = 0.0436 \text{ g O}$$

We now know the *mass* ratio of K to O, but we need the *mole* ratio to make an empirical formula. We thus need the mass-to-moles tools provided by the atomic masses of K and O, 39.10 for K and 16.00 for O. When you carry out the operation on each mass, you find that there is 0.005442 mol K and 0.00272 mol O. To work our way toward a ratio in *whole* numbers, we divide both by the smaller, 0.00272 and get a ratio of 2.00 : 1.00. The formula is thus K_2O.

3   No problem in reaction stoichiometry can be worked without a balanced equation, so this is the first step.

$$2SO_2 + 3O_2 \rightarrow 2SO_3$$

Now we know (from the coefficients) that the ratio of moles of SO_2 to moles of O_2 is 2:3. So,

$$4.5 \text{ mol } O_2 \times \frac{2 \text{ mol } SO_2}{3 \text{ mol } O_2} = 3.0 \text{ mol } SO_2$$

The coefficients tell us that for each mole of SO_2 used, one mole of SO_3 will form; the mole ratio is 1:1. So 3.0 mol of SO_3 can form.

4   We recognize this as a limiting reactant problem because the masses of *two* reactants are given. But first we need tools. One is the balanced equation, which we can write from the information given as

$$2Al + 3S \rightarrow Al_2S_3$$

We must convert the masses given into moles, using the grams–to–moles tool, so that we can compare the *mole* ratios of the reactants as they were taken.

$$5.65 \text{ g Al} = ? \text{ mol Al.}$$

We need the atomic mass of Al, 26.98

$$5.65 \text{ g Al} \times \frac{1 \text{ mol Al}}{26.98 \text{ g Al}} = 0.209 \text{ mol Al}$$

Doing the same kind of calculation on 12.4 g S (atomic mass 32.07) tells us that 12.4 g S = 0.387 mol S. Now we have to do the comparison. The mole ratio of Al to S is supposed to be 2 Al : 3 S. Arbitrarily beginning with Al, we can see how many moles of S are needed for 0.209 mol of Al.

$$0.209 \text{ mol Al} \times \frac{3 \text{ mol S}}{2 \text{ mol Al}} = 0.314 \text{ mol S}$$

More than 0.314 mol of S was taken, so all of the Al will be used up. Al limits the yield, so we calculate the mass of Al_2S_3 based on Al.

$$0.209 \text{ mol Al} \times \frac{1 \text{ mol Al}_2S_3}{2 \text{ Al}} = 0.104 \text{ mol Al}_2S_3$$

Finally, using a mole-to-mass tool—from the formula mass of Al_2S_3 (150.17) on 0.104 mol of Al_2S_3 would show that a mass of 15.6 g of Al_2S_3 can be obtained.

5   We first need the equation.

$$2Al + 3Br_2 \rightarrow 2AlBr_3$$

Because we're given the masses of *two* reactants, we have to do a limiting reactant calculation first. We find the following moles of reactants using their formula masses.

$$10.00 \text{ g Al} = 0.3706 \text{ mol Al}$$

$$80.00 \text{ g Br}_2 = 0.5006 \text{ mol Br}_2$$

We can tell using the equation's coefficients that 0.3706 mol Al needs 3/2 times as many moles of Br_2 or 0.5559 mol of Br_2. But this much was *not* used, so there will be aluminum left over. It's the Br_2 that limits. Now we see how much $AlBr_3$ we could obtain in theory from 0.5006 mol of Br_2.

$$0.5006 \text{ mol Br}_2 \times \frac{2 \text{ mol AlBr}_3}{3 \text{ mol Br}_2} = 0.3337 \text{ mol AlBr}_3$$

Because the formula mass of $AlBr_3$ calculates to be 166.68, we find the mass of $AlBr_3$ as follows.

$$0.3337 \text{ mol AlBr}_3 \times \frac{166.68 \text{ g AlBr}_3}{1 \text{ mol AlBr}_3} = 88.99 \text{ g AlBr}_3$$

There was obtained only 80.00 g $AlBr_3$, so the percentage yield is

$$\frac{80.00 \text{ g}}{88.99 \text{ g}} \times 100\% = 89.90\%$$

Answers to Self-Test Questions

1.   9848.5 u
2. 1.90×10^{23} atom Ag
3.   6 molecules CO_2
4.   58.1 u
5. 1.22×10^{23} molecules
6.   Formula mass = 203.14. $\dfrac{1 \text{ mol } (NH_4)_3PO_4 \cdot 3H_2O}{203.14 \text{ g } (NH_4)_3PO_4 \cdot 3H_2O}$ and $\dfrac{203.14 \text{ g } (NH_4)_3PO_4 \cdot 3H_2O}{1 \text{ mol } (NH_4)_3PO_4 \cdot 3H_2O}$
7. $\dfrac{6.02 \times 10^{23} \text{ molecules } NH_3}{17.0 \text{ g } NH_3}$ and $\dfrac{17.0 \text{ g } NH_3}{6.02 \times 10^{23} \text{ molecules } NH_3}$
8. 2.41×10^{23} atoms Fe
9.   49.04 g H_2SO_4
10.   2.50 mol CO_2
11.   40.078 g Ca
12.   0.0323 mol P
13.   58.1 g $C_{12}H_{12}N_2O_3$
14. 3.75×10^{-3} mol C_6HCl_5O
15. 3.13×10^{23} atom Fe
16. 4.52×10^{23} atom H

17. 9.36×10^{-21} g Na
18. 2.53×10^{21} atom U
19. 1.7×10^{21} molecules H_2O
20. 0.800 mol O; 1.60 mol H
21. 0.600 mol Ca; 1.20 mol O; 1.20 mol H
22. 12.8 g O
23. 1.21 g H
24. 13.52 g P; 10.48 g O. (Note the sum of these; it's 24.00 g.)
25. 1.4 mol O
26. 39.70% Cr; 17.55% Na; 47.76% O
27. The formula mass of $Na_2S_2O_3$ is 158.12. The calculated percentages are Na, 29.08; S, 40.56; and O, 30.36, which are very close to the experimental percentages, so the formula is correct.
28. C_5H_7
29. C_5H_8
30. (a) 12.5 mol O_2, (b) 0.500 mol P_4O_{10}, (c) 20.0 mol O_2
     (d) 14,4 mol P, 18.0 mol O_2
31. (a) 7.33 g As, (b) 2.16 g O_2, (c) 7.72 g As_2O_3
32. (a) $2Li + H_2 \rightarrow 2LiH$
     (b) $4Na + O_2 \rightarrow 2Na_2O$
     (c) $2Sr + O_2 \rightarrow 2SrO$
     (d) $2HCl + SrO \rightarrow SrCl_2 + H_2O$
     (e) $2HBr + Na_2O \rightarrow 2NaBr + H_2O$
     (f) $2Al + 3S \rightarrow Al_2S_3$
     (g) $CH_4 + 4F_2 \rightarrow CF_4 + 4HF$
     (h) $CO_2 + 4H_2 \rightarrow CH_4 + 2H_2O$
33. (a) Al limits, (b) 30.9 g $AlBr_3$
     (c) 1 g Br_2 left over (rounded from 0.959 g)
34. 87.9%
35. 15.5 g HCl not consumed

Tools you have learned

Consider removing this chart from the Study Guide so you can have it handy when tackling homework problems.

Tool	*How it Works*
Avogadro's number	We use Avogadro's number when we have to relate macroscopic lab-sized quantities (e.g., moles) to numbers of individual atomic-sized particles such as atoms, molecules, or ions.
Chemical formula	We use the subscripts in a formula to establish atom ratios and mole ratios between the elements in the substance.
Atomic mass	We use atomic masses to form conversion factors to calculate mass from moles of an element, or moles from the mass of an element.
Formula mass; molecular mass	We use these to form conversion factors to calculate mass from moles of a compound, or moles from the mass of a compound.
Percentage composition	This gives a way to represent the composition of a compound. It can be used as the basis for computing the empirical formula. Comparing experimental and theoretical percentage compositions can help establish the identity of a compound.
Balanced chemical equation	When we need to relate amounts of substances in a chemical reaction, we use the coefficients of a balanced equation to establish stoichiometric equivalencies that relate moles of one substance to moles of another. The coefficients also establish the ratios by formula units among the reactants and products.
Theoretical, actual, and percentage yields	These are needed when we wish to estimate the efficiency of a reaction. Remember that the theoretical yield is calculated from the limiting reactant using the balanced chemical equation.

Chapter **5**

Reactions between Ions in Aqueous Solutions

Solutions of compounds are often used for carrying out reactions, and we begin this chapter with the basic terms used to describe solutions. The principal focus of the chapter is on reactions that involve ionic substances as well as substances that form ions when dissolved in water. The latter are usually acids or bases, and we introduce you here to these important kinds of chemicals and the reactions they undergo with each other. You will learn about several classes of reactions for which it is possible to predict the products. These are common reactions often encountered in the laboratory, so you should learn how to deal with them.

To deal with the stoichiometry of reactions in solutions, the concentration unit *moles per liter* or *molarity* is studied. If you know a solution's molarity, you can obtain the number of moles of its solute required for an experiment by dispensing a certain volume of the solution instead of weighing the reactant. You will learn how to prepare solutions of a desired molarity, how to dilute solutions quantitatively, and how to use molarity in stoichiometric calculations when substances are in solution.

Learning Objectives

In this chapter, you should keep in mind the following goals.

1   To learn the meaning of the terms solute, solvent, concentration, concentrated solution, dilute solution, saturated solution, unsaturated solution, supersaturated solution, and solubility.

2   To learn what happens to ionic solutes when they dissolve in water and to learn how to write chemical equations to represent the changes that take place.

3   To learn how to convert a molecular equation into ionic and net ionic equations.

4   To learn how to use the solubility rules to predict precipitation reactions.

5   To learn the properties of acids and bases.

6   To learn how molecular acids and bases form ions by reacting with water and how a dynamic equilibrium is able to account for the low concentrations of ions in solutions of weak acids and bases.

7   To learn how to name acids and bases.

8   To learn how to write ionic and net ionic equations for neutralization reactions.

9   To learn how to write ionic and net ionic equations for reactions that produce gases.

10   To learn which factors cause a metathesis reaction and how to predict the products of metathesis reactions.

11   To become familiar with molarity as a concentration unit and how to prepare solutions with a specified molarity.

12   To learn how to do the calculations needed to prepare a dilute solution from a more concentrated solution of known molarity.

13   To learn how to work problems that deal with the stoichiometry of ionic reactions in solution, especially those that deal with the experimental technique called titration.

5.1   Special terminology applies to solutions

Review

The chief reason that reactions are carried out using *solutions* of the reactants is to let their molecules or ions have the freedom to move around and find each other. Several new terms are given in this section for describing solutions. Be sure to learn their meanings, because they will be used often in the classroom and the lab.

Self-Test

1.   When sugar dissolves in water, which is the *solvent* and which is the *solute*?

    Solvent _____   Solute   _____

2.   When a crystal of sodium acetate was added to a solution of sodium acetate in water, the crystal dissolved. What term would be used to describe this solution (saturated, unsaturated, or supersaturated)?

    _____

3.   When a crystal of sodium acetate was added to a different solution of sodium acetate in water, considerable crystalline sodium acetate separated from the solution. What term should be used to describe the sodium acetate solution before more solute was introduced?

    _____

New Terms

Write the definitions of the following terms, which were introduced in this section. If necessary, refer to the Glossary at the end of the text.

solution	solvent	solute
saturated solution	unsaturated solution	supersaturated solution
solubility	concentration	percentage concentration
concentrated	dilute	precipitate
precipitation reaction		

5.2   Ionic compounds conduct electricity when dissolved in water

Review

Ionic compounds *dissociate* (break apart) into their individual ions when they dissolve in water. These ions become surrounded by water molecules and are said to be *hydrated*. Because the solute yields charged particles in the solution, the solution conducts electricity, which is why such solutes are called *electrolytes*. Ionic compounds are essentially 100% dissociated in water, so their solutions conduct electricity well and ionic compounds are called strong electrolytes. Even ionic compounds that have very low solubilities in water are called strong electrolytes because they are effectively completely dissociated.

Compounds such as sugar that are not able to undergo dissociation and do not yield ions in solution are called *nonelectrolytes*; their solutions do not conduct electricity. You should be sure that you can write equations for the *dissociation* of an ionic compound in water. To do this, it is necessary that you know the formulas and the charges of the ions. If the formulas of polyatomic ions are still difficult to remember, review them in Table 2.5 on page 62 of the text.

In studying this section, notice how the formula of the salt determines the number of each kind of ion that is found in the solution. For example, the dissociation of chromium(III) sulfate, $Cr_2(SO_4)_3$, in water produces two Cr^{3+} ions and three SO_4^{2-} ions for each formula unit of the salt.

$$Cr_2(SO_4)_3(s) \rightarrow 2Cr^{3+}(aq) + 3SO_4^{2-}(aq)$$

Self-Test

4.   What are the formulas of the ions that would be found in aqueous solutions of the following salts?

(a) $AgC_2H_3O_2$ _____

(b) $(NH_4)_2Cr_2O_7$ _____

(c) $Ba(OH)_2$ _____

5.   Write chemical equations for the dissociation of the compounds in Question 4 when they are dissolved in water.

(a)   _____

(b)   _____

(c)   _____

New Terms

Write the definitions of the following terms, which were introduced in this section. If necessary, refer to the Glossary at the end of the text.

dissociation	nonelectrolyte	electrolyte
strong electrolyte	hydrated	

5.3   Equations for ionic reactions can be written in different ways

Review

Reactions between ionic compounds in aqueous solutions are really reactions between their ions because ionic compounds exist in solution in dissociated form. There are three ways that we normally represent these reactions by chemical equations. A *molecular equation* shows complete formulas for all reactants and products.

Generally we indicate whether the substances are soluble or insoluble by writing (*aq*) or (*s*) following their formulas. In an *ionic equation*, we write the formulas for all soluble strong electrolytes in dissociated form. In a *net ionic equation*, we show only those ions that are actually involved in the chemical reaction. We obtain the net ionic equation by dropping *spectator ions* from the ionic equation.

As an example, consider the reaction of solutions of silver nitrate ($AgNO_3$) and sodium chloride (NaCl) to give a precipitate of silver chloride (AgCl) and a solution that contains sodium nitrate ($NaNO_3$). Here are the molecular, ionic, and net ionic equations for the reaction.

Molecular equation:

$$AgNO_3(aq) + NaCl(aq) \rightarrow AgCl(s) + NaNO_3(aq)$$

Ionic Equation:

$$Ag^+(aq) + NO_3^-(aq) + Na^+(aq) + Cl^-(aq) \rightarrow AgCl(s) + Na^+(aq) + NO_3^-(aq)$$

Net Ionic Equation:

$$Ag^+(aq) + Cl^-(aq) \rightarrow AgCl(s)$$

Notice that to obtain the net ionic equation, we have dropped the Na^+ and NO_3^- ions from the ionic equation. Because these ions do not change during the reaction, they are called spectator ions.

When you write ionic and net ionic equations, be sure you follow the two criteria for having them balanced: (1) There must be the same number of atoms of each kind on both sides of the equation, and (2) the net electrical charge must be the same on both sides.

Self-Test

6. Balance the following molecular equations and then write their ionic and net ionic equations:

(a) ___ $Cu(NO_3)_2(aq) +$ ___ $KOH(aq) \rightarrow$ ___ $Cu(OH)_2(s) +$ ___ $KNO_3(aq)$

Ionic equation:

_____

Net ionic equation:

_____

(b) ___ $NiCl_2(aq) +$ ___ $AgNO_3(aq) \rightarrow$ ___ $Ni(NO_3)_2(aq) +$ ___ $AgCl(s)$

Ionic equation:

_____

Net ionic equation:

_____

(c) ___ $Cr_2(SO_4)_3(aq) +$ ___ $BaCl_2(aq) \rightarrow$ ___ $CrCl_3(aq) +$ ___ $BaSO_4(s)$

Ionic equation:

_____

Net ionic equation:

New Terms

Write the definitions of the following terms, which were introduced in this section. If necessary, refer to the Glossary at the end of the text.

ionic reaction   molecular equation

ionic equation   net ionic equation

spectator ion

5.4   Reactions that produce precipitates can be predicted

Review

In this section you learn how to predict whether a *metathesis reaction* will occur between a pair of reactants to form a precipitate. To do this, you will apply the Solubility Rules on page 164. The first step is to write a molecular equation for the reaction. To do this, it is necessary to determine what the products are. This is done by exchanging the anions between the two cations. However, *we must be very careful to write the correct formulas of the products*. That is, we must be sure that the formulas represent electrically neutral formula units. For example, in the reaction between $Cr_2(SO_4)_3$ and $BaCl_2$, we have to be sure to check the changes on the anions and cations before we write the products. The anions are SO_4^{2-} and Cl^-; the cations are Cr^{3+} and Ba^{2+}. When we exchange the anions, the Cl^- goes with the Cr^{3+} to give $CrCl_3$ and the SO_4^{2-} goes with the Ba^{2+} to give $BaSO_4$. Therefore, the equation before balancing is

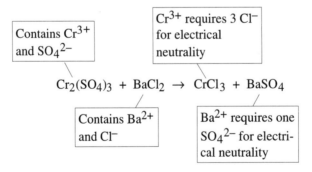

To balance the equation, we place a 3 in front of $BaCl_2$, a 2 in front of $CrCl_3$, and a 3 in front of $BaSO_4$ to give

$$Cr_2(SO_4)_3 + 3BaCl_2 \rightarrow 2CrCl_3 + 3BaSO_4$$

The next step is to write the ionic equation for the reaction. To do this we must know whether the reactants and products are soluble in water. This is where we apply the solubility rules. After we've identified any precipitate that might form, we can then divide the equation into ionic and net ionic equations. For this reaction, the molecular, ionic, and net ionic equations are, respectively,

$$Cr_2(SO_4)_3(aq) + 3BaCl_2(aq) \rightarrow 2CrCl_3(aq) + 3BaSO_4(s)$$

$$2Cr^{3+}(aq) + 3SO_4^{2-}(aq) + 3Ba^{2+}(aq) + 6Cl^-(aq) \rightarrow 2Cr^{3+}(aq) + 6Cl^-(aq) + 3BaSO_4(s)$$

$$3SO_4^{2-}(aq) + 3Ba^{2+}(aq) \rightarrow 3BaSO_4(s)$$

Self-Test

7. Test your knowledge of the solubility rules by applying them to the salts listed below. *Circle* the formulas of those that are soluble in water:

    Na_2SO_4 $CuCO_3$ $Ni(NO_3)_2$ Hg_2Cl_2 $PbBr_2$ Cr_2O_3 $Ca_3(PO_4)_2$

    $(NH_4)_2CO_2$ $Ba(ClO_4)_2$ AgI ZnS $MgSO_3$ $Sr(C_2H_3O_2)_2$ $(NH_4)_2S$

8. Write molecular, ionic, and net ionic equations for the reactions that occur between the following compounds.

    (a) $AgC_2H_3O_2 + (NH_4)_2S \rightarrow$

    Molecular equation:

    _____

    Ionic equation:

    _____

    Net ionic equation:

    _____

    (b) $Cr(ClO_4)_3 + KOH \rightarrow$

    Molecular equation:

    _____

    Ionic equation:

    _____

    Net ionic equation:

    _____

    (c) $Pb(NO_3)_2 + K_2SO_4 \rightarrow$

    Molecular equation:

    _____

    Ionic equation:

    _____

    Net ionic equation:

    _____

New Terms

Write the definitions of the following terms, which were introduced in this section. If necessary, refer to the Glossary at the end of the text.

double replacement reaction   precipitation reaction

metathesis   solubility rules

5.5   Acids and bases are classes of compounds with special properties

Review

Many substances can be categorized as acids or bases because they possess certain characteristic properties and because they undergo predictable reactions with each other called *neutralization*. Be sure to review the general properties of acids and bases described on pages 167 and 168.

According to the description of acids and bases presented in this chapter (a modern version of the Arrhenius definition) an *acid* is a substance that produces H_3O^+ by transferring H^+ ions to water molecules. A base produces OH^- in water. The neutralization reaction of an acid and a base generally produces water plus a *salt*. We use the term salt to mean any ionic compound not containing OH^- or O^{2-}.

In general, acids are molecular substances. Their reaction with water is called an *ionization reaction* because ions are formed from neutral molecules—before the reaction there are no ions and afterwards there are. Study the general equation for the reaction given on page 169. Acids are classified according to the number of hydrogens of the parent molecule that are capable of ionizing (e.g., monoprotic, diprotic, triprotic, polyprotic). Polyprotic acids ionize stepwise, losing one hydrogen ion at a time. Some oxides of nonmetals, such as SO_3, N_2O_5, and CO_2, react with water to give acid molecules, which subsequently undergo ionization in water. Such oxides are called *acidic anhydrides*, meaning "acids without water."

There are two kinds of Arrhenius bases. One consists of the metal hydroxides—compounds such as NaOH and $Ca(OH)_2$. These are ionic substances and simply dissociate in water to give the metal ion and the hydroxide ion. Because they are ionic compounds, they are completely dissociated. Soluble metal oxides react with water to form hydroxides and are said to be *basic anhydrides*. (Notice that nonmetal oxides are acidic anhydrides and metal oxides are basic anhydrides; this is a chemical distinction between metals and nonmetals.)

The second kind of base consists of molecules that react with water and release OH^- ions. An example is ammonia. When a molecular base ionizes, an H^+ ion is transferred from a water molecule to a molecule of the base. Study the general equation for the reaction of a molecular base with water on page 172.

Thinking It Through

For the following, identify the information needed to solve the problem and show (or explain) what must be done with it.

1   A white solid readily dissolves in water to form a solution that turns red litmus paper blue. Which of the following compounds could this solid be?

CO_2, Na_2O, HNO_3, KOH, SO_3, $HC_2H_3O_2$

2   Consider the oxide of an element in group IIA of the periodic table. Let Z be the symbol of this
    element. (a) Write the formula of this oxide using the symbol Z. (b) Will a solution of this oxide in
    water give a blue or a red color to litmus paper?

Self-Test

$$ZO + H_2O \rightarrow ZOH^+ + OH^-$$

9. Write chemical equations for the ionization of the following acids in water.

    (a) $HClO_2$ $HClO_2 + H_2O \rightarrow ClO_2^- + H_3O^+$

    (b) H_2SeO_3 $H_2SeO_3 + H_2O \rightarrow H_3O^+ + HSeO_3^-$

    $HSeO_3^- + H_2O \rightarrow H_3O^+ + SeO_3^{2-}$

    (c) H_3AsO_4 $H_3AsO_4 + H_2O \rightarrow H_3O^+ + H_2AsO_4^-$

    $H_2AsO_4^- + H_2O \rightarrow H_3O^+ + HAsO_4^{2-}$

    $HAsO_4^{2-} + H_2O \rightarrow H_3O^+ + AsO_4^{3-}$

10. Tetraphosphorus decaoxide is the acidic anhydride of phosphoric acid, H_3PO_4. Write the chemical
     equation showing the reaction of this oxide with water to form the acid.

     _____

11. Write an equation for the ionization reaction of the base hydrazine, N_2H_4, in water.

     _____

12. The radioactive element radium forms a soluble oxide with the formula RaO. Write an equation for
     its reaction with water.

     _____

13. Why is Na_2O called a *basic anhydride*? Write the equation for its reaction with water.

     _____

     _____

14. Why is SO_3 called an *acidic anhydride*? Write the equation for its reaction with water.

     _____

     _____

New Terms

Write the definitions of the following terms, which were introduced in this section. If necessary, refer to the
Glossary at the end of the text.

acid	monoprotic acid	polyprotic acid
base	diprotic acid	acidic anhydride
reagent	triprotic acid	basic anhydride
ionization reaction		

5.6 Naming acids and bases follows a system

Review

In this section, we extend the system of chemical nomenclature introduced in Chapter 2 to cover acids and bases.

Binary acids and their salts

The binary acids are the binary hydrogen compounds of elements in Groups VIA and VIIA. Examples include hydrogen chloride, HCl, and hydrogen sulfide, H_2S.

$$HCl(g) + H_2O \rightarrow H_3O^+(aq) + Cl^-(aq)$$

$$H_2S(g) + H_2O \rightarrow H_3O^+(aq) + HS^-(aq)$$

These two acids are called binary acids because they contain only one element besides hydrogen, and so are binary compounds in their pure states. In naming the acids, we add the prefix *hydro-* and the suffix *-ic acid* to the stem of the nonmetal name. Thus, in water, hydrogen chloride gives *hydro*chlor*ic acid* and hydrogen sulfide gives *hydro*sulfur*ic acid.*

As mentioned earlier, acids react with bases in a reaction called neutralization. The products are water and an ionic compound. Salts formed by neutralizing binary acids contain a monatomic anion of the acid and have the *-ide* ending. Thus, hydro*chlor*ic acid gives *chlor*ide salts.

Oxoacids and their salts

An oxoacid contains hydrogen, oxygen, plus a third element. Examples include HNO_3 (nitric acid) and H_2SO_4 (sulfuric acid). Many nonmetals form more than one oxoacid. If only two of them are possible, the name of the one having the larger number of oxygen atoms ends in *-ic* and the name of the acid with the smaller number of oxygens ends in *-ous*. The following examples were given in the text.

H_2SO_4 sulfuric acid		HNO_3 nitric acid	
H_2SO_3 sulfurous acid		HNO_2 nitrous acid	

When there are more than two oxoacids for a given nonmetal (as there are for the halogens), the following prefixes are used (in the order of increasing numbers of oxygen atoms).

hypo ... ous acid	(For example, hypo*chlor*ous acid — HClO)
... ous acid	(For example, *chlor*ous acid — $HClO_2$)
... ic acid	(For example, *chlor*ic acid — $HClO_3$)
per ... ic acid	(For example, per*chlor*ic acid — $HClO_4$)

Remember that ...*ic* acids give anions that end in *-ate*, and that ...*ous* acids give anions that end in *-ite*.

Acid salts

Monoprotic acids such as HCl have just one hydrogen to be neutralized, so they form only one kind of salt. However, polyprotic acids, such as H_2SO_4 and H_3PO_4, have more than one hydrogen to be neutralized, and the neutralization can be incomplete to give anions that still contain "acidic hydrogens." Compounds formed by these anions are called acid salts. In naming them, the hydrogen is specified either as "hydrogen" or, when only one acid salt is possible, by the prefix *bi-* before the name of the anion (for example, HSO_4^- is named as hydrogen sulfate ion or bisulfate ion).

Bases

Ionic bases contain hydroxide ions and are named just as any other ionic compound by specifying the cation first, followed by the anion (hydroxide). Molecular bases are named just by giving the name of the molecule.

Self-Test

15. Potassium arsenite has the formula K_3AsO_3. Name the following.

    (a) H_3AsO_3 _____

    (b) H_3AsO_4 _____

    (c) $Ca_3(AsO_4)_2$ _____

16. Name the following;

    (a) $H_2Se(g)$ _____

    (b) $H_2Se(aq)$ _____

    (c) $HF(g)$ _____

    (d) $HF(aq)$ _____

    (e) $KHSO_3$ _____

    (f) HIO_3 _____

    (g) $HBrO$ _____

    (h) $CuH_2AsO_4^{..}$ _____

17. Write formulas for the following.

    (a) sodium perbromate _____

    (b) periodic acid _____

    (c) sodium bioxalate _____

    (d) perbromic acid _____

    (e) hydrotelluric acid _____

    (f) barium hydroxide _____

New Terms

Write the definitions of the following terms, which were introduced in this section. If necessary, refer to the Glossary at the end of the text.

binary acid   oxoacid

acid salt

5.7   Acids and bases are classified as strong or weak

Review

Strong electrolytes are substances that are completely divided into ions in aqueous solution. Salts are examples. Some acids are also strong electrolytes and are classified as *strong acids*; they are 100% ionized in water. A list of six strong acids is given on page 175. You should memorize them because if you encounter an acid and it's *not* on the list, you can be fairly confident that it is *not* a strong acid.

The *strong bases* are the soluble metal hydroxides. These are the Group IA hydroxides and the hydroxides of the Group IIA metals from calcium on down. (Actually, you don't need to memorize them because they are covered by the solubility rules you learned in Section 5.4.) Metal hydroxides with low solubilities, such as $Mg(OH)_2$, are also strong electrolytes, but their solutions contain so little hydroxide ion that they are not strongly basic.

For a *weak acid,* only a small fraction of the molecules are ionized. Most of the weak acid is present in the solution as non-ionized molecules, and there is an equilibrium between the ions of the acid and its molecules. Similarly, in solutions of *weak bases* (which are the molecular bases), only a small fraction of the base exists in ionized form. Most of the base exists in solution as molecules.

Dynamic equilibrium in solutions of weak acids and bases

When describing the breakup into ions of weak acids or bases, an equation with oppositely pointing arrows, $\rightleftharpoons$, is usually used to describe the *dynamic equilibrium* in the solution. The *forward reaction* (going from left to right) is the ionization and the *reverse reaction* (going right to left) is the reconversion of the ions to the un-ionized or undissociated species. The double arrows symbolize the dynamic nature of an equilibrium—both forward and reverse reactions occur in the solution continuously and at equal rates when there is equilibrium. But there is no net change in the concentrations of either the ions or the un-ionized acid or base. For a weak acid or base, the extent to which the forward reaction proceeds toward completion is small and we say that the *position of equilibrium* lies to the left in the ionization reaction. (For a strong acid, the ionization reaction is essentially complete, so if we were to look at the reaction as an equilibrium, the position of equilibrium would lie far to the right.)

Thinking It Through

For the following, identify the information needed to solve the problem and show (or explain) what must be done with it.

3   Is butyric acid a strong or a weak acid? How can you tell without even knowing its formula and without having access to a table?

Self-Test

18.   Hydrogen cyanide, HCN, is a weak acid when dissolved in water. Write the chemical equation that illustrates the equilibrium that exists in the aqueous solution.

$$HCN + H_2O \rightleftharpoons CN^-_{(aq)} + H_3O^+_{(aq)}$$

Which reaction, the forward or the reverse, is far from completion?

forward

19. Chloric acid, $HClO_3$ is a strong acid. Write a chemical equation that shows the reaction to this acid with water.

$$HClO_3 + H_2O \rightarrow H_3O^+ + ClO_3^-{}_{(aq)}$$

20. Aniline, $C_6H_5NH_2$, is a weak base in water. Write a chemical equation that illustrates the equilibrium that exists in aqueous aniline solutions.

$$C_6H_5NH_2 + H_2O \rightleftharpoons C_6H_5NH_3^+{}_{(aq)} + OH^-{}_{(aq)}$$

21. From memory, write the names and the formulas of the six strong acids.

$$HCl, \quad H_2SO_4, \quad HClO_4, \quad HClO_3, \quad HBr,$$
$$HI$$

New Terms

Write the definitions of the following terms, which were introduced in this section. If necessary, refer to the Glossary at the end of the text.

strong acid	weak acid	weak electrolyte
strong base	weak base	strong electrolyte
dynamic equilibrium		

5.8   Neutralization occurs when acids and bases react

Review

Strong acids and strong bases react with each other to form a salt and water in a reaction we call *neutralization*. If both are soluble, the net reaction is

$$H^+ + OH^- \rightarrow H_2O$$

The anion of the acid and the cation of the base are spectator ions in the reaction.

Polyprotic acids can be completely neutralized or they can be partially neutralized to form acid salts.

When constructing an ionic equation, weak acids and bases are written in molecular form because in a solution most of the weak acid or base is present as molecules, not ions. This changes the nature of the spectator ions and the form of the net ionic equation. Study the examples that are shown on pages 180 through 181.

Insoluble oxides and hydroxides react with both strong and weak acids. In constructing the ionic equation, remember that insoluble solids and weak acids are written in "molecular" form.

Self-Test

22. Write the net ionic equation for the reaction of

   (a)   KOH with H_2SO_4 (complete neutralization)

$$KOH_{(aq)} + H_2SO_4{}_{(aq)} \longrightarrow$$
$$OH^-{}_{(aq)} + H_2^+{}_{(aq)} \rightarrow H_2O$$

(b) $Ca(OH)_2$ with $HCHO_2$ (a soluble weak acid)

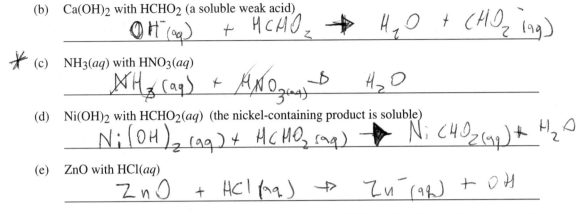

$$OH^-_{(aq)} + HCHO_2 \rightarrow H_2O + CHO_2^-_{(aq)}$$

✱ (c) $NH_3(aq)$ with $HNO_3(aq)$

$$NH_3(aq) + HNO_{3(aq)} \rightarrow H_2O$$

(d) $Ni(OH)_2$ with $HCHO_2(aq)$ (the nickel-containing product is soluble)

$$Ni(OH)_{2\,(aq)} + HCHO_2\,_{(aq)} \rightarrow NiCHO_{2(aq)} + H_2O$$

(e) ZnO with $HCl(aq)$

$$ZnO + HCl_{(aq)} \rightarrow Zn^-_{(aq)} + OH$$

23.   What salts are formed in the neutralization of H_3PO_4 by $Ca(OH)_2$.? Consider both complete and partial neutralization of the acid, but assume complete reaction of the $Ca(OH)_2$.

24.   Write the molecular, ionic, and net ionic equations for the complete neutralization of a solution of citric acid by aqueous ammonia. Citric acid has the formula $H_3C_6H_5O_7$.

New Terms

None

5.9   Gases are formed in some metathesis reactions.

Review

In this section we describe four kinds of reactions in which gases are formed.

1.   Acids react with soluble sulfides and many insoluble sulfides to give gaseous hydrogen sulfide and a salt.

   Because H_2S has a low solubility in water, it leaves a solution when it is formed as a product in a metathesis reaction.

2.   Cyanides react with acids to give gaseous hydrogen cyanide.

   HCN has a low solubility in water, and will be released when a strong acid reacts with a metal cyanide.

2.   Acids (strong and weak) react with soluble *and* insoluble carbonates and bicarbonates to give carbon dioxide and water.

   When you write a metathesis equation and find H_2CO_3 as one of the products, remember that it decomposes to give CO_2 and H_2O.

3.   Acids react with sulfites and bisulfites to give sulfur dioxide and water.

   When you write a metathesis equation and find H_2SO_3 as one of the products, remember that it decomposes to give SO_2 and H_2O.

4. Bases react with ammonium salts to give gaseous ammonia.

When you write a metathesis equation and find NH_4OH as one of the products, you should rewrite it as NH_3 plus H_2O.

Be sure to study Table 5.3 on page 185. Ask your instructor whether you are expected to learn all the substances released as gases that are described in this table.

Self-Test

[handwritten: $BaSO_3(s) + HCl \rightarrow BaCl + HSO_3 \rightarrow$]

25. What gas is formed in the reaction of HCl with

[handwritten: $K_2CO_3(aq) + HCl(aq)$]

(a) $K_2CO_3(aq)$ _____ *[handwritten: CO_2]*

(b) $BaSO_3(s)$ _____ *[handwritten: SO_2]*

(c) $(NH_4)_2S(aq)$ _____ *[handwritten: H_2S]*   *[handwritten: $(NH_4)_2S + HCl \rightarrow$]*

(d) $CaCO_3(s)$ _____ *[handwritten: CO_2]*

(e) $KHSO_3(aq)$ _____ *[handwritten: SO_2]*

(f) $NH_4HCO_3(aq)$ _____ *[handwritten: CO_2]*

[handwritten: $NH_4HCO_3 + HCl \rightarrow NH_4Cl + H_2CO_3$ $\rightarrow CO_2 + H_2O$]

26. Which of the substances in the preceding question would release a gas if treated with concentrated NaOH solution?

[handwritten: $NaOH(aq) + (NH_4)_2S(aq) \rightarrow$]

New Terms

[handwritten: $NH_4HCO_3(aq) + NaOH \rightarrow NaNH_4 + CO_2 H_2O \rightarrow H_2O + CO_2$]

None

[handwritten: $(NH_4)_2S + NaOH \rightarrow NaNH_4 + SOH HSO$]

5.10   Predicting Metathesis Reactions; A Summary

Review

In this section you learn that if a net ionic equation exists for a metathesis reaction, then a reaction does occur. However, if all the ions are spectator ions, and therefore cancel, there is no net reaction. As noted on page 187, a net ionic equation will be expected to exist under the following conditions.

1. A precipitate forms from a solution of soluble reactants.
2. Water forms in the reaction of an acid and a base.
3. A weak electrolyte forms from a solution of strong electrolytes. *[handwritten: ✗]*
4. A gas forms that escapes from the reaction mixture.

The preceding provides some general guidelines, but the real test is to write the equation as a metathesis using entire formulas for reactants and products (i.e., the molecular equation). Then write the ionic equation taking into account that (1) weak electrolytes are written in molecular form, (2) insoluble substances are written in molecular form, and (3) gases may be produced as described in Section 5.9. After writing the ionic equation, cancel any spectator ions. If an equation remains, there is a net reaction. If all the ions cancel, there is no net reaction. Study Examples 5.7 through 5.9.

Self-Test

27. Write net ionic equations for the reactions, if any, that occur between the following compounds. If there is no net reaction, state so by writing N.R. Assume any weak electrolytes are soluble in water, unless they are insoluble gases.

(a) $Pb(C_2H_3O_2)_2 + (NH_4)_2S \rightarrow$

_____

(b) $Cr(ClO_4)_3 + MgSO_4 \rightarrow$

_____

(c) $Zn(NO_3)_2 + KOH \rightarrow$

_____

(d) $Cr_2O_3 + HClO_4 \rightarrow$

_____

(e) $BaSO_3 + HC_2H_3O_2 \rightarrow$

_____

(f) $(NH_4)_2SO_4 + Ba(OH)_2 \rightarrow$

_____

(g) $Ca(C_3H_5O_2)_2 + HCl \rightarrow$

_____

New Terms

None

5.11   The composition of a solution is described by its concentration

Review

The composition of a solution is variable, meaning the proportions of solute to solvent can vary. To express the composition of a solution, we specify the *concentration* of the solute, and for purposes of stoichiometry, the most convenient expression of concentration is molarity. **Molarity** is defined as the ratio of the number of moles of solute to the volume of the solution expressed in liters.

$$\text{molarity } (M) = \frac{\text{moles of solute (mol)}}{\text{liters of solution (L)}}$$

Thus, a solution that contains 0.25 mole of solute per liter of solution has a concentration that we can express as 0.25 mol L^{-1}, or 0.25 M. The symbol M stands for the units mol L^{-1}. For example, a solution that is labeled 0.10 M HCl has a concentration of HCl equal to 0.10 mol L^{-1}.

To calculate the molarity of a solution, we need only the number of moles of solute (or the mass of solute, which we can convert to moles) and the volume of the solution. For instance, suppose we had 275 mL of a solution in which is dissolved 0.0264 mol of $CaCl_2$. To calculate the molarity, we divide the number of moles by the volume expressed in liters.

$$\text{molarity} = \frac{0.0264 \text{ mol } CaCl_2}{0.275 \text{ L solution}} = 0.0960 \text{ } M \text{ } CaCl_2$$

> Converting from milliliters to liters involves moving the decimal three places to the left.

Notice how we converted milliliters to liters. You should practice converting between these two units (from mL to L and from L to mL) until it becomes effortless. It is a skill you will use often.

Equivalent ways of expressing molar concentration

Because molarity is a ratio, it is possible to express the ratio in units other than moles and liters, and sometimes it's convenient to do so. One alternative is to convert the unit liter in the denominator to milliliters. Because 1 L = 1000 mL, we can express molarity in the units "mol/1000 mL". Thus, the concentration 0.20 M HCl can be expressed as follows:

$$0.20 \text{ } M \text{ HCl} = \frac{0.20 \text{ mol HCl}}{1 \text{ L solution}} = \frac{0.20 \text{ mol HCl}}{1000 \text{ mL solution}}$$

Another alternative is to convert the numerator to millimoles and the denominator to milliliters using: 1 mmol = 10^{-3} mol and 1 mL = 10^{-3} L. Let's do this for the 0.20 M HCl solution.

$$0.20 \text{ } M = \frac{0.20 \text{ mol HCl} \times \left(\dfrac{1 \text{ mmol HCl}}{10^{-3} \text{ mol HCl}} \right)}{1 \text{ L solution} \times \left(\dfrac{1 \text{ mL solution}}{10^{-3} \text{ L solution}} \right)} = \frac{0.20 \times \dfrac{1}{10^{-3}} \text{ mmol HCl}}{1 \times \dfrac{1}{10^{-3}} \text{ mL solution}}$$

Notice that the result gives the quantity $1/10^{-3}$ in both numerator and denominator. When we cancel this, we obtain 0.20 mmol HCl/1 mL solution

$$0.20 \text{ } M = \frac{0.20 \times \dfrac{1}{10^{-3}} \text{ mmol HCl}}{1 \times \dfrac{1}{10^{-3}} \text{ mL solution}} = \frac{0.20 \text{ mmol HCl}}{1 \text{ mL solution}}$$

Thus, the ratio of millimoles to milliliters is equal to the ratio of moles to liters. This means we can express molarity in either of these two sets of units. For example, if we had a 0.10 M solution of NaOH, we could write

$$0.10 \text{ } M \text{ NaOH} = \frac{0.10 \text{ mol NaOH}}{1 \text{ L solution}} \qquad \text{or} \qquad 0.10 \text{ } M \text{ NaOH} = \frac{0.10 \text{ mmol NaOH}}{1 \text{ mL solution}}$$

Molar concentration is the critical link between moles of solute and volume of solution

The reason molarity is so useful for stoichiometry is because it relates the amount of solute in a solution to the volume of solution. This makes it easy to dispense moles of solute simply by measuring out volumes of solution.

For calculations, molarity is the critical link that allows us to relate moles of solute and volumes of solution. Molarity provides the conversion factors needed to calculate the number of moles of solute in a given volume of a solution, or the volume of solution that contains a certain number of moles.

Let's see how we can use molarity to form conversion factors. Suppose we have a solution that is labeled 0.250 *M* HCl. The first thing we have to do is express this in units of moles and volume. For this solution we can write

$$0.250\ M = \frac{0.250\ \text{mol HCl}}{1\ \text{L solution}}$$

For convenience, we'll write the relationship between moles and volume as an equivalence

$$0.250\ \text{mol HCl} \Leftrightarrow 1.00\ \text{L solution}$$

(Notice that we've expressed the volume to 3 significant figures. For molarity, the precision is indicated by the numerator; the denominator can be written to the same number of significant figures.) From this relationship we can form two conversion factors useful in calculations.

$$\frac{0.250\ \text{mol HCl}}{1.00\ \text{L solution}} \quad \text{and} \quad \frac{1.00\ \text{L solution}}{0.250\ \text{mol HCl}}$$

A useful thing to remember is that when you multiply molar concentration by the volume of solution (expressed in liters), the result is the number of moles of solute. This is given as Equation 5.3 in the text.

$$\text{molarity x volume(L)} = \text{moles of solute}$$

If in a particular problem you are given both the volume and molarity of a solution, you are given information from which you can calculate the number of moles of solute in the solution. This is a critically important concept to remember.

Let's look at two examples.

Example 5.1   Working with molarity

How many grams of KOH are in 50.0 mL of 0.168 *M* KOH solution?

Analysis:

We can express the problem in equation form as follows:

$$50.0\ \text{mL KOH solution} \Leftrightarrow ?\ \text{g KOH}$$

Let's examine the data and see what we can calculate. We have both the molarity and volume of the solution. We know molarity relates moles and volume, so we should be able to use the molarity to calculate the number of moles of solute in the solution. We can write the molarity as

$$0.168\ M\ \text{KOH} = \frac{0.168\ \text{mol KOH}}{1\ \text{L solution}}$$

As written, this fraction is the conversion factor that will convert liters of solution to moles of KOH. However, to use it, we need the volume in liters, so we'll convert 50.0 mL to liters. In liters, the volume is 0.0500 L. Multiplying this volume by the molarity will give the number of moles of solute.

To convert moles of KOH to grams, the tool is the formula mass of KOH, which gives us the relationship,

$$1 \text{ mol KOH} = 56.10 \text{ g KOH}$$

Now we can see the path to the answer.

$$0.0500 \text{ L solution} \xrightarrow{\boxed{\text{tool}}} \underset{\text{molarity}}{} \text{mol KOH} \xrightarrow{\boxed{\text{tool}}} \underset{\substack{\text{formula} \\ \text{mass}}}{} \text{g KOH}$$

Solution:
Let's apply the conversion factors

$$0.0500 \text{ L KOH} \times \frac{0.168 \text{ mol KOH}}{1 \text{ L solution}} \times \frac{56.10 \text{ g KOH}}{1 \text{ mol KOH}} = 0.471 \text{ g KOH}$$

The 50.0 mL of solution contains 0.471 g of KOH.

Is the Answer Reasonable?
Let's do some proportional reasoning and approximate math. If we had an entire liter of the solution, it would contain 0.168 mol of KOH. One tenth (0.1) liter would contain 0.0168 mol, and half of that (0.05 L) would contain 0.0084 mol. This is a little less than 0.01 mol, so we'll use 0.01 as an approximate value. One mole of KOH weighs 56 g, so 0.01 mol would weigh 0.56 g. We have a little less than 0.01 mol, so the answer should be a little less than 0.56 g. The value we obtained (0.471 g) seems to be about right, so the answer is reasonable.

It is worth stating again that *whenever you have the molarity and volume of a solution, you can calculate the number of moles of solute*. Just multiply the molarity by the volume (in liters).

$$\text{L solution} \times \underset{\substack{\uparrow \\ \boxed{\text{volume (L)}} \quad \boxed{\text{molarity}}}}{\frac{\text{mol solute}}{\text{L solution}}} = \text{mol solute}$$

Let's look at another problem.

Example 5.2   Working with molarity

How many milliliters of 0.135 M Na_3PO_4 contains 12.5 g of the solute?

Analysis:
The problem, in equation form, is

$$12.5 \text{ g } Na_3PO_4 \Leftrightarrow \text{ ? mL solution}$$

The molarity (0.135 M or 0.135 mol L^{-1}) will allow us to relate volume of solution and moles of solute. The relationship is

$$0.135 \text{ mol } Na_3PO_4 \Leftrightarrow 1.00 \text{ L solution}$$

Before we can use it, however, we need to have the amount of solute expressed in moles. To convert grams to moles, we use the formula mass.

$$1 \text{ mol Na}_3\text{PO}_4 = 163.94 \text{ g Na}_3\text{PO}_4$$

We now have a way to go from g Na_3PO_4 to liters of solution. After we obtain this, we move the decimal point to change to milliliters.

Solution:

We begin by applying the unit conversions.

$$12.5 \text{ g Na}_3\text{PO}_4 \times \frac{1 \text{ mol Na}_3\text{PO}_4}{163.94 \text{ g Na}_3\text{PO}_4} \times \frac{1.00 \text{ L solution}}{0.135 \text{ mol Na}_3\text{PO}_4} = 0.565 \text{ L solution}$$

Finally, we convert the volume to milliliters to obtain 565 mL.

Is the Answer Reasonable?

One mole of Na_3PO_4 weighs about 160 g, so 0.1 mol would weigh about 16 g. Our amount (approximately 12 g) is about 3/4 of 16 g, or about 0.075 mol. If we round the molarity (0.135 M) to 0.15 M, then the amount of solution needed to have 0.075 mol would 0.5 L. Our solution is a little less concentrated than 0.15 M, so the volume required would be somewhat larger than 0.5 L. Our answer is "in the right ball park," so it's reasonable.

Preparing solutions of known molarity

If we want to prepare a solution of known molar concentration, we usually have in mind a particular final volume, so the calculation we have to do is to find first the moles of solute we need by using Equation 5.3. Then we convert moles to grams.

Preparing solutions by dilution

Often we prepare solutions by diluting a more concentrated solution with solvent. During this operation, the number of moles of solute remains constant. This leads to the Equation 5.4 in the text.

$$V_{\text{dil}}M_{\text{dil}} = V_{\text{concd}}M_{\text{concd}}$$

where the subscript "dil" refers to the more dilute solution and "concd" refers to the more concentrated one.

Self-Test

28. A solution is labeled 0.150 M HNO_3. Write the two conversion factors we can create from this information that would allow us to relate moles of HNO_3 and volume of this solution.

_____

29. For the solution in the preceding question, how many millimoles of HNO_3 would be found in 1 mL of the solution?

_____

How many micromoles of HNO_3 would be found in 1 μL of the solution?

_____

30. How would 500 mL of a 0.125 *M* CaCl$_2$ solution be prepared?

_____

31. How many moles and how many grams of NaCl are in 250 mL of 1.38 *M* NaCl?

_____

32. How many milliliters of 0.124 *M* HCl are needed to obtain 0.00244 mol HCl?

_____

33. How would you prepare 100 mL of 0.500 *M* H$_2$SO$_4$ from 0.800 *M* H$_2$SO$_4$?

_____

_____

34. How would you prepare 250 mL of 0.100 *M* HCl from 2.00 *M* HCl?

_____

_____

35. How many milliliters of water would have to be added to 50.0 mL of 0.250 *M* HCl to give a solution with a concentration of 0.100 *M*?

_____

New Terms

Write the definitions of the following terms, which were introduced in this section. If necessary, refer to the Glossary at the end of the text.

molar concentration   molarity

5.12   Molarity is used for problems in solution stoichiometry

Review

In previous problems dealing with reaction stoichiometry, you saw that the amounts of reactants and products could be expressed in grams or moles. Molarity and volume provide a third alternative, as illustrated in the following problem.

Example 5.3 Using molarity in a stoichiometry problem

How many grams of calcium carbonate will react with 75.0 mL of 0.250 *M* HCl? The equation for the reaction is

$$CaCO_3(s) + 2HCl(aq) \rightarrow CaCl_2(aq) + CO_2(g) + H_2O$$

Analysis:

The problem can be expressed as

$$75.0 \text{ mL HCl solution} \Leftrightarrow ? \text{ g CaCO}_3$$

We're dealing with a chemical reaction, so the critical link between HCl and $CaCO_3$ is provided by the coefficients of the equation. These give us the mole relationship between $CaCO_3$ and HCl.

$$1 \text{ mol CaCO}_3 \Leftrightarrow 2 \text{ mol HCl}$$

We have the volume and molarity of the HCl solution, which will allow us to calculate the number of moles of HCl. We just need to change 75.0 mL to liters.

We now know how we are going to go from the volume of the solution to moles of $CaCO_3$. To find grams of $CaCO_3$, the tool will be the formula mass (100.1 g mol^{-1})

We can diagram the flow of the problem as follows:

coefficients

$$\text{mol HCl} \longrightarrow \boxed{\text{tool}} \longrightarrow \text{mol CaCO}_3$$

molarity $\boxed{\text{tool}}$ $\boxed{\text{tool}}$ formula mass

75.0 mL HCl solution   ? g $CaCO_3$

Solution:

We'll start by changing the volume of the solution to liters: 75.0 mL = 0.0750 L. Then we apply the conversion factors following the path through the problem.

$$0.0750 \text{ L solution} \times \frac{0.250 \text{ mol HCl}}{1.00 \text{ L solution}} \times \frac{1 \text{ mol CaCO}_3}{2 \text{ mol HCl}} \times \frac{100.1 \text{ g CaCO}_3}{1 \text{ mol CaCO}_3} = 0.938 \text{ g CaCO}_3$$

Is the Answer Reasonable?

Let's suppose we had 100 mL (0.1 L) of the HCl solution. One liter of the HCl solution contains 0.25 mol HCl, so in 0.1 L, there would be 0.025 mol HCl. This would consume 0.0125 mol $CaCO_3$ which would weigh about 1.25 g. We have somewhat less than 100 mL, so we expect an answer somewhat less than 1.25 g. The answer we obtained, 0.938 g, seems to be reasonable.

Concentrations of ions in solutions of electrolytes

Multiply the molar concentration given for the salt by the number of ions per formula unit to obtain the concentration of the ion in the solution. For example, if the salt is $Al_2(SO_4)_3$ and its concentration is 0.10 M, the concentration of Al^{3+} is 2 x 0.10 M = 0.20 M, and the concentration of the SO_4^{2-} is 3 x 0.10 M = 0.30 M.

$$0.10 \text{ } M \text{ Al}_2(\text{SO}_4)_3$$

$\boxed{0.10 \times 2}$ $\boxed{0.10 \times 3}$

0.20 M Al^{3+} 0.30 M SO_4^{2-}

Study Examples 5.15 and 5.16 on pages 197 and 198; then work Practice Exercises 24 and 25. Additional exercises are found in the Self-Test below.

Using net ionic equations in stoichiometry calculations

Once you have learned to calculate the concentration of an ion using the formula of a salt and the salt concentration, you are ready to use a net ionic equation in a stoichiometry calculation. Follow Example 5.17 carefully.

Example 5.18 illustrates a limiting reactant problem and how we can keep track of each of the ions in an ionic reaction. Notice that we calculate the number of moles of each ion in the reaction mixture before and after the reaction, taking into account any reaction that takes place *as well as the change in total volume*.

Self-Test

36. What are the molar concentrations of the ions in

    (a)   0.15 M $Ba(OH)_2$ _____

    (b)   0.30 M Na_3PO_4 _____

37. In a certain solution of $(NH_4)_2SO_4$ the ammonium ion concentration was 0.42 M.

    (a)   What was the sulfate ion concentration in the solution?

                            _____

    (b)   What was the molar concentration of ammonium sulfate in the solution?

                            _____

38. How many moles of each ion are in

    (a)   50.0 mL of 0.250 M $(NH_4)_2SO_4$? _____

    (b)   80.0 mL of 0.60 M K_3PO_4?   _____

39. How many milliliters of 0.120 M $AgNO_3$ are needed to react with 18.0 mL of 0.0460 M $FeCl_3$ solution? The reaction follows the net ionic equation

$$Ag^+(aq) + Cl^-(aq) \rightarrow AgCl(s)$$

    Answer _____

40. A student mixed 30.0 mL of 0.25 M NaI solution with 45.0 mL of 0.10 M $Pb(NO_3)_2$ solution.

    (a)   Write the molecular equation for the reaction that occurred.

                _____

    (b)   How many moles of each of the ions were in the mixture before any reaction took place?

           Na^+ _____ I^- _____

           Pb^{2+} _____ NO_3^- _____

    (c)   How many moles of which ions reacted? _____

    (d)   How many moles of what compound were formed? _____

(e) How many moles of each of the ions were present in the solution after the reaction was over?

Na^+ _____ I^- _____

Pb^{2+} _____ NO_3^- _____

(f) What were the concentrations of each of the ions in the solution after the reaction was over?

Na^+ _____ I^- _____

Pb^{2+} _____ NO_3^- _____

New Terms

None

5.13 Chemical analysis and titration are applications of solution stoichiometry

Example 5.19 illustrates an important principle often used in chemical analyses. We begin with a mixture with an unknown composition (i.e., the *relative amounts* of the various components are unknown). A reaction is then carried out that transfers one component of the mixture quantitatively into a different compound, one with a known composition. By measuring the amount of this compound collected, the amount of the component in the original mixture can be calculated. By means of procedures of this sort, repeated for as many components as necessary, the composition of the entire mixture can be determined.

Titrations

This is a useful procedure for chemical analyses involving reactions in solution. In this section we describe acid-base titrations using appropriate indicators. Study Example 5.20 and be sure to work the Practice Exercises to test your knowledge. Then try the titration calculation in the Self-Test.

Thinking It Through

For the following, identify the information needed to solve the problem and show (or explain) what must be done with it.

4   You are hired to take charge of an analytical laboratory and one task is to analyze samples of a commercial product used to remove iron stains. It consists of NaHSO4, an acid salt that reacts with sodium hydroxide as follows:

$$NaHSO_4(aq) + NaOH(aq) \rightarrow Na_2SO_4(aq) + H_2O$$

The intent of the analysis is to ensure that the bottle labels show the actual percentage of $NaHSO_4$. The legal standard is pure $NaHSO_4$. You propose to analyze the samples by titrating 0.300 g portions of the product (dissolved in water) with 0.100 M NaOH. How many milliliters of the NaOH solution will be needed if the sample is pure $NaHSO_4$?

Self-Test

41. A sample of a weak acid weighing 0.256 g was dissolved in water and titrated with 0.200 *M* NaOH solution. The titration required 17.80 mL of the base.

    (a) How many moles of H$^+$ were neutralized in the titration?

    _____

    (b) If the acid is monoprotic, what is its molecular mass? (Hint: How many grams are there per mole of the acid?)

    _____

New Terms

Write the definitions of the following terms, which were introduced in this section. If necessary, refer to the Glossary at the end of the text.

titration	standard solution	acid–base indicator
buret	titrant	end point
stopcock		

Solutions to Thinking It Through

1   Go through the list and identify the kinds of substances and recall their general acid-base properties. Carbon dioxide is a gas and a nonmetal oxide that dissolves in water to make a weakly acidic solution. Sodium oxide is the oxide of a group IA element, all of which oxides react with water to give solutions of the hydroxides. HNO$_3$ is nitric acid, KOH is the hydroxide of a group IA metal, all of which are strongly basic. SO$_3$ is a nonmetal oxide that gives an acid in water (H$_2$SO$_4$). HC$_2$H$_3$O$_2$ is acetic acid,. Thus, only the metal oxide and the metal hydroxide are both solids and give a basic solution in water.

2   (a) ZO, because group IIA metals have charges of 2+, which balances the 2- charge on the O.

     (b) Blue; oxides of groups IA and IIA metals all do this.

3   We can tell that it is a weak acid because it is not on the list of strong acids.

4   The important fact is that the stoichiometric ratio is 1:1, meaning

$$1 \text{ mol NaHSO}_4 \Leftrightarrow 1 \text{ mol NaOH}$$

So we must convert 0.300 g NaHSO$_4$ into moles using the grams to moles tool (a conversion factor made from the formula mass of NaHSO$_4$, 120.07).

$$0.300 \text{ g NaHSO}_4 \times \frac{1 \text{ mol NaHSO}_4}{120.07 \text{ g NaHSO}_4} = 2.50 \times 10^{-3} \text{ mol NaHSO}_4$$

Therefore, we must take a volume of 0.100 *M* NaOH to provide 2.50 × 10^{-3} mol NaOH (because 1 mol NaHSO$_4$ ⇔ 1 mol NaOH).

$$2.50 \times 10^{-3} \text{ mol NaOH} \times \frac{1000 \text{ mL NaOH solution}}{0.100 \text{ mol NaOH}} = 25.0 \text{ mL NaOH solution}$$

Thus, the volume of the NaOH solution required = 25.0 mL.

Answers to Self-Tests

1. Solvent is water; solute is sugar.
2. unsaturated
3. supersaturated
4. (a) Ag^+ and $C_2H_3O_2^-$
    (b) NH_4^+ and $Cr_2O_7^{2-}$
    (c) Ba^{2+} and OH^-
5. (a) $AgC_2H_3O_2\ (aq) \rightarrow Ag^+(aq) + C_2H_3O_2^-(aq)$
    (b) $(NH_4)_2Cr_2O_7\ (aq) \rightarrow 2NH_4^+(aq) + Cr_2O_7^{2-}(aq)$
    (c) $Ba(OH)_2(aq) \rightarrow Ba^{2+}(aq) + 2OH^-(aq)$
6. (a) Coefficients are 1, 2, 1, 2
    $$Cu^{2+}(aq) + 2NO_3^-(aq) + 2K^+(aq) + 2OH^-(aq) \rightarrow$$
    $$Cu(OH)_2(s) + 2K^+(aq) + 2NO_3^-(aq)$$
    $Cu^{2+}(aq) + 2OH^-(aq) \rightarrow Cu(OH)_2(s)$
    (b) Coefficients are 1, 2, 1, 2
    $$Ni^{2+}(aq) + 2Cl^-(aq) + 2Ag^+(aq) + 2NO_3^-(aq) \rightarrow$$
    $$2AgCl(s) + Ni^{2+}aq) + 2NO_3^-(aq)$$
    $Cl^-(aq) + Ag^+(aq) \rightarrow AgCl(s)$
    (c) Coefficients are 1, 3, 1, 3
    $$2Cr^{3+}(aq) + 3SO_4^{2-}(aq) + 3Ba^{2+}(aq) + 6Cl^-(aq) \rightarrow$$
    $$2Cr^{3+}(aq) + 6Cl^-(aq) + 3BaSO_4(s)$$
7. Soluble: Na_2SO_4, $Ni(NO_3)_2$, $(NH_4)_2CO_3$, $Ba(ClO_4)_2$, $Sr(C_2H_3O_2)_2$, $(NH_4)_2S$
8. (a) $2AgC_2H_3O_2(aq) + (NH_4)_2S(aq) \rightarrow Ag_2S(s) + 2NH_4C_2H_3O_2(aq)$
    $2Ag^+(aq) + 2C_2H_3O_2^-(aq) + 2NH_4^+(aq) + S^{2-}(aq) \rightarrow Ag_2S(s) + 2NH_4^+(aq) + 2C_2H_3O_2^-(aq)$
    $2Ag^+(aq) + S^{2-}(aq) \rightarrow Ag_2S(s)$
    (b) $Cr(ClO_4)_3(aq) + 3KOH(aq) \rightarrow Cr(OH)_3(s) + 3KClO_4(aq)$
    $Cr^{3+}(aq) + 2ClO_4^-(aq) + 3K^+(aq) + 3OH^-(aq) \rightarrow Cr(OH)_3(s) + 3K^+(aq) + 3OH^-(aq)$
    $Cr^{3+}(aq) + 3OH^-(aq) \rightarrow Cr(OH)_3(s)$
    (c) $Pb(NO_3)_2(aq) + K_2SO_4(aq) \rightarrow PbSO_4(s) + 2KNO_3(aq)$
    $Pb^{2+}(aq) + 2NO_3^-(aq) + 2K^+(aq) + SO_4^{2-}(aq) \rightarrow PbSO_4(s) + 2K^+(aq) + 2NO_3^-(aq)$
    $Pb^{2+}(aq) + SO_4^{2-}(aq) \rightarrow PbSO_4(s)$
9. (a) $HClO_2 \rightarrow H^+ + ClO_2^-$
    (b) $H_2SeO_3 \rightarrow H^+ + HSeO_3^-$
    $HSeO_3^- \rightarrow H^+ + SeO_3^{2-}$
    (c) $H_3AsO_4 \rightarrow H^+ + H_2AsO_4^-$
    $H_2AsO_4^- \rightarrow H^+ + HAsO_4^{2-}$
    $HAsO_4^{2-} \rightarrow H^+ + AsO_4^{3-}$
10. $P_4O_{10} + 6H_2O \rightarrow 4H_3PO_4$
11. $N_2H_4 + H_2O \rightarrow N_2H_5^+ + OH^-$
12. $RaO + H_2O \rightarrow Ra(OH)_2$
13. Na_2O reacts with water to give sodium hydroxide. $Na_2O(s) + H_2O \rightarrow 2NaOH(aq)$
14. SO_3 reacts with water to give sulfuric acid. $SO_3(g) + H_2O \rightarrow H_2SO_4(aq)$
15. (a) arsenous acid, (b) arsenic acid, (c) calcium arsenate
16. (a) hydrogen selenide, (b) hydroselenic acid, (c) hydrogen fluoride, (d) hydrofluoric acid
    (e) potassium hydrogen sulfite (potassium bisulfite), (f) iodic acid, (g) bromous acid

(h) copper(II) dihydrogen arsenate

17. (a) $NaBrO_4$, (b) HIO_4, (c) $NaHC_2O_4$, (d) $HBrO_4$, (e) $H_2Te(aq)$, (f) $Ba(OH)_2$

18. $HCN(aq) + H_2O \rightleftharpoons H_3O^+(aq) + CN^-(aq)$; the forward reaction is far from completion.

19. $HClO_3(aq) + H_2O \rightarrow H_3O^+(aq) + ClO_3^-(aq)$

20. $C_6H_5NH_2(aq) + H_2O \rightleftharpoons C_6H_5NH_3^+(aq) + OH^-(aq)$

21. perchloric acid, $HClO_4$; chloric acid, $HClO_3$; hydrochloric acid, HCl; hydrobromic acid, HBr; hydriodic acid, HI; nitric acid, HNO_3; sulfuric acid, H_2SO_4

22. (a) $OH^-(aq) + H^+(aq) \rightarrow H_2O$
    (b) $OH^-(aq) + HCHO_2(aq) \rightarrow H_2O + CHO_2^-(aq)$
    (c) $NH_3(aq) + H^+(aq) \rightarrow NH_4^+(aq)$
    (d) $Ni(OH)_2(s) + 2HCHO_2(aq) \rightarrow Ni^{2+}(aq) + CHO_2^-(aq) + 2H_2O$
    (e) $ZnO(s) + 2H^+(aq) \rightarrow Zn^{2+}(aq) + H_2O$

23. $Ca(H_2PO_4)_2$, $CaHPO_4$, $Ca_3(PO_4)_2$

24. Molecular equation: $H_3C_6H_5O_7(aq) + 3NH_3(aq) \rightarrow (NH_4)_3C_6H_5O_7(aq)$
    Ionic and net ionic equations: $H_3C_6H_5O_7(aq) + 3NH_3(aq) \rightarrow 3NH_4^+(aq) + C_6H_5O_7^{3-}(aq)$

25. (a) CO_2 (b) SO_2 (c) H_2S (d) CO_2 (e) SO_2 (f) CO_2

26. $(NH_4)_2S$ and NH_4HCO_3

27. (a) $Pb^{2+}(aq) + S^{2-}(aq) \rightarrow PbS(s)$
    (b) N.R.
    (c) $Zn^{2+}(aq) + 2OH^-(aq) \rightarrow Zn(OH)_2(s)$
    (d) $Cr_2O_3(s) + 6H^+(aq) \rightarrow 2Cr^{3+}(aq) + 3H_2O$
    (e) $BaSO_3(s) + 2HC_2H_3O_2(aq) \rightarrow Ba^{2+}(aq) + 2C_2H_3O_2^- + SO_2(g) + H_2O$
    (f) $NH_4^+(aq) + OH^-(aq) \rightarrow NH_3(g) + H_2O$
    (g) $C_3H_5O_2^-(aq) + H^+(aq) \rightarrow HC_3H_5O_2(aq)$

28. $\dfrac{0.150 \text{ mol } HNO_3}{1 \text{ L solution}}$ and $\dfrac{1 \text{ L solution}}{0.150 \text{ mol } HNO_3}$

29. 0.150 mmol HNO_3

30. Dissolve 6.94 g $CaCl_2$ in water and make the final volume 500 mL.

31. 0.345 mol NaCl; 20.2 g NaCl

32. 19.7 mL HCl solution

33. Add water to 62.5 mL of 0.800 M H_2SO_4 to a final volume of 100 mL.

34. Add water to 12.5 mL of 2.00 M HCl to a final volume of 250 mL.

35. Add 75.0 mL of water to give a total volume of 125 mL.

36. (a) 0.15 M Ba^{2+}, 0.30 M OH^-, (b) 0.90 M Na^+, 0.30 M PO_4^{3-}

37. (a) 0.21 M SO_4^{2-}, (b) 0.21 M $(NH_4)_2SO_4$

38. 0.0250 mol NH_4^+, 0.0125 mol SO_4^{2-}

39. 20.7 mL $AgNO_3$ solution

40. (a) $Pb(NO_3)_2(aq) + 2NaI(aq) \rightarrow PbI_2(s) + 2NaNO_3(aq)$,
    (b) 0.0075 mol Na^+, 0.0075 mol I^-, 0.0045 mol Pb^{2+}, 0.0090 mol NO_3^-
    (c) 0.0075 mol I^- and 0.0038 mol Pb^{2+}
    (d) 0.0038 mol PbI_2
    (e) 0.0075 mol Na^+, 0.0 mol I^-, 0.0007 mol Pb^{2+}, 0.0090 mol NO_3^-
    (f) 0.10 M Na^+, 0.0 M I^-, 0.009 M Pb^{2+}, 0.12 M NO_3^-

41. (a) 3.56×10^{-3} mol H^+, (b) 71.9 g mol^{-1}

Tools you have learned

Consider removing this chart from the Study Guide so you can have it handy when tackling homework problems.

Tool	*How it Works*
List of strong acids	When you encounter the formula for an acid and need to know whether the acid is strong or weak, you use this list. If the acid is on the list, it's a strong acid; if not, it's probably a weak acid. Strong acids are written in dissociated form in an ionic equation.
Criteria for Balanced Ionic Equations Atoms must be in balance. Charge must be in balance.	If both the charge and number of atoms are the same on both sides of an equation, the equation is balanced.
Solubility rules See below.	You use the rules to tell if a given salt is soluble in water. Knowing that soluble salts dissociate fully in water enables you to write ionic equations correctly. The rules help you predict the course of metathesis reactions.
Types of substances that react with either acids or bases to give gases	This list enables you to predict when to expect H_2S, HCN, CO_2, SO_2, or NH_3 to form in a metathesis reaction.
Molarity $\text{molarity} = \dfrac{\text{moles of solute}}{\text{liters of solution}}$	Molarity is the critical link in relating moles of solute to volume of solution. We use it to construct conversion factors to use in calculations.

Solubility Rules

Soluble Compounds

1. All compounds of the alkali metals (Group IA) are soluble.
2. All salts containing NH_4^+, NO_3^-, ClO_4^-, ClO_3^-, and $C_2H_3O_2^-$ are soluble..
3. All chlorides, bromides and iodides (salts containing Cl^-, Br^-, and I^-) are soluble *except* when combined with Ag^+, Pb^{2+}, and Hg_2^{2+} (note the subscript "2").
4. All sulfates (salts containing SO_4^{2-}) are soluble *except* those of Pb^{2+}, Ca^{2+}, Sr^{2+}, Hg_2^{2+}, and Ba^{2+}.

Insoluble Compounds

5. All metal hydroxides (ionic compounds containing OH^-) and all metal oxides (ionic compounds containing O^{2-}) are insoluble *except* those of Group IA and of Ca^{2+}, Sr^{2+}, and Ba^{2+}.

     When metal oxides do dissolve, they react with water to form hydroxides. The oxide ion, O^{2-}, does not exist in water. For example: $Na_2O(s) + H_2O \rightarrow 2NaOH(aq)$

6. All salts that contain PO_4^{3-}, CO_3^{2-}, SO_3^{2-}, and S^{2-} are insoluble, *except* those of Group IA and NH_4^+.

Chapter **6**

Oxidation–Reduction Reactions

In Chapter 5 we discussed various kinds of chemical reactions, including metathesis and acid/base reactions. These reactions have certain features that make discussing them together convenient. Now we turn our attention to another class of chemical reactions. These are often characterized by the transfer of electrons from one species to another, and they include many of our most important chemical changes, as you will learn in this chapter.

Learning Objectives

As you study this chapter, keep in mind the following goals:

1   To learn the criteria that we use in classifying reactions as oxidation-reduction reactions.

2   To learn a bookkeeping method called oxidation numbers that we use to follow the course of redox reactions.

3   To learn how to balance redox reactions in aqueous solution by dividing the reaction into parts, balancing the parts separately, and then recombining them to give the balanced equation.

4   To learn how acids are able to attack metals, and to learn what the products are in such reactions.

5   To learn how to predict reactions in which one metal displaces another from its compounds.

6   To learn about the reactions that molecular oxygen undergoes with organic compounds, with metals, and with nonmetals.

7   To learn how redox reactions can be used in the laboratory and to explore the stoichiometry of these reactions.

6.1   Oxidation–reduction reactions involve electron transfer

Review

Oxidation-reduction reactions (often called *redox* reactions) are very common, so it is important that you learn how to identify them and know the terminology used in discussing them. These reactions can be thought of as involving the transfer of electrons from one substance to another. (Sometimes an electron transfer actually takes place, as when Na and Cl_2 react to form ions, but for many reactions it is just a convenience to think of them as involving electron transfer.)

Remember the following definitions:

Oxidation can be viewed as a loss of electrons.

Reduction can be viewed as a gain of electrons.

Oxidation and reduction always occur simultaneously during a chemical reaction. If one substance loses electrons and is oxidized, then another substance must gain electrons and be reduced.

The terms *oxidizing agent* and *reducing agent* often cause confusion. The basis for assigning these terms is given on page 216, but the simplest way to remember how to apply the terms properly is to find which substance is oxidized and which is reduced. Then, switch words—if a certain substance is oxidized, then it's the reducing *agent*; if the substance is reduced, then it's the oxidizing *agent*.

When you write separate equations showing electron gain or loss, as in Example 6.1, keep in mind that electrons are negatively charged particles. That way, you will be sure to write the electrons on the correct side. For instance, in Example 6.1 we find that magnesium atoms become magnesium ions.

Charge is becoming more positive, so Mg
must be losing *negative* electrons.

$$Mg \rightarrow Mg^{2+}$$

The magnesium is becoming more positive, so it must be losing electrons. Electrons are written on the product side of the equation to show that they have been separated from the magnesium atom, which has become a magnesium ion.

$$Mg \rightarrow Mg^{2+} + 2e^-$$

Notice that the net charge on both sides of the arrow are the same. Recall that this is a requirement for balanced equations involving ions, so the balanced charge helps assure us that we've placed the electrons on the correct side.

Oxidation numbers

Oxidation numbers are a bookkeeping device. We assign them following the rules given on page 218, which you should study carefully. The best way to learn how to apply the rules is to practice assigning oxidation numbers, so study Examples 6.2 through 6.7, work the Practice Exercises, and then work the questions in the Self-Test here in the Study Guide. As you apply these rules, remember that if you encounter a conflict between two rules, the one with the lower number (the one higher-up on the list) is the one that applies and we ignore the rule with the higher number. This is illustrated in Examples 6.4 and 6.5.

For binary compounds of a metal and a nonmetal (such as $NaCl$, $CaCl_2$, or Al_2O_3), the ions are simple monatomic ions and their oxidation numbers are equal to their charges. The formulas of the ions formed by the representative elements were given in Table 2.3 on page 59. If necessary, review them and learn how to use the periodic table to obtain their correct charges.

If you recognize a polyatomic ion in a formula, for example, SO_4^{2-} in $Cr_2(SO_4)_3$, you can use its charge as the *net* oxidation number of the ion. For the "SO_4" in this example, we can assign it a net oxidation number of -2, which means that the chromium must have an oxidation number of $+3$ (applying Rule 3, the summation rule: $[2 \times (+3)] + [3 \times (-2)] = 0$).

Redox is redefined on page 222 in terms of changes in oxidation number.

Oxidation: an increase in oxidation number (oxidation state)

Reduction: a decrease in oxidation number (oxidation state)

Study Example 6.8 to review how we use oxidation numbers to identify oxidation and reduction.

Self-Test

1. Consider the reaction of magnesium with fluorine to give the ionic compound MgF_2.

$$Mg + F_2 \rightarrow MgF_2$$

   (a) Write separate equations showing the gain and loss of electrons.

   _____

   _____

   (b) Which substance is oxidized?   _____

   (c) Which substance is reduced?   _____

   (d) Which reactant is the oxidizing agent?   _____

   (e) Which reactant is the reducing agent?   _____

2. Assign oxidation numbers to each atom in the following:

   (a) NO_3^- _____

   (b) $SbCl_5$ _____

   (c) $CaHAsO_4$ _____

   (d) ClF_3 _____

   (e) I_3^- _____

   (f) S_8 _____

3. Determine whether the following changes are oxidation, reduction, or neither oxidation nor reduction.

   (a) SO_3^{2-} to SO_4^{2-} _____

   (b) Cl_2 to ClO_3^- _____

   (c) N_2O_4 to NH_3 _____

   (d) PbO to $PbCl_4^{2-}$ _____

   (e) Ag to Ag_2S _____

4. Consider the balanced equation,

$$3H_2O + 3Cl_2 + NaI \rightarrow 6HCl + NaIO_3$$

   (a) Which substance is oxidized?   _____

   (b) Which substance is reduced?   _____

   (c) Which substance is the oxidizing agent? _____

   (d) Which substance is the reducing agent? _____

New Terms

Write the definitions of the following terms, which were introduced in this section. If necessary, refer to the Glossary at the end of the text.

oxidation	oxidation-reduction reaction
reduction	oxidizing agent
redox reaction	reducing agent
oxidation number	oxidation state

6.2   The ion-electron method creates balanced net ionic equations for redox reactions

Review

In applying the ion-electron method, we divide a redox equation into two half-reactions, balance the half-reactions separately, and then combine the balanced half-reactions to give the balanced overall net ionic equation. The method is not difficult to apply if you proceed in a stepwise fashion, without skipping any of the steps. If you can't obtain a balanced equation, it is probably because you haven't remembered to do things in sequence. Be sure you learn the seven steps summarized on page 227 for reactions that occur in acidic solutions.

In applying the ion-electron method, one of the most frequent causes for error is not remembering to write the correct charges on the formulas. You need these charges to get the correct numbers of electrons, so if you forget to write the charges on the ions, you will surely get wrong answers.

Another common problem students have is computing the net charge on each side of a half-reaction. Suppose we have reached the following stage in balancing a half-reaction

$$6H_2O + N_2H_5^+ \rightarrow 2NO_3^- + 17H^+$$

To obtain the charge contributed by each substance, multiply its coefficient by its charge.

$$6H_2O + N_2H_5^+ \rightarrow 2NO_3^- + 17H^+$$

$$\boxed{2 \times (1-) = 2-} \qquad \boxed{17 \times 1+ = 17+}$$

$$\boxed{(2-) + (17+) = 15+}$$

Thus, on the right side of this half-reaction the net charge is

$$[2 \times (1-)] + [17 \times (1+)] = 15+$$

The 2 is the coefficient of the NO_3^- and the 1– is the charge on the NO_3^-. Similarly, 17 is the coefficient of H^+ and 1+ is the charge on H^+.

In determining the number of electrons that must be added to balance a half-reaction, be especially careful when the charges on opposite sides of the half-reaction have *opposite* algebraic signs. For example, consider the following half-reaction just before we add electrons to it.

Net charge is 5+   Net charge is 1–

$$7H^+ + SO_3^{2-} \rightarrow HS^- + 3H_2O$$

The net charge on the left is 5+, the net charge on the right is 1–. The number of electrons we must add equals the *algebraic difference* between them.

$$\text{number of electrons to be added} = (5+) - (1-) = 6$$

The electrons are always added to the more positive (or less negative) side, so the half-reaction properly balanced is

$$6e^- + 7H^+ + SO_3^{2-} \rightarrow HS^- + 3H_2O$$

Notice that the net charge is the same on both sides.

Reactions in basic solutions

To balance an equation for basic solution, first balance it as if the solution were acidic. Then follow the three-step conversion to basic solution described on page 229. Notice that the result of the change-over is to convert all the H^+ to H_2O and to add an equal number of OH^- to the other side. For example,

These become
$7H_2O$.

We add $7OH^-$ to the right side.

$$6e^- + 7H^+ + SO_3^{2-} \rightarrow HS^- + 3H_2O$$

This gives

$$6e^- + 7H_2O + SO_3^{2-} \rightarrow HS^- + 3H_2O + 7OH^-$$

Finally, we delete three water molecules from each side to give the balanced equation (a half-reaction, in this instance).

$$6e^- + 4H_2O + SO_3^{2-} \rightarrow HS^- + 7OH^-$$

Self-Test

5. Balance the following half-reactions for an acidic solution.

   (a) $NO \rightarrow NO_3^-$ _____

   (b) $Br_2 \rightarrow BrO_3^{2-}$ _____

   (c) $P_4 \rightarrow HPO_3^{2-}$ _____

6. Balance the following half-reactions for a basic solution.

   (a) $Cl_2 \rightarrow OCl^-$ _____

   (b) $AsO_4^{3-} \rightarrow AsH_3$ _____

   (c) $S_2O_4^{2-} \rightarrow SO_4^{2-}$ _____

7.   Balance the following reaction that occurs in an acidic solution.

$$H_2SeO_3 + I^- \rightarrow I_2 + Se$$

8.   Balance this reaction for a basic solution.

$$SeO_3^{2-} + I^- \rightarrow I_2 + Se$$

New Terms

Write the definitions of the following terms, which were introduced in this section. If necessary, refer to the Glossary at the end of the text.

half-reaction   ion-electron method

6.3   Metals are oxidized when they react with acids

Review

Every acid releases hydrogen ions, H^+ (which, of course, actually exist in solution as H_3O^+). Hydrogen ion is a mild oxidizing agent and can oxidize many metals. The products are the metal ion and H_2 gas. For example, tin reacts with H^+ to give H_2 and Sn^{2+}.

$$Sn(s) + 2H^+(aq) \rightarrow Sn^{2+}(aq) + H_2(g)$$

Many metals dissolve in acids such as HCl. Some examples are iron, zinc, tin, aluminum, and magnesium. These metals are said to be *more active* than H^+. (In the next section, you will learn more about how you can tell whether a given metal will dissolve in acids.)

There are some metals that are not attacked by H^+, and they will not dissolve in acids that have H^+ as the strongest oxidizing agent. Such acids are called *nonoxidizing acids*. An example is HCl, which gives H^+ and Cl^- in solution. The Cl^- ion cannot gain any more electrons, so it cannot act as an oxidizing agent. Solutions of HCl therefore have H^+ as the only oxidizing agent.

Metals that won't dissolve in nonoxidizing acids often will dissolve in *oxidizing acids* such as HNO_3. Copper is the example that's given in the text on page 232. Notice that when the nitrate ion serves as the oxidizing agent, hydrogen gas is *not* among the products. Instead, the reduction product comes from the NO_3^- ion, which shows that NO_3^- is the oxidizing agent. Also, you should remember that when concentrated nitric acid is used as an oxidizing agent, the reduction product is usually NO_2, and when dilute nitric acid is used, the reduction product is usually NO. When a very strong reducing agent such as zinc reacts with NO_3^-, the nitrogen can be reduced all the way to the -3 oxidation state.

Hot concentrated H_2SO_4 is also an oxidizing agent in which SO_4^{2-} undergoes reduction to SO_2.

Self-Test

9.   Hydrobromic acid, HBr, is a nonoxidizing acid. Write the chemical equation for its reaction with magnesium.

10. Cadmium reacts with dilute solutions of sulfuric acid with the evolution of hydrogen. Write a chemical equation for the reaction.

11. In the reaction described in Question 10, which is the oxidizing agent and which is the reducing agent?

        Oxidizing agent _____

        Reducing agent _____

12. Mercury dissolves in concentrated nitric acid, with the evolution of a reddish-brown gas, just as in the reaction shown in Figure 6.2 for copper. The oxidation state of the mercury after reaction is +2. Write a balanced net ionic equation for the reaction.

13. Construct the balanced molecular equation for the reaction described in Question 12.

New Terms

Write the definitions of the following terms, which were introduced in this section. If necessary, refer to the Glossary at the end of the text.

    nonoxidizing acid   oxidizing acid

6.4   A more active metal will displace a less active one from its compounds

Review

If one metal is more active than another, then the more active metal will be able to reduce the ion of the less active metal. In the reaction, the more active metal is oxidized. This is what happens when metallic zinc is placed into a solution that contains a soluble copper salt. The zinc (which is the more active metal) is oxidized to zinc ion and the copper ion is reduced to metallic copper.

The activity series on page 235 lists metals in order of increasing activity (that is, in order of increasing ability to be oxidized and to serve as a reducing agent). Remember that any metal in this table is able to displace the ion of any metal above it from compounds. For instance, if you turn to Table 6.2 (page 235), you will see that aluminum can reduce the ions of manganese, zinc, chromium, iron, and ions of all the other metals above it in the table.

Table 6.2 can also be used to determine if a given metal will dissolve in a nonoxidizing acid. If the metal is *below* hydrogen in the table, a nonoxidizing acid will react with it to give the metal ion and hydrogen gas. If a metal is above hydrogen in this table, an oxidizing acid such as HNO_3 or hot, concentrated H_2SO_4 must be used to dissolve it.

Self-Test

14. Write molecular equations for any reaction that will occur when the following reactants are combined:

    (a) Manganese metal added to a solution of $CoCl_2$.

    _____

    (b) Iron metal added to a solution of $AuCl_3$.

    _____

    (c) Cadmium metal added to a solution of $MgCl_2$.

    _____

15. Which metal ions will be reduced if an excess amount of iron powder is added to a solution that contains a mixture of $Ca(NO_3)_2$, $Cu(NO_3)_2$, $AgNO_3$, KNO_3, $Pb(NO_3)_2$, and $Hg(NO_3)_2$?

    _____

16. Which of the following metals will dissolve in a solution of HBr: aluminum, tin, cobalt, gold, mercury, manganese?

    _____

New Terms

Write the definition of the following term, which was introduced in this section. If necessary, refer to the Glossary at the end of the text.

single replacement reaction   activity series

6.5   Molecular oxygen is a powerful oxidizing agent

Review

Molecular oxygen, O_2, is a very good oxidizing agent. In this section you learn about several kinds of reactions that are brought about by O_2. The purpose is to enable you to make reasonable predictions of the outcome of combustion reactions.

1. **Reactions of O_2 with organic compounds**. Organic compounds normally contain carbon, hydrogen, oxygen, and perhaps several other elements. When these compounds are burned in a plentiful supply of O_2, the carbon forms CO_2, the hydrogen forms H_2O, and if any sulfur is present in the compound, it is changed to SO_2. Any oxygen in the compound becomes incorporated in the other oxygen-containing products.

    If the combustion takes place in a limited supply of O_2, hydrogen still is changed to H_2O, but the carbon is only oxidized to CO. In extremely limited supplies of O_2, elemental carbon (soot) is formed.

2.   **Reactions of O_2 with metals**. Many metals combine with oxygen directly. Aluminum, iron, and magnesium are given as examples in the text. The products are metal oxides. When the reactions are slow, they are described by the term corrosion or tarnishing. If the reaction is rapid, it's combustion.

3.   **Reactions of O_2 with nonmetals**. These reactions give nonmetal oxides. An important nonmetal that does not react with O_2 at atmospheric pressure is nitrogen.

Self-Test

17.   Complete and balance the following equations:

(a) $C_2H_6 + O_2 \rightarrow$ _____

(b) $C_6H_{12}O_6 + O_2 \rightarrow$ _____

(c) $C_3H_6O + O_2 \rightarrow$ _____

(d) $C_2H_5SH + O_2 \rightarrow$ _____

(e) $Al + O_2 \rightarrow$ _____

(f) $S + O_2 \rightarrow$ _____

(g) $Ca + O_2 \rightarrow$ _____

18.   When burned in an excess supply of oxygen, phosphorus forms the compound P_4O_{10}. Write a chemical equation for the reaction.

_____

19.   Write a balanced chemical equation for the combustion of benzene, C_6H_6, when there is a severely limited supply of O_2.

_____

New Terms

Write the definitions of the following terms, which were introduced in this section. If necessary, refer to the Glossary at the end of the text.

organic compound	hydrocarbon	corrosion
tarnishing	combustion	

6.6   Redox reactions follow the same stoichiometric principles as other reactions

Review

The general principles of stoichiometry discussed earlier apply to redox reactions as well. However, redox reactions are usually more complex than metathesis and their equations are similarly more complex.

The use of potassium permanganate, $KMnO_4$, as a reactant in redox titrations is discussed. The advantage of this substance is that the MnO_4^- ion is deeply colored, but its reduction product is almost colorless. Permanganate ion serves as its own indicator in a titration. When the last of the reducing agent has reacted, the next drop of the MnO_4^- solution causes the reaction mixture to take on a pink color. This signals the end point in the titration. Study Examples 6.12 and 6.13 to review the principles of stoichiometry, and then work Practice Exercise 14. To test your understanding, work the Thinking It Through problem and the Self-Test below.

Thinking It Through

For the following, identify the information needed to solve the problem and show (or explain) what must be done with it.

1   A sample of tin ore with a mass of 0.225 g was dissolved in acid and all the tin converted to Sn^{2+}. The solution required 24.33 mL of 0.0200 M $KMnO_4$ to reach an end point in a titration in which the Sn^{2+} was converted to Sn^{4+}. What was the percentage of SnO_2 in the original ore sample?

Self-Test

20. (a) Write a balanced net ionic equation for the reaction of sulfurous acid with permanganate ion in an acidic solution. The products of the reaction are sulfate ion and manganese(II) ion.

   _____

   (b) How many milliliters of 0.150 M $KMnO_4$ are required to react with a solution that contains 0.440 g of H_2SO_3?

   _____

21. The reaction of MnO_4^- with Sn^{2+} in acidic solution follows the equation:

$$2MnO_4^- + 5Sn^{2+} + 16H^+ \rightarrow 2Mn^{2+} + 5Sn^{4+} + 8H_2O$$

   (a) How many grams of $KMnO_4$ must be used to react with 35.0 g of $SnCl_2$?

   _____

   (b) How many milliliters of 0.0500 M $KMnO_4$ solution would be needed to react with all the tin in 25.0 mL of 0.250 M $SnCl_2$ solution?

   _____

   (c) A 0.500 g sample of solder, which is an alloy containing lead and tin, was dissolved in acid and all the tin was converted to Sn^{2+}. The solution was then titrated with 0.0200 M $KMnO_4$ solution. The titration required 27.73 mL of the $KMnO_4$ solution. What is the percentage by mass of tin in the solder?

   _____

22. Sodium chlorite, $NaClO_2$, is used as a bleaching agent. A 2.00 g sample of a bleach containing this compound was dissolved in an acidic solution and treated with excess NaI. The net ionic equation for the reaction that took place is

$$4H^+ + ClO_2^- + 6I^- \rightarrow Cl^- + 2I_3^- + 2H_2O$$

The solution containing the I_3^- was then titrated with 28.80 mL of 0.200 M $Na_2S_2O_3$ solution. The reaction that took place during the titration was

$$I_3^- + 2S_2O_3{}^{2-} \rightarrow 3I^- + S_4O_6{}^{2-}$$

What is the percentage by mass of $NaClO_2$ in the bleach?

_____

New Terms

None

Solution to Thinking It Through

1   First, we need to write a balanced chemical equation for the reaction of Sn^{2+} with MnO_4^- to give Sn^{4+} and Mn^{2+}. The product of molarity and volume for the permanganate solution gives us the moles of MnO_4^- used. The coefficients of the equation are then used to calculate the number of moles of Sn^{2+} that reacted. Since 1 mol of Sn comes from 1 mol of SnO_2, the number of moles of Sn^{2+} obtained is equal to the number of moles of SnO_2 that were in the ore sample. Multiply this value by the formula mass of SnO_2 to obtain the mass of SnO_2 in the ore sample. The mass of SnO_2 multiplied by 100% and then divided by 0.225 g gives the percentage of SnO_2 in the ore sample. [The answer is 81.5 % SnO_2.]

Answers to Self-Test Questions

1.   (a) $Mg \rightarrow Mg^{2+} + 2e^-$; $F_2 + 2e^- \rightarrow 2F^-$ (b) Mg , (c) F_2, (d) F_2, (e) Mg

2.   (a) N, +5; O, –2 (b) Sb, +5; Cl, –1 (c) Ca, +2; H, +1; As, +5; O, –2

    (d) Cl, +3; F, –1 (e) I, $-\frac{1}{3}$ (f) S, zero

3.   (a) oxidation, (b) oxidation, (c) reduction, (d) neither, (e) oxidation

4.   (a) NaI, (b) Cl_2, (c) Cl_2, (d) NaI

5.   (a) $2H_2O + NO \rightarrow NO_3^- + 4H^+ + 3e^-$

    (b) $6H_2O + Br_2 \rightarrow 2BrO_3^- + 12H^+ + 10e^-$

    (c) $12H_2O + P_4 \rightarrow 4H_3PO_3 + 12H^+ + 12e^-$

6.   (a) $4OH^- + Cl_2 \rightarrow 2OCl^- + 2H_2O + 2e^-$

    (b) $8e^- + 7H_2O + AsO_4{}^{3-} \rightarrow AsH_3 + 11OH^-$

    (c) $8OH^- + S_2O_4{}^{2-} \rightarrow 2SO_4{}^{2-} + 4H_2O + 6e^-$

7. $4H^+ + 4I^- + H_2SeO_3 \rightarrow Se + 2I_2 + 3H_2O$

8. $3H_2O + 4I^- + SeO_3{}^{2-} \rightarrow Se + 2I_2 + 6OH^-$

9. $Mg + 2HBr \rightarrow MgBr_2 + H_2$

10. $Cd + H_2SO_4 \rightarrow CdSO_4 + H_2$

11. Cd is the reducing agent, H^+ is the oxidizing agent.

12. $Hg + 2NO_3^- + 4H^+ \rightarrow Hg^{2+} + 2NO_2 + 2H_2O$

13. $Hg + 4HNO_3 \rightarrow Hg(NO_3)_2 + 2NO_2 + 2H_2O$

14. (a) $Mn + CoCl_2 \rightarrow MnCl_2 + Co$, (b) $2AuCl_3 + 3Fe \rightarrow 3FeCl_2 + 2Au$

    (c) $Cd + MgCl_2 \rightarrow$ no reaction

15. Cu^{2+}, Ag^+, Pb^{2+}, Hg^{2+}

16. Al, Sn, Co, Mn

17. (a) $2C_2H_6 + 15O_2 \rightarrow 12CO_2 + 6H_2O$ (b) $C_6H_{12}O_6 + 6O_2 \rightarrow 6CO_2 + 6H_2O$

    (c) $C_3H_6O + 4O_2 \rightarrow 3CO_2 + 3H_2O$

    (d) $2C_2H_5SH + 9O_2 \rightarrow 4CO_2 + 6H_2O + 2SO_2$

    (e) $2Al + 3O_2 \rightarrow Al_2O_3$ (f) $S + O_2 \rightarrow SO_2$

    (g) $2Ca + O_2 \rightarrow 2CaO$

18. $4P + 5O_2 \rightarrow P_4O_{10}$

19. $2C_6H_6 + 3O_2 \rightarrow 12C + 6H_2O$

20. (a) $5H_2SO_3 + 2MnO_4^- \rightarrow 5SO_4^{2-} + 2Mn^{2+} + 4H^+ + 3H_2O$ (b) 14.3 mL

21. (a) 23.3 g $KMnO_4$, (b) 50.0 mL, (c) 32.9% Sn

22. 6.51% $NaClO_2$

Tools you have learned

Consider removing this chart from the Study Guide so you can have it handy when tackling homework problems.

Tool	*How It Works*
Rules for assigning oxidation numbers (See below.)	Oxidation numbers can be used to determine whether redox changes take place in a reaction. We use the rules to assign oxidation numbers to the atoms in a chemical formula.
Ion-electron method (See below.)	Use this method when you need to have a balanced net ionic equation for a redox reaction.
Table of oxidizing and nonoxidizing acids	The information in the table enables you to determine the products of reactions of acids with metals.
Activity series	The activity series enables you to determine whether one metal will displace another from its compounds in solution.

Summary of Useful Information

Rules for Assigning Oxidation Numbers

1. The oxidation number of any free element is zero, regardless of how complex its molecules might be.
2. The oxidation number of any simple, monatomic ion is equal to the charge on the ion.
3. The sum of all the oxidation numbers of the atoms in a molecule or ion must equal the charge on the particle.
4. In its compounds, fluorine has an oxidation number of -1.
5. In its compounds, hydrogen has an oxidation number of $+1$.
6. In its compounds, oxygen has an oxidation number of -2.

Ion-Electron Method — Acidic Solution

Step 1. Divide the equation into two half-reactions.

Step 2. Balance atoms other than H and O.

Step 3. Balance O by adding H_2O.

Step 4. Balance H by adding H^+.

Step 5. Balance net charge by adding e^-.

Step 6. Make e^- gain equal e^- loss; then add half-reactions.

Step 7. Cancel anything that's the same on both sides.

Additional Steps for Basic Solution

Step 8. Add to *both* sides of the equation the same number of OH^- as there are H^+.

Step 9. Combine H^+ and OH^- to form H_2O.

Step 10. Cancel any H_2O that you can.

Chapter 7

Energy and Chemical Change: Breaking and Making Bonds

We turn from the materials budgets of chemical reactions (Chapters 4, 5, and 6) to their energy budgets. Some chemical reactions will not occur until the reactants have received energy—usually heat energy but sometimes electrical energy, and sometimes other forms. Other reactions occur and spontaneously give off energy as heat, light, sound, electricity, or as mechanical energy (as from an explosion). We here introduce the important concepts needed to describe such energy changes. We also introduce some very important topics in thermodynamics.

Learning Objectives

In this chapter, keep in mind the following goals.

1   To learn what energy is and what it isn't.

2   To learn the distinctions between *kinetic* and *potential energy* and to understand the roles played by attractions and repulsions in potential energy.

3   To learn the units of energy, the joule and the calorie, and how they are related.

4   To understand the concepts of thermal equilibrium, internal energy, and molecular kinetic energy.

5   To learn how the kinetic theory of matter and the concept of molecular kinetic energy help us understand heat, temperature, and how heat transfers.

6   To learn how bond making and bond breaking affect a system's potential energy.

7   To learn what is meant by the state of a system and a state function.

8   To learn the distinctions among open, closed, and isolate systems.

9   To learn how the heat absorbed or released by an object is related to its temperature change.

10   To learn how the heat capacity of an object can be determined and how it can be used along with temperature changes to measure heat gain or loss.

11   To learn the relationship between heat capacity and specific heat.

12   To learn how to use specific heat in calculations.

13   To learn how the making and breaking of chemical bonds are related to changes in potential energy (chemical energy).

14   To understand the meaning of the terms endothermic and exothermic.

15   To learn the concept of pressure to enable a study of pressure–volume work.

16   To learn how the terms work (w) and heat (q) are used to describe the exchange of energy between a system and its surroundings.

17   To learn why neither heat nor work is necessarily a state function, but that the algebraic sum of the two, the change in internal energy of a system (ΔE), is a state function.

18   To learn how algebraic sign conventions are used for values of heat and work to indicate whether energy is flowing into a system or out of a system.

19   To learn that the heat of reaction at constant volume (q_v) equals ΔE when the system does no pressure–volume work, as in a bomb calorimeter.

20   To learn that the heat of reaction at constant pressure (q_p) is not necessarily the same as q_v, but introduces a new energy term called the enthalpy change, ΔH.

21   To learn how the sign of ΔH determines whether a change is exothermic or endothermic.

22   To learn how data obtained with calorimeters can be used to calculate thermal properties.

23   To learn what is meant by a standard heat of reaction and to write thermochemical equations.

24   To learn the conditions used to define standard states.

25   To manipulate and interpret thermochemical equations and enthalpy diagrams.

26   To use Hess's law and enthalpy diagrams to calculate enthalpy changes.

27   To use standard enthalpies of formation and Hess's law to calculate enthalpy changes.

7.1   Energy is the ability to do work or supply heat

Review

Energy is the ability to do work and supply *heat*. *Work* is motion accomplished against an opposing force. For example, work is accomplished when you lift an object against the opposing force of gravity, or when you squeeze a spring that offers resistance.

The energy of a moving object is called its *kinetic energy (KE)* and is defined by the equation KE. = $(1/2)mv^2$, where m is the mass and v is the velocity. *Potential energy* is stored energy. Potential energy exists in a stretched spring, a compressed spring, or in a book raised above the desk. These are examples that illustrate some of the ways by which potential energy can be stored. There are two important situations where we can tell that changes in potential energy have occurred.

1.   Potential energy increases by pulling apart objects that attract each other. (An example is lifting a book above a desk against the force of gravitational attraction.)

2.   Potential energy increases by pushing together objects that push back or repel each other (for example, pushing objects together that are held apart by a spring, which involves compressing the spring).

The existence of positively and negatively charged particles in atoms, ions, and molecules means that there are repelling and attracting forces in substances. Motions of these particles lead to changes in both kinetic and potential energies.

The SI unit of energy is the *joule (J)*, although we will often express energies in the larger unit, *kilojoule (kJ)*. Older energy units are the calorie (cal) and kilocalorie (kcal). The nutritional Calorie unit is really the kilocalorie. Study Example 7.1 in the text to be sure you understand how to apply the conversions between joules and calories, and between kilojoules and kilocalories. Be sure to learn the following conversion relationships between the energy units.

$$4.184 \text{ J} = 1 \text{ cal}$$

$$4,184 \text{ kJ} = 1 \text{ kcal}$$

The original definition of the calorie was based on how the absorption of heat affects the temperature of water; 1 cal is enough heat to raise the temperature of 1 g of water by 1 °C. Study Example 7.2 to gain an appreciation for the meaning of this definition.

Self-Test

1.   Describe how the kinetic energy and the potential energy of a stone thrown upward change during its upward motion, at the top of its trajectory, and during its downward flight.

_____

2.   Which actions decrease the potential energies of the systems?

(a)   Two electrons move away from each other.

(b)   A stretched rubber band is released.

(c)   Coal burns.

(d)   All of the above.   _____

3.   How much heat, measured in calories, is needed to raise the temperature of 35.0 g of water from 15.0 °C to 40.0 °C?

_____

4.   Express the answer to the preceding question in units of joules.

_____

5.   How many kilocalories corresponds to an energy of 45.8 kJ?

_____

6.   How many joules are possessed by a 12.0 kg object traveling at a speed of 120 m/s?

_____

New Terms

Write the definitions of the following terms, which were introduced in this section. If necessary, refer to the Glossary at the end of the text.

kinetic energy	potential energy
joule	kilocalorie
calorie	kilojoule

7.2   Internal energy is the total energy of an object's molecules

Review

There are several key concepts introduced in this section. One is that the atoms and molecules of a substance are in constant motion and possess molecular kinetic energy. The sum of all the kinetic and potential energies of the particles in a substance is its internal energy, E. We use the Greek letter Δ (delta) to signify a change, so a change in internal energy is given the symbol ΔE. Changes in quantities are defined as "final" minus "initial." Therefore,

$$\Delta E = E_{final} - E_{initial} \quad \text{or} \quad \Delta E = E_{products} - E_{reactants}$$

The algebraic sign of the internal energy change defines the direction of energy flow. If ΔE is positive for a chemical reaction, then the products have more internal energy than the reactants.

Another important concept is that the temperature of an object is related to the average molecular kinetic energy of its particles (atoms, molecules, or ions). Figure 7.5 is particularly important because it illustrates how the molecular kinetic energies are distributed among the molecules, and how this distribution changes with temperature. Notice how the shapes of the curves change with changing temperature. We will used these graphs in later chapters to help us understand properties of liquids and solids and how temperature affects the speeds of chemical reactions.

The concept of a *state function* is introduced in this section as well. The change in the value of a state function depends only on the initial and final conditions or *state*, not on how the change occurs. For example, temperature qualifies as a state function because a change in the temperature of an object depends only on its initial and final temperatures. If the final temperature is 30 °C and the initial temperature is 25 °C, the temperature changes, Δt, equals +5 °C. The quantity Δt doesn't depend on how the temperature went from 25 to 30 °C. State functions are useful quantities because we don't have to worry about the path from start to finish, only on the values in the initial and final states.

Finally, you learn in this section how molecular kinetic energy accounts for the transfer of heat by collisions between molecules.

Self-Test

7. Define internal energy.

8. Considering the equation for kinetic energy, why does kinetic energy never have a negative value?

9. Why is it impossible for a system to have a temperature lower than 0 K?

10. When a substance cools, what happens to the average kinetic energy of its atoms, molecules, or ions?

11. What is a state function? Give two examples.

_____

_____

12. Suppose that in a chemical reaction, the products have 25.6 kJ less energy than the reactants. What is the value (with appropriate algebraic sign) of ΔE?

_____

13. On a separate sheet of paper, make a graph that illustrates the distribution of molecular kinetic energies (i.e., how molecular kinetic energy varies with the fraction of molecules with that kinetic energy) for two different temperatures. Indicate on the graph the average kinetic energy for each distribution and indicate which distribution corresponds to the higher temperature. (Refer to Figure 7.5 to check your answer.)

New Terms

Write the definitions of the following terms, which were introduced in this section. If necessary, refer to the Glossary at the end of the text.

thermal equilibrium	internal energy	molecular kinetic energy
temperature	state	state function

7.3   Heat can be determined by measuring temperature changes

Review

Two important terms that you must be sure to understand are *system* and *surroundings*, because almost everything we do in thermochemistry and thermodynamics relates to these concepts. The *system* consists of that part of the universe we are studying. All the rest is the surroundings. Separating the system and the surroundings is a *boundary*, which can be visible (like the walls of a beaker) or invisible (like the arbitrary boundary that separates the earth's atmosphere from outer space).

Be sure to understand the differences among open, closed, and isolated systems given on page 263 of the text. In this chapter we will be most interested in closed systems and will study the flow of energy between systems and their surroundings.

Heat flow is specified by the symbol q and is given a positive sign if heat enters a system and a negative sign if heat leaves. When heat moves across the boundary between system and surroundings, the sign of q for the system is just the opposite of the sign of q for the surroundings. Thus, if the system absorbs 30 J of heat, it must absorb the heat from the surroundings and the surroundings will lose 30 J of heat. For the system, $q = +30$ J and for the surroundings, $q = -30$ J. (When energy flows across the boundary between the system and surroundings, the amount of energy absorbed by one is always equal in magnitude to the amount of energy released by the other. The signs are opposite to indicate the direction of energy flow.)

The amount of heat that enters or leaves an object is directly proportional to the temperature change.

$$q \propto \Delta t$$

The proportionality constant is called the *heat capacity* and is given the symbol C. Inserting this constant into the expression above gives the equation

$$q = C\Delta t$$

This is Equation 7.4 on page 265 of the text. Study Example 7.3.

Heat capacity is a property that depends on the size (mass) of the sample. If we divide the heat capacity C by the mass m (both extensive properties), we obtain a quantity that is independent of sample size. We call it the specific heat, s. (This is similar to dividing mass by volume to obtain the intensive property density.)

$$s = C/m$$

Solving for C gives $C = s\,m$, and substituting into the equation for heat gives

$$q = s\,m\,\Delta t$$

This is Equation 7.6 in the textbook. Study Example 7.4 in the text to see how we use specific heat in a calculation.

Self-Test

14. What is the difference between an open and a closed system?

   _____

15. Suppose that for a beaker of water, $q = +45$ J.

   (a) Will the temperature of the water increase or decrease? _____

   (b) What is the value of q for the surroundings? _____

16. A 48.6 g sample of nickel required 298 J to raise its temperature from 35.8 °C to 49.6 °C.

   (a) What is the heat capacity of the nickel sample (with correct units)? _____

   (b) What is the specific heat of nickel (with correct units)? _____

   (c) How many joules would be needed to raise the temperature of 155 g of nickel from 35.0 °C to 45.0 °C?

      _____

   (d) If 355 J were removed from a 62.0 g sample of nickel initially at 40.0 °C, what would the temperature change to?

      _____

New Terms

Write the definitions of the following terms, which were introduced in this section. If necessary, refer to the Glossary at the end of the text.

system	surroundings	boundary
state of a system	open system	closed system
isolated system	heat capacity	specific heat
specific heat capacity		

7.4   Bond breaking requires energy; bond formation releases energy

Review

The net attractive force that exists between two atoms in a molecule is called a *chemical bond*. The nature of these attractions will be studied in detail later. For now, our attention is on the potential energy associated with chemical bonds, because when bonds form and break, potential energy is lost and gained, respectively. This kind of potential energy, which we associate with chemical bonds in molecules, is often called *chemical energy*. Remember the following:

• Breaking a chemical bond requires energy, so it is accompanied by an increase in the potential energy (chemical energy) of the atoms.

• Forming a chemical bond releases energy, so it is accompanied by a decrease in the potential energy (chemical energy) of the atoms.

When a chemical reaction involves both the breaking and forming of bonds, there is almost always a net change in potential energy. Overall, if there is a net increase in attractions, there will be a lowering of the potential energy. Let's see what this does to the temperature of the system.

If the potential energy decreases, it can't just disappear; that would violate the law of conservation of energy. Instead, the PE that's lost changes to molecular KE. This causes the *total* molecular KE to rise and also leads to an increase in the *average molecular KE*. As we noted earlier, when the average molecular KE increases, so does the temperature. The conclusion, then, is that forming stronger bonds (with greater attractions between atoms) leads to a rise in temperature. This is what is described for the reaction of methane with oxygen on page 272.

When the temperature of a reacting system rises, we can view it as resulting from an increase in the amount of heat in the system. This heat comes from the reaction that's taking place, so we can view the heat as a product of the reaction. Chemical reactions in which heat is released by the chemicals are said to be *exothermic*; those in which heat is consumed and changed to potential energy are said to be *endothermic*.

It is easy to tell whether a reaction is exothermic or endothermic. If the reaction mixture becomes warmer as the reaction occurs, the reaction is exothermic. If the mixture becomes cooler, the reaction is endothermic.

Self-Test

17.   How does the average molecular kinetic energy of the molecules of a system change during an exothermic reaction among them?

_____

18.   When the products of a reaction have less potential (chemical) energy than the reactants, will the reaction be exothermic or endothermic?

_____

19.   When potassium iodide dissolves in water, it dissociates and the ions become hydrated. We could write the dissociation reaction as

$$KI(s) + nH_2O \rightarrow K^+(aq) + I^-(aq)$$

where the water that surrounds the ions is written as a reactant. Forming a solution of KI in water is endothermic.

(a)   Does the mixture become warmer or cooler as the KI dissolves?

_____

(b)   Rewrite the equation showing heat as a reactant or product, which ever is appropriate.

_____

(c)   Overall, do the attractions increase or decrease as the dissociation takes place?

_____

(d)   What happens to the molecular KE and chemical energy during the dissociation process? (increase or decrease).

_____

New Terms

Write the definitions of the following terms, which were introduced in this section. If necessary, refer to the Glossary at the end of the text.

chemical bond   chemical energy

endothermic   exothermic

7.5   Heats of reaction are measured at constant volume or at constant pressure

Review

The heat absorbed or given off during a chemical reaction is called the heat of reaction and is measured using a calorimeter either under conditions of constant volume (using a rigid container that cannot change in volume) or at constant pressure.

Pressure is a ratio of force to area. The gases of the atmosphere exert a pressure (atmospheric pressure) equal to approximately 14.7 lb in.$^{-2}$. This pressure is very close to the pressure units called the *standard atmosphere* (*atm*) and the *bar*.

Heats of reaction measured at constant volume are indicated by q_v and those measured at constant pressure are indicated by q_p. Study Example 7.6, which illustrates that q_p may not equal q_v, especially for reactions that involve gases.

The difference between q_p and q_v arises from pressure volume work, which is given by Equation 7.7 on page 275. Pressure-volume work is another way for a system to transfer energy. According to the *first law of thermodynamics*, the change in internal energy of a system is a function of the heat, q, added to the system and the work, w, done on the system. (Compressing a gaseous system is an example of work done on or to a system.)

$$\Delta E = q + w$$

It's important to keep the algebraic signs of q and w straight. Both are positive when energy is being added to the system (i.e., when heat is added or the system has work done on it). Be sure to study the sign conventions about q and w given in Section 6.4 in the text.

The values of q and w depend on how a change is carried out, as the example of the discharging battery of Figure 7.10 in the text explains. Therefore, q and w are not state functions. Nevertheless, their algebraic sum, ΔE, is a state function.

If a system cannot change in volume during a reaction, then the heat of reaction at constant volume, q_V equals ΔE for the change.

$$\Delta E = q_V \text{ (where } q_V \text{ is the heat of reaction at constant volume)}$$

Study Example 7.7, which describes measuring q_V using a bomb calorimeter.

Enthalpy and enthalpy changes

The "corrected" internal energy of a system at constant pressure is called the system's *enthalpy*. Like internal energy, enthalpy is a state function. By definition, $H = E + PV$, so when P is a constant, a change in H can be represented by

$$\Delta H = \Delta E + P\Delta V$$

When the only kind of work a system can do is expansion work (i.e., pressure-volume work) against the opposing atmospheric pressure, then ΔH equals the heat of reaction at constant pressure, q_p.

$$\Delta H = q_p \text{ (where } q_p \text{ is the heat of reaction at constant pressure)}$$

For most reactions, particularly those that do not involve gases, the difference between ΔE and ΔH is very small and can be neglected. However, as shown in Example 7.6, the difference can be significant for reactions involving gases.

As with the internal energy, a change in enthalpy is the difference between the enthalpies of the initial and final states.

$$\Delta H = H_{final} - H_{initial}$$

or

$$\Delta H = H_{products} - H_{reactants}$$

The value of ΔH is positive for endothermic reactions and negative for exothermic ones. Study Example 7.8, which illustrates how ΔH for a reaction can be measured using a constant pressure calorimeter.

Thinking It Through

1   When 0.1000 mol of diethyl ether, an anesthetic, was burned in a calorimeter, the temperature of the system rose from 24.000 °C to 26.332 °C. The heat capacity of the calorimeter was 118.0 kJ °C^{-1}. The equation for the reaction is

$$C_4H_{10}O(l) + 6O_2(g) \rightarrow 4CO_2(g) + 5H_2O(l)$$

(a)   How many kilojoules were liberated by the combustion of this sample?

(b)   How many kilojoules were liberated per mole of diethyl ether?

Self-Test

20. To what state functions do q_p and q_v each belong? _____

21. Give the definition of enthalpy in terms of internal energy, pressure, and volume.

    _____

22. When a reacting system in an insulated container becomes warmer, does its total energy change? Explain.

    _____

23. ΔH_{system} is defined as

    (a) $H_{reactants} - H_{products}$

    (b) $H_{products} - H_{reactants}$

    (c) $\Delta H_{products} - \Delta H_{reactants}$

    (d) the change in the heat capacity of the system.

    _____

24. If we had some way of knowing that $H_{products} = 400$ kcal and that $H_{reactants} = 600$ kcal, then $\Delta H_{reaction} =$

    (a) 1000 kcal

    (c) −200 kcal

    (b) 200 kcal

    (d) −1000 kcal   _____

25. If the value of $\Delta H_{system} = -200$ J, and the system's temperature before and after the change is the same, then the value of $\Delta H_{surroundings}$ is

    (a) −200 J

    (c) 0

    (b) 200 J

    (d) not measurable   _____

26. A 6.48 kg mass of water in a well-insulated vat was warmed from 20.15 °C to 21.39 °C. How much energy entered the water? Give the answer in J, kJ, cal, and kcal. The specific heat of water in this temperature range is 4.179 J g^{-1} °C^{-1}.

    _____

27. Hydrochloric acid, HCl(aq), can be made by bubbling hydrogen chloride, HCl(g), into water. When HCl(g) was bubbled into 225.00 g of water at 25.00 °C until the solution had a mass of 226.16 g, the temperature rose to 27.51 °C. Assume that the specific heat of the water is 4.184 J g^{-1} °C^{-1} throughout the reaction and the change in temperature. Calculate $\Delta H_{solution}$ for this change in kJ/mol HCl.

    _____

28. The heat of the reaction between KOH and HBr solutions was determined using a coffee cup calorimeter. The reaction was

    $$KOH(aq) + HBr(aq) \rightarrow KBr(aq) + H_2O$$

    A solution of 45.0 mL of 1.00 *M* HBr at 23.5 °C was mixed with a solution of 45.0 mL of 1.00 *M* KOH, also at 23.5 °C. The temperature quickly rose to 30.2 °C. Assume that the specific heats of each solution is 4.18 J g^{-1} °C^{-1}, that the densities of the solutions are 1.00 g mL^{-1}, and that the system loses no heat to its surroundings. Calculate the heat of reaction in kilojoules per mole of HCl.

New Terms

Write the definitions of the following terms, which were introduced in this section. If necessary, refer to the Glossary at the end of the text.

heat of reaction	calorimeter
calorimetry	pressure
atmospheric pressure	expansion work
pressure-volume work	first law of thermodynamics
heat of reaction at constant volume, q_v	heat of reaction at constant pressure, q_p
bomb calorimeter	enthalpy, H
enthalpy change, ΔH	

7.6   Thermochemical equations are chemical equations that quantitatively include heat

Review

The major concept in this section is the *thermochemical equation,* what it is and how to write one.

The value of ΔH for a given reaction depends on both the temperature and the pressure at which the system is kept. To compare ΔH values for different reactions, they must be measured under the same conditions. For enthalpy experiments, the *standard conditions* are 25 °C and 1 atm of pressure (and 1 *M* concentrations of reactants in solution). Values of ΔH measured under these conditions are called *standard heats of reaction* (or *standard enthalpies of reaction*), and the associated symbol is $\Delta H°$.

The actual value of $\Delta H°$ depends on the *scale* of the reaction—the actual mole quantities of substances as specified by the coefficients in the balanced equation. An equation that includes $\Delta H°$ is called a *thermochemical equation.* The following are all valid thermochemical equations for the reaction of hydrogen and oxygen that gives water.

(1)	$2H_2(g) + O_2(g) \rightarrow 2H_2O(l)$	$\Delta H° = -571.5$ kJ
(2)	$3H_2(g) + \frac{3}{2} O_2(g) \rightarrow 3H_2O(l)$	$\Delta H° = -857.3$ kJ
(3)	$4H_2(g) + 2O_2(g) \rightarrow 4H_2O(l)$	$\Delta H° = -1143$ kJ

The value of $\Delta H°$ differs for each because, as the coefficients indicate, the scale changes.   Always be careful to notice the physical states of the substances involved (*s, l,* or *g*), because enthalpy data depend on these, too. Notice also that the coefficients in a thermochemical equation do not have to be whole numbers and that they are not necessarily the smallest set of coefficients needed to balance the equation. This is because the coefficients specify numbers of *moles,* not individual molecules, and it is possible to have fractions of moles.

Self-Test

29. What are the experimental conditions to which standard enthalpy data refer?

_____

30. The thermochemical equation for the combustion of 2 mol of $CO(g)$ is

$$2CO(g) + O_2(g) \rightarrow 2CO_2(g) \qquad \Delta H^\circ = -566 \text{ kJ}$$

Write the thermochemical equation for the combustion of 4 mol of $CO(g)$.

_____

New Terms

Write the definitions of the following terms, which were introduced in this section. If necessary, refer to the Glossary at the end of the text.

standard state   standard heat of reaction

thermochemical equation

7.7   Thermochemical equations can be added because enthalpy is a state function

Review

Because enthalpy is a state function, it doesn't matter how we proceed from an initial state to a final state. The enthalpy change, ΔH, will be the same along any path as long as we begin and end at the same initial and final states. This interesting fact allows us to calculate an unknown ΔH for a reaction by combining known ΔH values for a different set of reactions that take us from the same reactants to the same products.

Suppose, for example, that we wished to determine ΔH° for the reaction

$$2NO(g) + O_2(g) \rightarrow 2NO_2(g)$$

We could use the following two thermochemical equations to accomplish this.

$$(1) \qquad N_2(g) + O_2(g) \rightarrow 2NO(g) \qquad \Delta H^\circ = +180.7 \text{ kJ}$$
$$(2) \qquad N_2(g) + 2O_2(g) \rightarrow 2NO_2(g) \qquad \Delta H^\circ = +67.6 \text{ kJ}$$

An *enthalpy diagram* can be used to obtain a graphical solution to the problem, as illustrated at the top of the next page.

The vertical axis is standard enthalpy, H°, as indicated by the heavy arrow pointing upward at the left. The line at the bottom represents the enthalpy of $N_2(g)$ plus $2O_2(g)$, which we've written as $O_2(g) + O_2(g)$. This will make it a bit easier to see what's happening.

Combining $N_2(g)$ with $O_2(g)$ to form $2NO(g)$ produces an enthalpy increase of +180.7 kJ, so the enthalpy of these products are shown higher on the enthalpy scale. (Recall that a positive value of ΔH means the products have a higher enthalpy than the reactants.) Notice that we've carried along an extra mole of O_2 that hasn't reacted. We'll use that shortly.

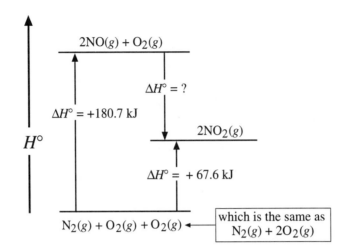

The second reaction combines $N_2(g)$ with $2O_2(g)$ to give $2NO_2(g)$, which gives an enthalpy increase of +67.6 kJ. Therefore, the $2NO_2(g)$ is shown higher on the enthalpy scale, too [but not as high as $2NO(g)$ + $O_2(g)$].

Examining the diagram, we can see that to go from $2NO(g)$ + $O_2(g)$, at the top of the diagram, to $2NO_2(g)$ we will move downward on the enthalpy scale. This means $H°$ is decreasing, so $\Delta H°$ will be negative. The *size* of the difference is also easy to determine from the graph; we just take the difference between 180.7 kJ and 67.6 kJ, or 113.1 kJ. Therefore, for the reaction $2NO(g)$ + $O_2(g)$ → $2NO_2(g)$, $\Delta H° = -113.1$ kJ.

Combining thermochemical equations without an enthalpy diagram

Hess's law, given on page 285, tells us we can combine Equations 1 and 2 above to obtain the same results as with the enthalpy diagram. To accomplish this, we set up the given thermochemical equations so that when we add them together we obtain the desired equation (let's call it the *target equation*). *If the sum of the chemical equations gives the target equation, then the sum of their $\Delta H°$ values will equal the $\Delta H°$ of the target reaction.*

Before we can add the equations, we have to adjust them so that, when added together, the sum is the target equation. We will use the rules for manipulating thermochemical equations given on pages 285 and 286 of the text. In this case, we will reverse Equation 1, which will change the sign of $\Delta H°$. Let's do that and then add the two equations.

(1 reversed)	$2NO(g)$ → $N_2(g)$ + $O_2(g)$	$\Delta H° = -180.7$ kJ
(2)	$N_2(g)$ + $2O_2(g)$ → $2NO_2(g)$	$\Delta H° = +67.6$ kJ

$$2NO(g) + \cancel{N_2(g)} + \cancel{2}O_2(g) \rightarrow \cancel{N_2(g)} + \cancel{O_2(g)} + 2NO_2(g) \qquad \Delta H° = -113.1 \text{ kJ}$$

Notice that we've canceled $N_2(g)$ as well as one $O_2(g)$ from each side of the final equation. What's left is our target equation. Also notice that we've obtained the same value of $\Delta H°$ as when we used the enthalpy diagram.

Applying Hess's law can sometimes seem like a daunting project, but it's not so bad if you proceed slowly. The following Example illustrates a strategy you can follow. We suggest you read through the Solution from beginning to end and then go back and study it carefully.

Example 7.1   Manipulating Thermochemical Equations

Calculate $\Delta H°$ for the reaction of copper(II) sulfide, $CuS(s)$, with gaseous hydrogen to give metallic copper and hydrogen sulfide.

$$CuS(s) + H_2(g) \rightarrow Cu(s) + H_2S(g)$$

Ust the thermochemical equations below.

(1)	$2CuS(s) + 3O_2(g) \rightarrow 2CuO(s) + 2SO_2(g)$	$\Delta H° = -807.18$ kJ
(2)	$2CuO(s) + C(s) \rightarrow 2Cu(s) + CO_2(g)$	$\Delta H° = -83.05$ kJ
(3)	$3H_2(g) + SO_2(g) \rightarrow H_2S(g) + 2H_2O(l)$	$\Delta H° = -292.32$ kJ
(4)	$C(s) + O_2(g) \rightarrow CO_2(g)$	$\Delta H° = -393.51$ kJ
(5)	$2H_2(g) + O_2(g) \rightarrow 2H_2O(l)$	$\Delta H° = -571.70$ kJ

Analysis:

We have two goals as we proceed. The first is to rearrange the listed equations so that we obtain each substance in the target equation (the one we want to work towards) on the correct side of the arrow with the correct coefficient. The second goal is to be sure anything that should *not* be in the target equation will cancel when we add the adjusted equations.

Before you get experience with these kinds of problems, the biggest difficulty is knowing where to start. There aren't any pat rules, but probably the best advice is to pick one of the compounds in the target equation *that appears in only one of the listed equations.* If we have more than one such choice, then just select one of them and begin.

Solution:

First, let's scan the equations to find compounds in the target equation that appear in only one of the listed equations. The formulas $CuS(s)$, $Cu(s)$, and $H_2S(g)$ fit this requirement. Notice, however, that $H_2(g)$ does not; it appears in both Equations (3) and (5), so we have no way of knowing which equation to adjust and in what way. We will have to leave $H_2(g)$ until near the end.

Let's begin with $CuS(s)$. In the target, it appears on the left with a coefficient of 1. In Equation (1) $CuS(s)$ is on the left, but its coefficient is 2. We'll divide all the coefficients of Equation (1) by 2 and we'll divide its $\Delta H°$ by 2 as well. We suggest you follow along by listing the equations on a piece of scrap paper as we adjust them. Then, in the end, you'll have them arranged so you can simply add them.

$$\text{(1-adjusted)} \quad \boxed{CuS(s)} + \tfrac{3}{2} O_2(g) \rightarrow CuO(s) + SO_2(g) \qquad \Delta H° = -403.59 \text{ kJ}$$

We've placed a box around the CuS so we know we have it where we want it with the correct coefficient.

Next, let's work on $Cu(s)$. This appears in the second equation, but its coefficient has to be divided by 2 also. Therefore, we divide all the coefficients of Equation (2) by 2, including the value of $\Delta H°$.

$$\text{(2-adjusted)} \quad CuO(s) + \tfrac{1}{2} C(s) \rightarrow \boxed{Cu(s)} + \tfrac{1}{2} CO_2(g) \qquad \Delta H° = -41.52 \text{ kJ}$$

Now we'll work on $H_2S(g)$, which is in Equation (3). This compound is on the correct side and its coefficient is what we want, so we don't have to do anything to this equation. Let's just rewrite it, placing a box around the H_2S.

$$\text{(3)} \quad 3H_2(g) + SO_2(g) \rightarrow \boxed{H_2S(g)} + 2H_2O(l) \qquad \Delta H° = -292.32 \text{ kJ}$$

Let's combine these three equations, ignoring their $\Delta H°$ for the time being, so we can see where we stand.

$$CuS(s) + \tfrac{3}{2}O_2(g) \rightarrow CuO(s) + SO_2(g)$$

$$CuO(s) + \tfrac{1}{2}C(s) \rightarrow Cu(s) + \tfrac{1}{2}CO_2(g)$$

$$3H_2(g) + SO_2(g) \rightarrow H_2S(g) + 2H_2O(l)$$

$$CuS(s) + \tfrac{3}{2}O_2(g) + \tfrac{1}{2}C(s) + 3H_2(g) \rightarrow Cu(s) + \tfrac{1}{2}CO_2(g) + H_2S(g) + 2H_2O(l)$$

Notice that the SO_2 and CuO cancel, but there is still work to do. We need to cancel $3/2 O_2(g)$, $1/2 C(s)$, and $2H_2(g)$ from the left, and $1/2 CO_2(g)$ and $2H_2O(l)$ from the right. We'll do this by using the remaining thermochemical equations.

We can eliminate the $\tfrac{1}{2}C(s)$, $\tfrac{1}{2}O_2(g)$, and $\tfrac{1}{2}CO_2(g)$ by reversing Equation (4) and dividing its coefficients by 2.

(4-adjusted) $$\tfrac{1}{2}CO_2(g) \rightarrow \tfrac{1}{2}C(s) + \tfrac{1}{2}O_2(g) \quad \Delta H° = +196.76 \text{ kJ}$$

Notice that this places $\tfrac{1}{2}CO_2(g)$ on the left, so it will cancel with the $\tfrac{1}{2}CO_2(g)$ on the right. We similarly will be able to cancel $\tfrac{1}{2}C(s)$ and $\tfrac{1}{2}O_2(g)$. This still leaves $\tfrac{2}{2}O_2(g)$ to be removed, as well as $2H_2(g)$. We can finally clear these up by reversing Equation (5).

(5-reversed) $$2H_2O(l) \rightarrow 2H_2(g) + O_2(g) \quad \Delta H° = +571.70 \text{ kJ}$$

Let's check to be sure everything cancels to give the target equation. Then we can add the $\Delta H°$ values to get $\Delta H°$ for the target equation.

$$CuS(s) + \tfrac{3}{2}O_2(g) + \tfrac{1}{2}C(s) + 3H_2(g) \rightarrow Cu(s) + \tfrac{1}{2}CO_2(g) + H_2S(g) + 2H_2O(l)$$

$$\tfrac{1}{2}CO_2(g) \rightarrow \tfrac{1}{2}C(s) + \tfrac{1}{2}O_2(g)$$

$$2H_2O(l) \rightarrow 2H_2(g) + O_2(g)$$

These combine to give $\tfrac{3}{2}O_2(g)$

Target equation $$CuS(s) + H_2(g) \rightarrow Cu(s) + H_2S(g)$$

Spend some time to study this so you can see how everything cancels except the formulas of the compounds in the target equation.

Equation Used	$\Delta H°$
(1) adjusted by dividing coefficients by 2	−403.59 kJ
(2) adjusted by dividing coefficients by 2	−41.52 kJ
(3) as is	−292.32 kJ
(4) reversed and coefficients divided by 2	+196.76 kJ
(5) reversed +571.70 kJ	
Net $\Delta H°$ =	+31.03 kJ

Therefore, for the target equation we can write

$$CuS(s) + H_2(g) \rightarrow Cu(s) + H_2S(g) \quad \Delta H° = 31.02 \text{ kJ}$$

Is the answer reasonable?

The best check is to note that when we add the adjusted equations, we obtain the target equation. We can check to be sure we've followed the rules for adjusting the $\Delta H°$ values when manipulating thermochemical equations. Once we've done that, we need to check to be sure we've done the arithmetic correctly. After all that, we can feel confident in our answer.

The Example we've just worked out is about as complicated as they get. If you're careful and follow the procedure we've outlined above, you will get correct answers. Just be sure the adjusted equations add to give the target equation, and keep your eye on the target as you make decisions as to how to manipulate the equations.

Self-Test

31. The direct reaction of methane (CH_4) with oxygen to give carbon monoxide is hard to accomplish without also producing carbon dioxide. However, we can use a Hess's law calculation to calculate $\Delta H°$, anyway. Calculate $\Delta H°$ in kJ for the following reaction from the thermochemical equations given.

$$2CH_4(g) + 3O_2(g) \rightarrow 2CO(g) + 4H_2O(l)$$

Use the following thermochemical equations:

(1) $2C(s) + O_2(g) \rightarrow 2CO(g)$ $\qquad\qquad\qquad$ $\Delta H° = -221.08$ kJ

(2) $CH_4(g) + 2O_2(g) \rightarrow CO_2(g) + 2H_2O(l)$ $\quad$ $\Delta H° = -890.4$ kJ

(3) $C(s) + O_2(g) \rightarrow CO_2(g)$ $\qquad\qquad\qquad$ $\Delta H° = -393.51$ kJ

---

32. Acetic acid can be made from acetylene by the following overall reaction:
$$2C_2H_2(g) + 2H_2O(l) + O_2(g) \rightarrow 2HC_2H_3O_2$$
$\qquad\quad$ acetylene $\qquad\qquad\qquad\qquad\quad$ acetic acid

Calculate $\Delta H°$ for the formation of *one* mole of acetic acid by this reaction using the following thermochemical equations:

(1) $C_2H_4(g) \rightarrow H_2(g) + C_2H_2(g)$ $\qquad\qquad\qquad$ $\Delta H° = 174.464$ kJ
$\quad$ ethylene

(2) $C_2H_5OH(l) \rightarrow C_2H_4(g) + H_2O(l)$ $\qquad\qquad$ $\Delta H°\ \ 44.066$ kJ
$\quad$ ethyl alcohol

(3) $C_2H_5OH(l) + O_2(g) \rightarrow HC_2H_3O_2(l) + H_2O(l)$ $\qquad$ $\Delta H° = -495.22$ kJ

(4) $2H_2(g) + O_2(g) \rightarrow 2H_2O(l)$ $\qquad\qquad\qquad$ $\Delta H° = -571.70$ kJ

New Terms

Write the definitions of the following terms, which were introduced in this section. If necessary, refer to the Glossary at the end of the text.

enthalpy diagram   Hess's law of heat summation (Hess's law)

7.8   Heats of reaction can be predicted using standard enthalpies of formation

Review

Thermochemical data is available in several formats, including heats of combustion. The standard heat of combustion, ΔH_c°, is the energy released in the complete combustion of 1 mol of a compound in oxygen, all substances being in their standard states. For example, for butane, ΔH_c° corresponds to ΔH° for the reaction

$$C_4H_{10}(g) + \frac{13}{2}\,O_2(g) \rightarrow 4CO_2(g) + 5H_2O(l)$$

Study Example 7.12 on page 187 to see how we can use ΔH_c° in calculations.

The *standard enthalpy of formation* (also called the *standard heat of formation*), ΔH_f°, is the amount of heat absorbed or released in the formation of 1 mol of a compound at 25 °C and a pressure of 1 bar from its elements in their standard states. The key phrase here is "from its elements in their standard states," which means from the elements as we normally find them at 25 °C and ordinary atmospheric pressure. Be sure to examine the thermochemical equations on page ♦♦♦, which illustrate this point. Notice that ΔH_f° has units of energy per mole (e.g., kJ mol^{-1}).

Heats of formation are useful for calculating heats of reaction using the *Hess's law equation* (Equation 7.12 on page 189).

$$\Delta H^\circ = \left(\text{sum of } \Delta H_f^\circ \text{ of products}\right) - \left(\text{sum of } \Delta H_f^\circ \text{ of reactants}\right)$$

Study Example 7.13 on page 289 to see how we apply the Hess's law equation. Problems of this kind are not difficult as long as you are careful about the algebraic signs of the energy terms. Take your time to avoid errors.

Thinking It Through

2   Calculate ΔH° for the target reaction in Example 7.1 which we worked out in the preceding section of the Study Guide. The reaction is

$$CuS(s) + H_2(g) \rightarrow Cu(s) + H_2S(g)$$

Use the following ΔH_f° data.

$$CuS(s) \quad \Delta H_f^\circ = -48.53 \text{ kJ mol}^{-1}$$

$$H_2S(g) \quad \Delta H_f^\circ = -17.506 \text{ kJ mol}^{-1}$$

Does your result agree with the answer obtained in the worked example (making allowances for errors caused by rounding or in experimental data)?

Self–Test

33. Any substance in its most stable chemical form at 25 °C and 1 bar is said to be in its

    (a)   thermochemical state   (b)   standard state

    (c)   STP state   (d)   zero H state   _____

34. Write the thermochemical equations together with ΔH_f° values (Table 7.2 in the textbook) for the formation of each of the following compounds from its elements. Use units of kJ/mol.
(a) $H_2SO_4(l)$ (b) $Al_2O_3(s)$ (c) $CO(NH_2)_2(s)$

    _____

    _____

    _____

35. Write the thermochemical equations, including correct values of ΔH° in kJ, for the formation of the specified amount of each of the following compounds from its elements. Use Table 7.2 in the textbook for necessary data.

    (a)   2 mol of Fe_2O_3

    _____

    (b)   3 mol of $C_2H_6(g)$

    _____

36. Draw an enthalpy diagram for the formation of one mole of H_2O_2 from its elements. Use data in Table 7.2 of the textbook in units of kJ as needed.

37. Use the Hess law equation and ΔH_f° data in Table 7.2 of the textbook to calculate ΔH° for the reaction of methane with oxygen given in Problem 31. (Having done 31 the long way you'll see the powerful advantage of the Hess law equation.)

    _____

38. Use the Hess law equation, data in Table 7.2 of the textbook, and the results of the calculations in Problem 32 to calculate ΔH_f° for acetic acid in kJ mol^{-1}.

                      _____

39. Calculate ΔH_c° for the combustion of acetylene, $C_2H_2(g)$, in kJ mol^{-1} from ΔH_f° data from Table 7.2 of the textbook. The equation is

$$2C_2H_2(g) + 5O_2(g) \rightarrow 4CO_2(g) + 2H_2O(l)$$

New Terms

Write the definitions of the following terms, which were introduced in this section. If necessary, refer to the Glossary at the end of the text.

Heat of combustion	standard heat of combustion
Hess's law equation	standard state
standard enthalpy of formation	standard heat of formation

Solutions to Thinking It Through

1   (a)   What we first want is the relationship between the temperature change and the amount of heat released:

$$\Delta t \Leftrightarrow ? \text{ kilojoules}$$

The connection between Δt and kilojoules is the heat capacity of the calorimeter, 118.0 kJ °C^{-1}.

$$\Delta t = 26.332 \text{ °C} - 24.000 \text{ °C} = 2.332 \text{ °C}$$

$$118 \frac{\text{kJ}}{\text{°C}} \times 2.332 \text{ °C} = 275.2 \text{ kJ}$$

(b)   Because 0.1000 mol of diethyl ether liberated 275.2 kJ, the heat liberated per mole is the ratio:

$$\frac{275.2 \text{ kJ}}{0.1000 \text{ mol}} = 2751 \text{ kJ mol}^{-1}$$

2   Hess's law equation is applicable:

$$\Delta H° = \left(\text{sum of } H_f° \text{ of products} \right) - \left(\text{sum of } H_f° \text{ of reactants} \right)$$

$$= \left(-17.506 \text{ kJ mol}^{-1} \times 1 \text{ mol (H}_2\text{S)} \right) - \left(-\left(-48.53 \text{ kJ mol}^{-1} \times 1 \text{ mol (CuS)} \right) \right)$$

$$= +31.02 \text{ kJ}$$

Answers to Self-Test Questions

1.   On the upward flight, kinetic energy changes to potential energy. At the top, where the stone is motionless, all of the initial kinetic energy has become potential energy. On the way down, potential energy changes to kinetic energy as the stone picks up speed.

2.   d

3.   875 cal

4. 3.66×10^3 J

5.   10.9 kcal

6. 1.73×10^5 J

7.   E equals the sum of all KE and PE of atoms, molecules, and ions in a sample.

8. KE = $1/mv^2$; mass (m) is never negative and even if velocity (v) were negative, the square of the velocity must be a positive number. Hence, KE is always positive (or zero). (Note. As you may have already studied in a physics course, velocity can be negative because it is a vector quantity.)

9. Because there is no such thing as negative kinetic energy and because 0 K corresponds to the cessation of all motions associated with molecular kinetic energy.

10. The average molecular kinetic energy decreases.

11. A change in a state function depends only on the initial and final states, not on the path taken by the system as it changes from one state to the other. Examples are temperature and internal energy.

12. $\Delta E = 25.6$ kJ

13. See Figure 7.5 in the textbook.

14. Open system: exchanges matter and energy with surroundings. Closed system: only exchanges energy.

15. (a) increase, (b) –45 J

16. (a) 21.6 J $°C^{-1}$, (b) 0.444 J g^{-1} $°C^{-1}$, (c) 688 J, (d) 27.1 °C

17. It increases.

18. exothermic

19. (a) cooler, (b) heat + KI(s) + nH$_2$O $\rightarrow$ K$^+$(aq) + I$^-$(aq), (c) decrease, (d) KE decrease, PE increase

20. $q_v = \Delta E$ at constant volume; $q_p = \Delta H$ at constant pressure

21. $H = E + PV$

22. No. Law of conservation of energy.

23. b

24. c

25. b

26. 33.6×10^3 J, 33.6 kJ, 8.03×10^3 cal, 8.03 kcal

27. 74.9 kJ/mol HCl

28. 55 kJ/mol HCl

29. 25 °C and 1 atm

30. $4CO(g) + 2O_2(g) \rightarrow 4CO_2(g)$ $\Delta H° = 1132$ kJ

31. –1214.9 kJ

32. –427.90 kJ

33. b

34. (a) $H_2(g) + S(s) + 2O_2(g) \rightarrow H_2SO_4(l)$ $\qquad$ $\Delta H_f° = -811.32$ kJ mol^{-1}

    (b) $2Al(s) + \frac{3}{2}O_2(g) \rightarrow Al_2O_3(s)$ $\qquad$ $\Delta H_f° = -1669.8$ kJ mol^{-1}

    (c) $C(s) + \frac{1}{2}O_2(g) + N_2(g) + 2H_2(g) \rightarrow CO(NH_2)_2(s)$ $\Delta H° = -333.19$ kJ mol^{-1}

35. (a) $4Fe(s) + 3O_2(g) \rightarrow 2Fe_2O_3(s)$ $\qquad$ $\Delta H° = -1644.4$ kJ

    (b) $6C(s) + 9H_2(g) \rightarrow 3C_2H_6(g)$ $\qquad$ $\Delta H° = -254.001$ kJ

36.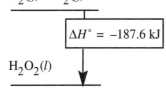

    Because $\Delta H°$ is negative, the product is at a lower enthalpy than the reactants.

37. –1214.9 kJ

38. –487.05 kJ

39. 1299.7 kJ mol^{-1}

Tools you have learned

Consider removing this chart from the Study Guide so you can have it handy when tackling homework problems.

Tool	How It Works
Heat Capacity and Specific Heat	We use the heat capacity and a measured temperature change to determine the amount of heat gained or lost by a system. To use specific heat to calculate energy transfer, we need the mass of the sample and the temperature change.
Thermochemical Equations	We use thermochemical equations for one set of reactions to write a thermochemical equation for some net reaction.
Enthalpy Diagrams	These help us visualize enthalpy changes in multistep reactions.
Hess's law equation	We use this equation and standard heats of formation to calculate the enthalpy of a reaction.

Summary of Important Equations

Heat capacity

$$\text{Heat capacity} = \frac{\text{heat absorbed}}{\Delta t}$$

Specific heat

$$\text{Specific heat} = \frac{\text{heat capacity}}{\text{mass}(g)}$$

Using specific heat

$$\text{Heat energy} = \text{specific heat} \times \text{mass} \times \Delta t$$

Hess's law equation

$$\Delta H^\circ = \left(\text{sum of } \Delta H_f^\circ \text{ of products}\right) - \left(\text{sum of } \Delta H_f^\circ \text{ of reactants}\right)$$

Chapter **8**

The Quantum Mechanical Atom

In Chapter 2 you learned that atoms are not the simplest particles of matter. They are composed of still simpler parts called protons, neutrons, and electrons. The protons and neutrons of an atom are found in the nucleus and the electrons surround the nucleus in the atom's remaining volume. Because the nucleus is so small and so far from the outer parts of an atom, it has very little direct influence on ordinary chemical and physical properties. Its indirect role is in determining the number of electrons a neutral atom will have. As you will learn in this chapter, it is the number of electrons and their energy distribution (which we call the atom's *electronic structure*) that is the primary factor in determining an atom's chemical and physical properties.

An important fact about electrons in atoms is that they do not obey the classical laws of physics, which would predict that an orbiting electron should quickly crash into the nucleus. In fact, tiny particles like the electron sometimes behave like particles and at other times they behave as waves, a phenomenon we call *wave/particle duality*. The understanding of atomic structure came about through the development of *quantum theory*, also called *quantum mechanics*. We introduce you to some of the concepts and results of the theory, but (mercifully!) we cannot delve too deeply into it because of its highly mathematical nature.

You will find that much of this chapter deals with theory, and you may find that a lot of it seems remote from everyday experience. If you find it difficult to understand at first, don't be discouraged. Reread the discussions, study the worked examples, and do the exercises. It may take some time, but it should all fit together eventually.

Learning Objectives

Throughout your study of this chapter, keep in mind the following objectives:

1   To learn about the properties of light, and to learn how the light emitted by an atom when it is excited, or energized, gives clues to how the electrons are distributed in the space outside the atom's nucleus.

2   To see how the theoretical model of the electronic structure of the atom developed, historically.

3   To see that matter behaves not only as though it were composed of discrete particles, but also as waves.

4   To learn the conditions under which the wave properties of electrons are most noticeable.

5   To learn how waves interact with each other to produce a diffraction pattern.

6   To learn how the theoretical study of matter waves leads to a more sophisticated explanation of atomic structure.

7   To learn the names, symbols, and allowed values for the three quantum numbers associated with electron waves in atoms.

8   To examine the relationship between the quantum numbers and the energies of the electron waves in atoms.

9   To see how the magnetic behavior of the electron leads to the idea that the electron is spinning, and to see how this electron spin affects the electronic structure of an atom.

10   To see how electrons are arranged among the orbitals of an atom, and to predict such arrangements using Figure 6.15.

11   To see how the structure of the periodic table correlates with predicted electron configurations and to learn how to use the periodic table as a tool for predicting electron configurations.

12   To learn the electron configurations of chromium and copper, which are not predicted by the rules for writing electron configuration of atoms.

13   To learn the shapes and directional properties (relative orientations) of *s*, *p*, and *d* orbitals.

14   To learn how electrons that are close to the nucleus shield more distant electrons from the full effects of the nuclear charge.

15   To see how atomic and ionic size are related to electronic structure and to learn how these properties vary within the periodic table.

16   To learn how the size of an atom changes when it gains or loses electrons to become an ion.

17   To learn how the energy needed to remove an electron from an atom can be related to the atom's electronic structure, and how this quantity varies according to the atom's position in the periodic table.

18   To see how the energy change associated with the addition of an electron to an atom varies according to the atom's position in the periodic table.

8.1   Electromagnetic radiation can be described as a wave or as a stream of photons

Review

Electromagnetic energy is energy that is carried by *electromagnetic radiation* (light waves). These waves are characterized by their *frequency*, ν (the number of oscillations per second), which is given in units called hertz (Hz). Remember,

$$1 \text{ Hz} = 1 \text{ s}^{-1}$$

(s^{-1} means *second* raised to the minus one power, which is the same as 1/second).

$$1 \text{ s}^{-1} = \frac{1}{s} = \frac{1}{\text{second}}$$

A wave is also characterized by its *wavelength*, λ, which is the distance between any two successive peaks. In the SI, wavelength is expressed in meters (or submultiples of meters, such as nanometers).

The product of wavelength and frequency is the speed of the wave, and for electromagnetic radiation traveling through a vacuum, this speed is a constant called the *speed of light* (symbol, c). Be sure you know the equation

$$\lambda \times \nu = c$$

Check with your teacher to find out whether you are expected to know the value of c, 3.00×10^8 m/s (or 3.00×10^8 m s^{-1}). Study Examples 8.1 and 8.2 to be sure you can change wavelength to frequency and frequency to wavelength.

The electromagnetic spectrum

Electromagnetic radiation comes in a range of frequencies or wavelengths that we call the electromagnetic spectrum. Visible light constitutes only a narrow band of wavelengths ranging from approximately 400 to 700 nm. Infrared, microwaves, and radio and TV broadcasts are waves of longer wavelength (lower frequency) than visible light. Ultraviolet, X rays, and gamma radiation are waves of shorter wavelength (higher frequency) than visible light.

The energy of light

Light often behaves as though it consists of a stream of tiny packets (also called *quanta*) of energy that we call *photons*. The energy of a photon is proportional to the radiation's frequency. The equation is

$$E = h\nu$$

where E is the energy of the photon, ν is its frequency, and h is a proportionality constant called Planck's constant. Ask your teacher whether you are expected to memorize the value of Planck's constant ($h = 6.62 \times 10^{34}$ J s). The relationship between energy and frequency tells us that light with high frequencies (short wavelengths) consist of photons of high energy, whereas light with low frequencies (long wavelengths) have photons of lower energy.

Thinking It Through

For the following, identify the information needed to solve the problem and show (or explain) what must be done with it.

1   How many microwave photons, each with a wavelength of 1.05 cm, are needed to raise the temperature of a cup of coffee (250 mL) from room temperature (25 °C) to 80 °C? Assume the coffee has a density of 1.0 g/mL and a specific heat of 4.18 J g^{-1} °C^{-1}.

Self-Test

1.   Radio station WGBB in New York broadcasts at a frequency of 1240 kHz on the AM radio band. What is the wavelength in meters of these radio waves?

   _____

2.   A certain compound is found to absorb infrared radiation strongly at a wavelength of 3.0×10^{-6} m. What is the frequency of this radiation?

   _____

3.   What is the wavelength in nanometers of electromagnetic radiation that has a frequency of 6.0×10^{15} Hz?

   _____

4.   Calculate the energy in joules of a photon of green light that has a wavelength of 546 nm.

   _____

5.   What is the energy in kilojoules of one mole of photons having a wavelength of 546 nm?

   _____

6.   Arrange the following kinds of electromagnetic radiation in order of increasing energy (lowest → highest): X rays, radio waves, infrared radiation, ultraviolet light, visible light, microwaves.

New Terms

Write the definitions of the following terms, which were introduced in the introduction to the chapter and in this section. If necessary, refer to the Glossary at the end of the text.

wave/particle duality	quantum mechanics	quantum theory
collapsing atom paradox	electromagnetic radiation	electromagnetic energy
frequency	hertz	wavelength
amplitude	electromagnetic spectrum	visible spectrum
quantum	photon	Planck's constant

8.2   Atomic line spectra are experimental evidence that electrons in atoms have quantized energies

Review

Atomic spectra

A *continuous spectrum* (Figure 8.8a) contains all wavelengths of light and is produced by the glow of a hot object, such as the filament in a light bulb, or the sun. The spectrum emitted by energized (*excited*) atoms is called an *atomic spectrum* or emission spectrum and is not continuous. It contains only a relatively few wavelengths, as illustrated in Figure 8.8b and c.

The atomic spectrum is different for each element, and can be used to identify the element; it's like the element's "fingerprint." Atomic spectra suggest that the energy of an electron in an atom is quantized, which means the electron can have only certain specific amounts of energy. These energies correspond to a characteristic set of *energy levels* possessed by atoms of the element. When an electron goes from one energy level to a lower one the difference in energy, ΔE, appears as light having a frequency determined by the equation $\Delta E = h\nu$.

Atomic spectrum of hydrogen and the Bohr model of the atom

The Rydberg equation is an *empirical equation*, which means it was derived from data obtained by experimental measurements. It can be used to calculate the wavelengths of the lines in the spectrum of hydrogen as illustrated in Example 8.3. (You probably don't need to memorize the Rydberg equation, but check with your instructor to be sure.)

Bohr developed his model of the atom to explain the spectrum of hydrogen. You should know the basic features of the model:

1   The electron was believed to travel in circular orbits.

2   The energy and size of an orbit can have only certain values, which are related to the value of the quantum number *n*.

3   When the atom absorbs energy, the electron is raised to a higher, more energetic orbit.

4   When the electron drops from a higher to a lower orbit, a photon is emitted whose energy equals the energy difference between the two orbits. The frequency of the photon is determined by the relationship $\Delta E = h\nu$.

Bohr was able to use his model to derive an equation that matched the Rydberg equation almost exactly. This was the theory's greatest success.

Although Bohr's model of the atom was later shown to be incorrect, he was the first to recognize the existence of quantized energy levels in atoms. He was also the first to introduce the idea of a *quantum number*—an integer related to the energy of an electron in an atom.

Self-Test

7.   According to Bohr's model of the hydrogen atom, what was the value of the quantum number for the lowest-energy orbit?

_____

8.   In terms of b in Equation 8.3, what is the energy of the electron in the first Bohr orbit ($n = 1$)?

_____

What is the energy of the electron in the second Bohr orbit ($n = 2$)?

_____

In terms of b, how much energy is emitted when the electron drops from the orbit with $n = 2$ to the orbit with $n = 1$?

_____

9.   Use the Rydberg equation to calculate the wavelength in nanometers of the spectral line produced when an electron drops from the 6th Bohr orbit to the second. What color is this spectral line?

_____

New Terms

Write the definitions of the following terms, which were introduced in this section. If necessary, refer to the Glossary at the end of the text.

continuous spectrum	quantum number	Rydberg equation
atomic spectrum	ground state	quantized energy
emission spectrum	energy level	

8.3   Electron waves in atoms are called orbitals

Review

De Broglie proposed that particles have wave properties, each with a wavelength, λ, that is inversely proportional to the product of the particle's mass and velocity.

$$\lambda = \frac{h}{mv}$$

The proportionality constant, h, is Planck's constant. Only very light particles, such as the electron, proton, and neutron, have wavelengths large enough to be observed experimentally. Diffraction—the constructive and destructive interference of waves—can be used to demonstrate the wave nature of these particles.

In this section, we examine two types of waves, traveling waves and standing waves (one in which the positions of the peaks, troughs, and nodes don't move). The discussion about notes on a guitar string shows that standing waves lead naturally to integer relationships concerning the wavelengths that can exist on a string. Bringing this together with the de Broglie relationship for an electron on a string shows that the energy of the electron, behaving like a wave, is restricted in its values. In other words, the energy of the electron is quantized. The significant point here is that this quantized behavior is a direct result of the wave characteristics of the electron.

In atoms, electrons behave as three-dimensional standing waves

Wave mechanics (quantum mechanics) is the name we give to the theory about the electronic structures of atoms and molecules. In this theory, the electron in an atom is considered to be a three-dimensional standing wave. Electron waves are described by a mathematical expression called a wave function, usually represented by the symbol ψ. Each individual wave function describes an *orbital,* which is another term used to describe an electron wave. Each orbital is identified by a set of values for three quantum numbers. You should learn the names of the quantum numbers and their allowed values (including the restrictions on their values).

The *principal quantum number, n,* can only have integer values that range from 1 to ∞.

$$n = 1 \text{ or } 2 \text{ or } 3 \text{ or } \ldots\infty$$

In categorizing the energies of the orbitals, n identifies the shell. An orbital with $n = 1$ is in the first shell, and so forth. The energy and size of an orbital is determined in large measure by its value of n.

The *secondary quantum number, l,* divides the shells into subshells. This quantum number determines the shapes of the orbitals, and to some degree, their energies as well. For a given n, allowed values of l range from 0 to $(n - 1)$. Remember the letter designations for the subshells.

value of l	0	1	2	3
letter	s	p	d	f

The *magnetic quantum number, m_l,* divides the subshells into individual orbitals and determines the orientations of the orbitals relative to each other. In a given subshell, the values of m_l range from $+l$ to $-l$. These values determine the number of orbitals in a given kind of subshell.

type of subshell	s	p	d	f
number of orbitals	1	3	5	7

These relationships are illustrated in Figure 1 on the next page, which should be studied in conjunction with Table 8.1 on page 323 of the text. The relative energies of the various subshells and orbitals are described by Figure 8.19 of the text. You need not memorize this figure, although we will use it in Section 8.5 to determine how the electrons in an atom are distributed among the various orbitals.

In Figure 8.19 of the text, there are several points to note. First, we see that every shell has an s subshell. Shells above the first shell each have a p subshell. Any shell above the second shell has a d subshell as well, and beyond the third shell, each also has an f subshell. Second, notice that the energies of the shells increase with increasing value of n. Third, notice that within a shell the energies of the subshells are in the order $s < p < d < f$, and that all orbitals of a given subshell are of equal energy. Finally, notice how the subshells of one shell begin to overlap with subshells of other shells as n becomes larger.

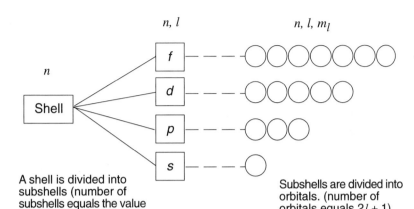

n, l   *n, l, m_l*

f

d

Shell

p

s

A shell is divided into subshells (number of subshells equals the value of *n* for the shell).

Subshells are divided into orbitals. (number of orbitals equals $2l + 1$)

Figure 1
The way the quantum numbers for electrons correspond to shells, subshells and individual orbitals in an atom. Each circle represents an orbital.

Self-Test

10. What do the terms *in phase* and *out of phase* mean?

   _____

   _____

11. What would be the subshell designation (e.g., $1s$) corresponding to the following sets of values of *n* and l?

   (a) $n = 2, l = 1$ _____

   (b) $n = 4, l = 0$ _____

   (c) $n = 3, l = 2$ _____

   (d) $n = 5, l = 3$ _____

12. Which of the following sets of quantum numbers are <u>unacceptable?</u>

   (a) $n = 2, l = 1, m_l = 0$ (b) $n = 2, l = 2, m_l = 1$

   (c) $n = 2, l = 1, m_l = -2$ (d) $n = 3, l = 2, m_l = -2$

   (e) $n = 0, l = 0, m_l = 0$ _____

13. What subshells are found in the fourth shell?

   _____

14. Which subshell is higher in energy?

   (a) $3s$ or $3p$ _____

   (b) $4p$ or $4d$ _____

   (c) $3p$ or $4p$ _____

New Terms

Write the definitions of the following terms, which were introduced in this section. If necessary, refer to the Glossary at the end of the text.

diffraction	interference fringes	diffraction pattern
traveling wave	standing wave	node
wave function	wave mechanics	quantum mechanics
principal quantum number (n)	secondary quantum number (l)	magnetic quantum number (m_l)
shell	subshell	orbital

8.4   Electron spin affects the distribution of electrons among orbitals in atoms

Review

The electron behaves like a tiny magnet. A way of explaining this is to imagine that it spins like a top. Two directions of spin are possible, which are identified by the values of the spin quantum number, m_s. The Pauli exclusion principle states that no two electrons in the same atom can have the same values for all four of their quantum numbers. The result is that no more than two electrons can occupy any one orbital. (No more than two electrons can have the same wave form.) When all of the electrons in an atom are not *paired*, a residual magnetism occurs and the atom is paramagnetic. Paramagnetic substances are attracted weakly to a magnetic field. On the other hand, if all the electrons are paired, the atom is diamagnetic. It is repelled weakly by a magnetic field.

Self-Test

15.   How many electrons can occupy

    (a)   a 2s subshell?   _____

    (b)   a 3d subshell?   _____

    (c)   the shell with $n = 3$?   _____

    (d)   the shell with $n = 4$?   _____

    (e)   the subshell with $n = 3$ and l $= 1$ _____

New Terms

Write the definitions of the following terms, which were introduced in this section. If necessary, refer to the Glossary at the end of the text.

electron spin	paramagnetism
spin quantum number (m_s)	diamagnetism
Pauli exclusion principle	

8.5   The ground state electron configuration is the lowest energy distribution of electrons among orbitals

Review

The distribution of electrons among an atom's orbitals is called the electronic structure, or electron configuration of the atom. Electrons always fill orbitals of lowest energy first (Figure 8.19). When filling a set of equal-energy orbitals, Hund's rule requires that the electrons spread out over the orbitals of a subshell as much as possible and that the number of unpaired electrons in the subshell be a maximum. This gives the ground state (lowest energy) configuration. In Section 8.6 you will see that the periodic table can also be used in place of Figure 8.19 to obtain an atom's electron configuration.

Learn to draw orbital diagrams. When indicating the orbitals of a p, d, or f subshell, be sure to show *all* of the orbitals of that subshell, even if some of them are unoccupied. For example, for boron we should write

B   (↑↓) (↑↓) (↑)○○   (correct)
         1s   2s   2p

The following is *incorrect*, because only one of the three $2p$ orbitals is shown.

B   (↑↓) (↑↓) (↑)   (incorrect)
         1s   2s   2p

Self-Test

16. Use Figure 8.19 to write the electron configuration for

    (a) Al _____

    (b) Br _____

    (c) V _____

17. Construct orbital diagrams for

    (a) phosphorus

_____

    (b) silicon

_____

New Terms

Write the definitions of the following terms, which were introduced in this section. If necessary, refer to the Glossary at the end of the text.

electronic structure	orbital diagram	Hund's rule
electron configuration	core electrons	aufbau principle
orbital diagram		

8.6   Electron configurations explain the structure of the periodic table

Review

Figure 8.22 illustrates how we can use the periodic table as a guide in writing electron configurations. Following the aufbau principle, we assign electrons to subshells beginning with the 1s (corresponding to Period 1 in the periodic table). Then, as we cross the periodic table row after row, we let the table tell us which subshells become occupied.

When we cross through Groups IA and IIA, we fill an s subshell; across Groups IIIA through Group VIII we fill a p subshell. For the s and p subshells, the value of n that goes with the s and p designations equals the period number. Crossing a row of transition elements corresponds to filling a d subshell with n equal to *one less than the period number*. Crossing a row of inner transition elements (the lanthanides or actinides) corresponds to filling an f subshell with n equal to *two less than the period number*. Be sure to study Example 8.4 on page 330 of the text.

Basis for the periodic recurrence of properties

The main point of this discussion is that atoms of a given group in the periodic table have similar electron configurations for their outer shells. Thus, all elements in Group IA have an outer shell with one electron in an s orbital. Similarly, the elements in Group IIA each have two electrons in their outer shell s orbital.

Abbreviated configurations

In the shorthand electron configuration of an element, we only show subshells above those that are filled in an atom of the preceding noble gas. For example, for potassium (atomic number $Z = 19$), we only show orbitals beyond those that are filled in argon, the preceding noble gas. Argon ($Z = 18$) has completed 1s, 2s, 2p, 3s, and 3p subshells ($2 + 2 + 6 + 2 + 6 = 18$ e$^-$). The next subshell is the 4s, so for potassium we write

$$\text{K}\ [\text{Ar}]\ 4s^1$$

where [Ar] stands for the "argon core" of a potassium atom. We write the abbreviated orbital diagram for potassium as

$$\text{K}\qquad [\text{Ar}]\quad \uparrow$$
$$4s$$

Following the same procedure, for the element nickel we write

$$\text{Ni}\ [\text{Ar}]\ 3d^8 4s^2$$

and give its orbital diagram as

$$\text{Ni}\qquad [\text{Ar}]\quad \uparrow\downarrow\ \uparrow\downarrow\ \uparrow\downarrow\ \uparrow\ \uparrow\qquad \uparrow\downarrow$$
$$3d\qquad\qquad\qquad 4s$$

Valence shell configurations

In writing the valence shell configuration for an element, we only list subshells that are in the *outer shell* of the atom. (Valence shell configurations are only of interest to us for the representative elements.) The valence shell configuration for the element sulfur, for example, is

$$\text{S}\qquad 3s^2 3p^4$$

Unexpected electron configurations sometimes occur for transition and inner transition elements

Half-filled subshells and (especially) filled subshells have extra stability that causes the electron configurations of some elements such as chromium, copper, silver, and gold to deviate from the configurations that we would predict by following the procedures described above. You should learn the electron configurations of these elements as exceptions (see page 333 in the text).

Self-Test

18. What is the shorthand electron configuration of iodine?

19. What is the shorthand electron configuration of cobalt?

20. How many unpaired *d* electrons are found in an atom of cobalt?

21. Use the periodic table to write the valence shell electron configuration of

    (a) oxygen

    (b) barium

    (c) indium

    (d) bismuth

22. Use the periodic table to predict the shorthand electron configuration for

    (a) Tc

    (b) Fe

    (c) Gd

23. From what you learned in this section, predict the electron configuration of europium, Eu (Z = 63).

New Terms

Write the definitions of the following terms, which were introduced in this section. If necessary, refer to the Glossary at the end of the text.

shorthand configuration   valence electrons

valence shell   valence shell configuration

outer shell   outer shell electrons

core electrons

8.7   Nodes in atomic orbitals affect their energies and their shapes

Review

The Heisenberg uncertainty principle comes from a mathematical proof that it is impossible to know or predict exactly where an electron is in an atom at any given moment. Wave mechanics, however, provides us with a way to predict the probability of finding the electron at any given location around the nucleus. This probability, which varies as we move from point to point, is given by the square of the wave function, ψ^2. The notion of a probability distribution for the electron in the space around the nucleus leads to the concept of the electron behaving as a blur or cloud with a greater *electron density* in some places than in others.

In general, the size of a given kind of orbital increases with increasing n. The s orbitals are all spherical in shape, meaning that if we examine a surface on which the probability of finding the electron is everywhere the same, the surface is a sphere. Study Figure 8.23, which illustrates the electron density distribution in a $1s$ orbital. In Figure 8.24, we see that the $2s$, $3s$, and higher s orbitals are also spherical, but spherical nodes, too. Nodes are a consequence of wave behavior, although they are not really important for our discussions.

The p orbitals are sometimes described as being dumbbell-shaped. Each p orbital has *two* regions of electron density located on opposite sides of the nucleus (Figures 8.25 and 8.26) and the three p orbitals of a p subshell are oriented perpendicular to each other (Figure 8.27).

The d orbitals are more complex, still, than the p orbitals (Figure 8.28). Four of them in a given subshell have the same shape, each consisting of four lobes of electron density. The fifth consists of a pair of lobes along the (arbitrary) z axis, plus a donut-shaped ring of electron density in the xy plane.

Self-Test

24.   On a separate piece of paper, practice sketching the shapes of s and p orbitals.

25.   On the axes below, sketch the shapes of d_{xy}, d_{xz}, d_{yz}, and $d_{x^2-y^2}$ orbitals.

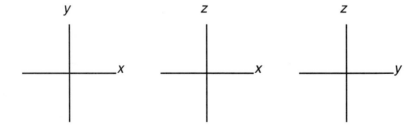

New Terms

Write the definitions of the following terms, which were introduced in this section. If necessary, refer to the Glossary at the end of the text.

uncertainty principle   electron density

electron cloud

8.8   Atomic properties correlate with an atom's electron configuration

Review

The key to understanding the way the properties discussed in this section vary within the periodic table is understanding the concept of effective nuclear charge. The notion here is that the outer electrons do not feel the full effect of the charge on the nucleus because the charge of the electrons in shells below the outer shell partially offset the nuclear charge. The positive charge that the outer electrons do feel is called the *effective nuclear charge*.

Sizes of atoms

Although, in a strict sense, atoms and ions have no true outer boundary, they often behave as if they have nearly constant sizes. The size of an atom or ion is generally given in terms of its radius and is expressed in units of picometers, nanometers, or angstroms. Even though the angstrom isn't an SI unit, it is still widely used for expressing small distances. Remember: 1 Å = 0.1 nm = 100 pm.

Within the periodic table, size *decreases* from left to right in a period and from bottom to top in a group. Going across a period, the amount of positive charge felt by the outer electrons increases because electrons in the same shell are not very effective at shielding each other from the nuclear charge. Going from top to bottom in a group, the outer-shell orbitals feel about a constant effective nuclear charge and become larger as *n* becomes larger. As a result, atoms become larger as we descend a group. Study Figure 8.30.

Sizes of ions

When ions are formed from atoms, their sizes increase as electrons are added and decrease as electrons are removed. Negative ions are always larger than the atoms from which they are formed. Positive ions are always smaller than the atoms from which they are formed.

Ionization energy

The *ionization energy* (IE) is the energy needed to pull an electron from an isolated atom or ion in its ground state. It is a measure of how tightly held the electrons are. Atoms with more than one electron have a series of ionization energies corresponding to the removal of the electrons one by one. For any given atom, successive ionization energies increase.

You should remember that, in general, the ionization energy increases from left to right across a period and decreases from top to bottom within a group. The increase across a period occurs because of the increasing effective nuclear charge felt by the outer electrons. The decrease down a group occurs because the outer electrons become farther from the nucleus (from which they are well shielded) and are held less tightly. You should also note the special stability of the noble gas configuration. It is also helpful to remember that as atomic radius becomes *smaller*, IE becomes *larger.*

Electron affinity

Energy is normally released when an electron is added to an isolated gaseous atom to form a negative ion. This energy is the *electron affinity* (EA). Most values are given with a negative sign because the process is usually exothermic.

The changes in EA within the periodic table parallel the changes in IE, and for the same reasons. Both increase from left to right in a period and from bottom to top in a group.

When more than one electron is added to an atom (i.e., when an ion with a charge of 2– or 3– is formed) the attachment of the second and third electron is always endothermic. Overall, formation of a negative ion with a charge larger than 1– is endothermic.

Self-Test

26. Which is the larger ion, Co^{2+} or Co^{3+}?   _____

27. Explain your answer to Question 25. _____

_____

28. The following particles each have the same number of electrons: N^{3-}, O^{2-}, F^-. Their atomic numbers increase from N to O to F. How would you expect their sizes to vary? Explain.

_____

_____

29. In each pair, choose the species with the larger IE.

     (a) K or Ca   _____   (d) Fe or Fe^{2+} _____

     (b) Ca or Sr   _____   (e) Ne or Na   _____

     (c) Al or C   _____

30. Which elements, metals or nonmetals, tend to have the larger ionization energies?

                                                        _____

31. Choose the species with the more exothermic EA.

     (a) S or Se   _____   (c) Te or Br   _____

     (b) S or Cl   _____   (d) S or S^- _____

32. From which ion, O^{2-} or O^-, would you expect it to be easier to remove an electron?

                                                        _____

New Terms

Write the definitions of the following terms, which were introduced in this section. If necessary, refer to the Glossary at the end of the text.

effective nuclear charge   ionization energy

angstrom (Å)   electron affinity

Solution to Thinking It Through

1   We can calculate the energy of one photon from the equation $E = h\nu$, but first we must convert the wavelength to frequency with the equation $\lambda\nu = c$. (Alternatively, we can combine these two equations to give $E = hc/\lambda$.) To find the number of such photons required, we have to calculate the total energy needed. We obtain this from the product of the specific heat of the coffee (4.18 J g^{-1} $°C^{-1}$), the mass of the coffee (250 g), and the temperature change that the coffee undergoes (55 °C).

$$\text{specific heat} \times \text{mass} \times \text{temp. change} = \text{energy}$$

Once we know the total amount of energy needed (in joules), we divide by the energy per photon to calculate the number of photons needed. (The answer is 3.04×10^{31} photons.)

Answers to Self-Test Questions

1. 242 m
2. 1.0×10^{14} Hz
3. 50 nm
4. 3.64×10^{-19} J
5. 219 kJ
6. radio waves, microwaves, infrared radiation, visible light, ultraviolet light, X rays
7. $n = 1$
8. $E = -b$ for $n = 1$, $E = -b/4$ for $n = 2$, change in energy is $0.75b$.
9. 410 nm, violet
10. *in phase*—amplitudes add to give a new wave with a larger amplitude.
    out of phase—amplitudes cancel to give a new wave with a smaller, or even zero amplitude.
11. (a) $2p$, (b) $4s$, (c) $3d$, (d) $5f$
12. (b), (c), (e) are unacceptable.
13. $4s, 4p, 4d, 4f$
14. (a) $3p$, (b) $4d$, (c) $4p$
15. (a) 2, (b) 10, (c) 18, (d) 32, (e) 6
16. (a) $1s^2 2s^2 2p^6 3s^2 3p^1$, (b) $1s^2 2s^2 2p^6 3s^2 3p^6 3d^{10} 4s^2 4p^5$
    (c) $1s^2 2s^2 2p^6 3s^2 3p^6 3d^3 4s^2$
17. (a) P
    (b) Si
18. [Kr] $4d^{10} 5s^2 5p^5$
19. Co [Ar] $3d^7 4s^2$
20. three
21. (a) $2s^2 2p^4$, (b) $6s^2$ (c) $5s^2 5p^1$, (d) $6s^2 6p^3$
22. (a) Tc [Kr] $4d^5 5s^2$, (b) Fe [Ar] $3d^6 4s^2$, (c) Gd [Xe] $4f^7 5d^1 6s^2$
23. Eu [Xe] $6s^2 4f^7$ (the $4f$ subshell becomes half-filled)
24. Check your sketches against Figures 8.24 and 8.26.

25.

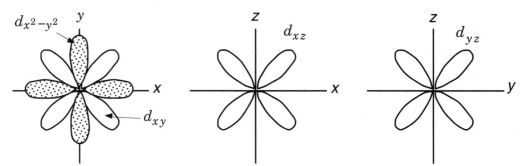

(*$d_{x^2-y^2}$* and *d_{xy}*, are concentrated in the *xy* plane, the *d_{xz}*, is concentrated in the *xz* plane, and the *d_{yz}*, is concentrated in the *yz* plane.)

26. Co^{2+}

27.   In Co^{3+} there is one less e^- than in Co^{2+} and therefore less inter-electron repulsion, allowing the e^- in Co^{3+} to be pulled closer to the nucleus.

28. $N^{3-} > O^{2-} > F^-$. As the nuclear charge increases, the electrons are pulled closer to the nucleus and the size decreases.

29.   (a) Ca (b) Ca (c) C (d) Fe^{2+} (e) Ne

30.   Nonmetals

31.   (a) S (b) Cl (c) Br (d) S

32. O^{2-}

Summary of Important Equations

The most important equations in this chapter involve the wavelength-frequency relationship and Planck's equation for the energy of a photon.

Wavelength-frequency relationship.

$$\lambda \times v = c = 3.00 \times 10^8 \text{ m s}^{-1}$$

Energy of a photon of frequency v.

$$E = hv$$

where h = Planck's constant (6.626×10^{34} J s)

Your teacher may also ask you to learn the following:

The Rydberg equation, where R_H = 109,687 cm^{-1}

$$\frac{1}{\lambda} = R_H \left(\frac{1}{n_1^2} - \frac{1}{n_2^2} \right)$$

De Broglie's equation for the wavelength of a matter wave,

$$\lambda = \frac{h}{mv}$$

were h is Planck's constant, m is the mass of the particle, and v is the velocity of the particle.

Tools you have learned

Consider removing this chart from the Study Guide so you can have it handy when tackling homework problems.

Tool	*How It Works*
Wavelength-frequency relationship $\lambda \nu = c$	We use this equation whenever we need to convert between wavelength and frequency.
Energy of a photon $E = h\nu$	We use this relationship when we need the energy carried by a photon of frequency ν.
Periodic Table Remember how the subshells become filled crossing row after row of the periodic table.	This is another use for the periodic table as a tool. It serves as a device to help us write electron configurations of the elements and in correlating properties of the elements such as atomic size, ionization energy, and electron affinity.
Periodic Trends in Atomic and Ionic Size Atomic size increases from top to bottom in a group and decreases from left to right in a period.	By learning the trends in atomic size, you can use the periodic table to compare the sizes of atoms and ions.
Periodic Trends in Ionization Energy IE becomes larger (more endothermic) from left to right in a period and from bottom to top in a group. For a given element, successive IEs become larger.	By learning the trends in this property, you can use the periodic table to compare the ionization energies of atoms and ions.
Periodic Trends in Electron Affinity EA becomes more exothermic from left to right in a period and from bottom to top in a group.	By learning the trends in this property, you can use the periodic table to compare the electron affinities of atoms and ions.

Chapter 9

Chemical Bonding: General Concepts

This is the first of two chapters that deal with the attractions that hold atoms to each other in chemical compounds. In this chapter we examine chemical bonds on a rather elementary level. Yet, even these simple explanations provide us with many useful tools for understanding chemical and physical properties.

Learning Objectives

Throughout your study of this chapter, keep in mind the following objectives:

1   To learn how ionic compounds are formed from their elements, what factors cause elements to form ionic compounds, and what determines the charges that atoms acquire when they form ions.

2   To learn to use a simple device called Lewis symbols to represent the valence electrons of an atom or ion of the representative elements.

3   To learn what happens to the energy of two atoms when they share electrons in a covalent bond, and to learn how we represent covalent bonds using Lewis symbols.

4   To learn how the tendency of atoms to acquire a noble gas electron configuration often determines the number of electrons they share with other atoms in covalent bonds, and how this leads to a general rule that many atoms tend to acquire eight electrons in their valence shell by electron sharing.

5   To learn some of the important classes of organic compounds.

6   To learn how electrons may be shared unequally in a covalent bond and to see how this affects the distribution of electric charge in a molecule.

7   To learn how electronegativity is related to the reactivities of metals and nonmetals, and how reactivity varies within the periodic table.

8   To learn how atoms in some molecules are able to be surrounded by more than or less than an octet of electrons.

9   To learn to draw Lewis structures for molecules and polyatomic ions.

10   To learn a method for selecting the best Lewis structure for a molecule when several are possible.

11   To learn how the structure of a molecule or ion is represented when a single Lewis structure does not adequately describe it.

12   To see how atoms can use unshared pairs of electrons to form coordinate covalent bonds.

9.1   Electron transfer leads to the formation of ionic compounds

Review

There are two broad categories of chemical bonds—*ionic bonds*, which occur when atoms transfer electrons between them, and *covalent bonds*, which occur when atoms share electrons. An ionic "bond" is really just the attraction that exists between oppositely charged ions.

For a bond to be formed between atoms, there must be a net lowering of the potential energy of the particles. For ionic bonding, the three most important factors that contribute to the overall potential energy change are (1) the ionization energy of the element that forms the cation, (2) the electron affinity of the element that forms the anion, and (3) the *lattice energy*, which is the potential energy lowering that is produced by the attractions between the ions. For most elements, (1) and (2) taken together produce a net increase in potential energy, so it is the stabilizing (energy-lowering) influence of the lattice energy that enables ionic compounds to exist. But the lattice energy can do this *only* if it is larger than the net PE increase caused by (1) and (2). This restriction leads to certain generalizations about the kinds of elements that form ionic compounds and the charges on the ions that are created in the reaction:

1   Ionic compounds tend to be formed between metals (low IE) and nonmetals (relatively large exothermic EA).

2   The representative metals of Groups IA and IIA plus aluminum lose electrons until they have achieved an electron configuration corresponding to that of a noble gas.

3   Nonmetals gain electrons until they have also achieved an electron configuration that is the same as that of a noble gas.

The tendency of certain elements to achieve a noble gas configuration gives rise to the octet rule, which states that *many elements tend to gain or lose electrons until they have achieved a valence shell that contains eight electrons.*

The transition metals and the metals that follow the transition elements in periods 4, 5, and 6 (the post-transition metals) do not follow any particular rule. They lose electrons, but often more than one cation is possible, depending on conditions. The charges on the ions of the common transition metals were given in Table 2.4 on page 61 of the text; if necessary, review them.

When electrons are lost by an atom, they always come first from the shell with largest n. For a transition element, the shell with largest n is an s subshell. After the s subshell is emptied, electrons are lost from the d subshell below the outer shell.

When electrons are lost from a given shell, they come from the highest energy subshell first. Because the energies of subshells vary in the order $s < p < d < f$, post-transition metals lose electrons from the outer shell p subshell before they lose them from the outer shell s subshell.

Self-Test

1.   Show what happens to the valence shells of Ba and I when these elements react to form an ionic compound. What is the formula of this compound?

_____

_____

_____

2.   What happens to the electron configuration of a manganese atom when it forms (a) the Mn^{2+} ion and (b) the Mn^{3+} ion?

_____

_____

3.   When Rb and Cl_2 react, why doesn't the compound $RbCl_2$ form? What compound does form?

_____

_____

_____

4.   Write the abbreviated electron configuration of

(a) Pb^{2+} _____   (b) Pb^{4+} _____

New Terms

Write the definitions of the following terms, which were introduced in this section. If necessary, refer to the Glossary at the end of the text.

chemical bond	lattice energy
ionic bond	octet rule
octet of electrons	

9.2   Lewis symbols help keep track of valence electrons

An element's Lewis symbol is constructed by placing one dot for each valence electron around the chemical symbol for the element. The number of valence electrons, and therefore the number of dots in the Lewis symbol, is equal to the element's group number. (In general, Lewis symbols are only used for the representative elements.)

A simple way to draw Lewis symbols is as follows. First write the chemical symbol. Then imagine a diamond with the symbol at its center. Place a dot at one of the corners of the diamond (it doesn't matter where you start). Move around the diamond from corner to corner as you place additional dots, until the required number are shown. For example, for sulfur (Group VIA), there must be six dots around the symbol S.

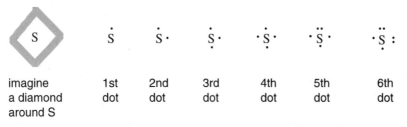

imagine a diamond around S	1st dot	2nd dot	3rd dot	4th dot	5th dot	6th dot

The Lewis symbol for sulfur is ·S: (or any other arrangement such as ·S: that shows two pairs of dots and two single dots).

Example 9.4 on page 362 illustrates how Lewis symbols can be used to diagram the transfer of electrons that takes place during the formation of an ionic substance. Notice that when we write the Lewis symbol for an anion, we enclose it in brackets to show that all the electrons belong to the ion.

Self-Test

5.   Write Lewis symbols for the following:

(a)   P   _____   (c)   Cl   _____

(b)   Te   _____   (d)   B   _____

6.   Diagram the reaction that occurs when Li reacts with Se to form the compound Li_2Se.

_____

New Terms

Write the definition of the following term, which was introduced in this section. If necessary, refer to the Glossary at the end of the text.

Lewis symbol

9.3   Covalent bonds are formed by electron sharing

Review

Covalent bonds are formed when ionic bonds are not energetically favored. In the formation of a covalent bond, there is also a potential energy lowering, but it is achieved by a different method than in ionic bonding.

When atoms approach each other to form a covalent bond, there is an attraction of the electrons of each atom toward both nuclei, which leads to an overall lowering of the potential energy. The two nuclei also repel each other because they are of the same charge, so the atoms cannot approach each other too closely. The *bond length* is determined by the balance of these attractive and repulsive forces, and the net decrease in PE that occurs when the bond is formed is called the *bond energy*, which is usually expressed in units of kilojoules per mole of bonds formed. (See Figure 9.3.) The bond energy is also the energy that would be necessary to break the bond.

One of the most important features of covalent bonding is the pairing of electrons that occurs; a covalent bond is the sharing of a *pair* of electrons with their spins in opposite directions. When Lewis symbols are used to represent a molecule, a shared pair of electrons is shown as a pair of dots or a dash between the chemical symbols. One dash stands for *two* paired electrons.

When counting electrons in the valence shells of atoms attached to each other by a covalent bond, we count both electrons of the shared pair as belonging to both atoms joined by the bond. When applied to covalent bonds, the octet rule states that *atoms tend to share sufficient electrons so as to obtain an outer shell with eight electrons.* An exception to the octet rule is hydrogen, which completes its valence shell when it has a share of two electrons (i.e., when it forms one covalent bond).

Atoms can share one, two, or three pairs of electrons to give single, double, and triple bonds. On page 366, notice how the numbers of electrons needed by atoms of carbon, nitrogen, and oxygen to achieve octets are related to the number of covalent bonds these atoms tend to form.

Self-Test

7. What happens to the electron density between the two atoms when they form a covalent bond?

    _____

    _____

8. On a separate piece of paper, sketch a diagram that shows how the potential energy varies with the distance between the nuclei of a pair of atoms becoming joined by a covalent bond. On the diagram indicate the bond energy and the bond length. Refer to Figure 9.3 to check your answer.

9. Predict the formulas of the simplest compound that might be formed between hydrogen and each of the following:

    (a)   Ge   _____

    (b)   Te   _____

    (c)   I   _____

10. Draw the Lewis structures for each of your answers to Question 8.

         (a)   (b)   (c)

11. Formaldehyde, used in preserving biological specimens, has the formula H_2CO. A molecule of this substance has two hydrogen atoms and an oxygen atom bonded to the carbon. Use Lewis symbols to write a structural formula for H_2CO. (Hint: There's a double bond in the structure.)

12. Count the number of electrons around the atoms in the following molecule (called urea).

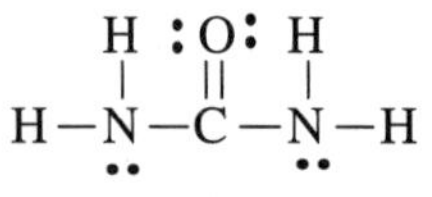

$$\begin{array}{ccc} H & :\!O\!: & H \\ | & || & | \\ H\!-\!\!\ddot{N}\!-\!C\!-\!\ddot{N}\!-\!H \end{array}$$

    _____

New Terms

Write the definitions of the following terms, which were introduced in this section. If necessary, refer to the Glossary at the end of the text.

octet rule	bond distance	triple bond
structural formula	bond energy	covalent bond
single bond	electron pair bond	bond length
double bond		

9.4   Carbon compounds illustrate the variety of structures possible with covalent bonds

Review

Carbon almost always forms four covalent bonds, either to other carbon atoms or to atoms of other elements such as hydrogen, oxygen, and nitrogen. Learn the structures of methane, ethane, and propane. Be sure you can write their structures in condensed form, as illustrated in the margin comment on page 368. You should also know the formulas and structures of ethylene and acetylene.

Three series of hydrocarbons are discussed in this section, alkanes (with general formula C_nH_{2n+2}), alkenes (C_nH_{2n}), and alkynes (C_nH_{2n-2}). Alkanes have only carbon-carbon single bonds, an alkene has one carbon-carbon double bond, and alkyne has one carbon-carbon triple bond. Notice that as the number of carbon atoms increases, more than one compound can have the same molecular formula. This phenomenon, called isomerism, is illustrated for butane, C_4H_{10}.

Oxygen-containing compounds

Alcohols, ketones, aldehydes, and organic acids are characterized by the following kinds of structures.

$$\underset{\text{alcohol}}{-\overset{|}{\underset{|}{C}}-OH} \qquad \underset{\text{ketone}}{-\overset{|}{\underset{|}{C}}-\overset{\overset{O}{\|}}{C}-\overset{|}{\underset{|}{C}}-} \qquad \underset{\text{aldehyde}}{-\overset{\overset{O}{\|}}{C}-H} \qquad \underset{\text{carboxylic aci}}{-\overset{\overset{O}{\|}}{C}-OH}$$

Ketones, aldehydes, and carboxylic acids contain the carbonyl group, $>C=O$. Acids that contain the carboxyl group are weak acids.

Nitrogen-containing organic compounds include amines which can be viewed as ammonia molecules in which hydrogens have been replaced by hydrocarbon groups. Amino acids contain both an amine group and a carboxyl group.

Self-Test

13. Write the formula for

    (a) An alkane with 7 carbon atoms. _____

    (b) An alkene with 4 carbon atoms. _____

    (c) An alkyne with 6 carbon atoms. _____

14. Consider the compound

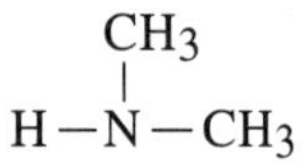

$$\begin{array}{c} CH_3 \\ | \\ H-N-CH_3 \end{array}$$

Which of the compounds below is one of its isomers?

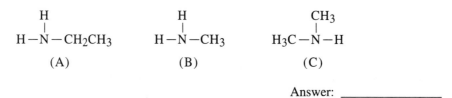

$$\begin{array}{ccc} \begin{array}{c} H \\ | \\ H-N-CH_2CH_3 \end{array} & \begin{array}{c} H \\ | \\ H-N-CH_3 \end{array} & \begin{array}{c} CH_3 \\ | \\ H_3C-N-H \end{array} \\ (A) & (B) & (C) \end{array}$$

Answer: _____

15. On a separate sheet of paper, draw structural formulas for

(a) propane   (b) methane   (c) ethane   (d) acetylene   (e) ethylene

(Check your answers by referring to the structures on pages 368.)

16. Consider the following molecules:

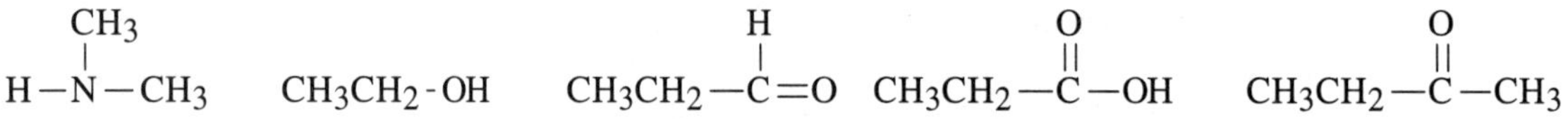

$$\begin{array}{lllll} \begin{array}{c} CH_3 \\ | \\ H-N-CH_3 \end{array} & CH_3CH_2\text{-}OH & CH_3CH_2-\overset{H}{\underset{}{C}}{=}O & CH_3CH_2-\overset{O}{\overset{||}{C}}-OH & CH_3CH_2-\overset{O}{\overset{||}{C}}-CH_3 \\ (A) & (B) & (C) & (D) & (E) \end{array}$$

(a) Which is an alcohol?   _____

(b) Which is an amine?   _____

(c) Which one is acidic?   _____

(d) Which one is basic?   _____

(e) Which is a ketone?   _____

(f) Which is an aldehyde?   _____

(g) Which contains a carboxyl group?   _____

New Terms

Write the definitions of the following terms, which were introduced in this section. If necessary, refer to the Glossary at the end of the text.

organic compound	alkane	alkene
alkyne	alcohol	ketone
amine	aldehyde	carboxyl group
carbonyl group	isomer	

9.5   Covalent bonds can have partial charges at opposite ends

Review

Because different atoms have different attractions for electrons, the electrons in a covalent bond can be shared unequally. When this happens, the electron density in the bond is shifted toward that atom with the greater attraction for electrons, so this atom acquires a partial negative charge, δ^-. The atom at the other end of the bond loses some electrical charge and carries a partial positive charge, $\delta+$. This produces an *electric dipole*, which is two equal but opposite electric charges separated by some distance. Bonds in which electrons are shared unequally are *polar bonds* and the bond is a *dipole* (two poles of equal but opposite charges separated by the bond distance). The degree of polarity of a bond is expressed quantitatively by the bond's *dipole moment* μ, which is the product of the charge on either end of the bond multiplied by the distance between the charges. Molecules as a whole can be polar and have dipole moments. Dipole moments are measured in units called *debeys* (symbol D).

Electronegativity is the attraction an atom has for the electrons in a bond. The greater the electronegativity *difference* between two atoms that are bonded to each other, the more polar the bond is and the greater is the partial negative charge on the more electronegative of the two atoms.

It is important to remember that there is no sharp dividing line between covalent and ionic bonding. If we arranged all sorts of bonds in order of increasing polarity, we would find a gradual transition from nonpolar covalent bonds to bonds that are essentially 100% ionic. All degrees of ionic character (polarity) are possible.

You should be sure to learn how electronegativity varies in the periodic table. This is illustrated in Figure 9.5 on page 373. Metals, in the lower left corner of the table, have low electronegativities and nonmetals, in the upper right, have high electronegativities.

Self-Test

17. Use the data in Table 9.3 of the textbook to calculate the amount of charge, in electronic charge units, on opposite ends of the HBr molecule. (1 D = 3.34×10^{-30} C m and the electronic charge unit = 1.60×10^{-19} C)

   _____

18. Use Figure 9.5 to arrange the following bonds in order of increasing percent ionic character: Li—I, Ba—F, N—O, As—Cl.

   _____

New Terms

Write the definitions of the following terms, which were introduced in this section. If necessary, refer to the Glossary at the end of the text.

partial charge	dipole
electric dipole	debey
polar bond	dipole moment
polar covalent bond	electronegativity
polar molecule	ionic character of a bond

9.6   The reactivities of metals and nonmetals can be related to their electronegativities

Review

For metals, reactivity refers to how easily oxidized the element is. Because oxidation involves the removal of electrons from a substance, a metal that is easily oxidized is one that does not hold its electrons very tightly. This corresponds to a metal with a low electronegativity. By this reasoning, we come to the generalization that the smaller the electronegativity of a metal, the more easily oxidized it should be. This correlation is reflected in Figure 9.7, where we see that the metals with the lowest electronegativities (Groups IA and IIA) are most easily oxidized. Study Figure 9.7 and note where in the periodic table the least reactive metals are found.

The reactivity of a nonmetal generally is associated with its ability to serve as an oxidizing agent and is directly proportional to the nonmetal's electronegativity, which increases from left to right in a period and decreases from top to bottom in a group.

Self-Test

19.   In each pair, select the element with the specified property.

(a)   The better reducing agent: K or Be   _____

(b)   The more easily oxidized: Mg or Fe   _____

(c)   The more reactive metal: Ti or Au   _____

(d)   The better oxidizing agent: P or Cl   _____

(e)   The more easily reduced: O or Se   _____

(f)   The more easily oxidized: Ga or Se   _____

New Terms

None

9.7   Drawing Lewis structures is a necessary skill

Review

Although the octet rule is often a useful tool in constructing the Lewis structure for a molecule or polyatomic ion, there are some instances in which the rule is not obeyed. Whenever an atom is bonded to more than four other atoms, the octet rule cannot be obeyed. Examples are PCl_5 and SF_6 illustrated on page 376. Remember that only atoms below Period 2 are able to exceed an octet; their valence shells are able to accommodate more than eight electrons. (Period 2 elements never go beyond eight electrons in the outer shell because the second shell has only *s* and *p* subshells, which together can hold a maximum of 8 e^-.) Beryllium and boron are unusual elements because they sometimes have *less* than an octet in compounds.

Procedure for drawing Lewis structures

To draw a Lewis structure for a molecule or ion, follow the guidelines in Figure 9.8 on page 377. The first step in the procedure is to decide on the skeletal structure.

Remember:

- In oxoacids, hydrogen is bonded to oxygen, which in turn is bonded to the other nonmetal.
- The central atom in a molecule is usually the least electronegative atom. In the formulas for most simple molecules and polyatomic ions, the central atom is written first.
- When all else fails, the most symmetrical arrangement of atoms is the "best guess."

Next, count all of the valence electrons. Remember that the number of the group in which a representative element (A-group element) is found is the same as the number of valence electrons that it contributes. Sulfur, for example, is in Group VI, so a sulfur atom contributes six valence electrons. Also remember that if the species whose Lewis structure you are working on is a negative ion, add one electron for each negative charge. If it is a positive ion, subtract one electron for each positive charge.

Once you know how many electrons are to be represented in the structure, place them in the structure as described in Figure 9.8. Study Examples 9.5 through 9.8, work the Practice Exercises, and then answer the Self-Test Questions below.

Self-Test

20. If an atom forms more than four bonds, it definitely does not obey the octet rule. Why?

_____

21. What is the maximum number of electrons that can be held in the orbitals in the valence shell of

    (a) nitrogen   _____

    (b) phosphorus   _____

    (c) What kind of orbitals does the valence shell of phosphorus have that the valence shell of nitrogen does not?

    _____

22. How many valence electrons are in each of the following?

    (a) PO_4^{3-} _____   (d) IF_5 _____

    (b) IF_4^- _____   (e) CO_3^{2-} _____

    (c) NO_2^+ _____   (f) C_2^{2-} _____

23. Write the Lewis structures for (a) $HClO_2$ and (b) H_2CO_3 following the procedure in Figure 9.8.

    (a)   (b)

24.   Construct Lewis structures for each substance in Question 22 following the procedure in Figure 9.8.

   (a)   (b)   (c)

   (d)   (e)   (f)

New Terms

None

9.8   Formal charges help select correct Lewis structures

Review

The bond length between a given pair of atoms *decreases* as the bond order increases (i.e., on going from a single to a double to a triple bond). At the same time, the bond energy increases. These trends, along with experimentally measured values of these bond properties, help us "fine tune" our descriptions of the bonding in various molecules. One of the observations we can make is that the best Lewis structure for a molecule or ion is one in which the *formal charges* on the atoms are a minimum.

To determine the formal charge on an atom in a Lewis structure, add up the number of bonds it forms plus the number of unshared electrons. Subtract this number from the number of electrons an isolated atom of the element has. The difference is the formal charge. (This is what Equation 9.2 tells us to do.) To select the preferred ("best") Lewis structure for a molecule or ion, we seek to minimize the number of formal charges. Study Examples 9.8 through 9.10 as well as the additional worked example below.

Example 9.1   Using formal charges to select Lewis structures

What is the preferred Lewis structure for chlorous acid, $HClO_2$?

Analysis:
We know that the preferred Lewis structure is the one that has the minimum number of formal charges. Therefore, the first step is to construct the Lewis structure following the general procedure given on page 377. Then we calculate the formal charge on each atom. Since the best Lewis structure is the one with the minimum number of formal charges, we attempt to reduce the formal charges by moving electrons.

Solution:

Below we have the Lewis structure for $HClO_2$ drawn according to the procedure outlined earlier. Notice that we have counted the number of bonds and unshared electrons, which we need to know to calculate the formal charges according to the formula

$$\text{Formal charge} = \left(\begin{array}{c}\text{number of electrons}\\ \text{an isolated atom}\\ \text{of the element has}\end{array}\right) - \left(\begin{array}{c}\text{number of bonds}\\ \text{to the atom}\end{array} + \begin{array}{c}\text{number of}\\ \text{unshared } e^-\end{array}\right)$$

$$H\overset{\cdot\cdot}{\underset{\cdot\cdot}{O}}\overset{\cdot\cdot}{\underset{\cdot\cdot}{Cl}}\overset{\cdot\cdot}{\underset{\cdot\cdot}{O}}:$$

2 bonds plus   1 bond plus
4 unshared e^- 6 unshared e^-

For the oxygen between the H and Cl, the sum of 2 bonds and 4 unshared e^- gives a total of 6. An isolated oxygen atom has 6 valence electrons, so the formal charge on this oxygen is zero. For the chlorine, the sum of bonds and unshared e^- is also 6, and when subtracted from 7 (the number of valence e^- an isolated Cl atom has) we obtain +1. This is the formal charge on the Cl. For the oxygen at the right, the sum of bonds and unshared e^- is 7. When 7 is subtracted from 6 (the number of valence e^- an oxygen atom has), we get –1. This is the formal charge on the oxygen at the right. Placing these formal charges in circles alongside the atomic symbols gives us

$$H\overset{\cdot\cdot}{\underset{\cdot\cdot}{O}}\overset{\oplus}{\underset{\cdot\cdot}{Cl}}\overset{\cdot\cdot}{\underset{\cdot\cdot}{O}}:{}^{\ominus}$$

The next step is to attempt to reduce formal charges to obtain a better structure. We do this by shifting an unshared pair of electrons on the more negative atom into the bond it forms to the more positive atom.

$$H\overset{\cdot\cdot}{\underset{\cdot\cdot}{O}}\overset{\oplus}{\underset{\cdot\cdot}{Cl}}\overset{\cdot\cdot}{\underset{\cdot\cdot}{O}}:{}^{\ominus} \quad - - \rightarrow \quad H\overset{\cdot\cdot}{\underset{\cdot\cdot}{O}}\overset{\cdot\cdot}{Cl}=\overset{\cdot\cdot}{\underset{\cdot\cdot}{O}}:$$

A lone pair of electrons is moved from the oxygen into the Cl–O bond. This effectively transfers an electron from O to Cl, thereby reducing the formal charges to zero.

This gives a Lewis structure with no formal charges, so we select it as the preferred Lewis structure for the molecule.

Is the Answer Reasonable?

First, we can check the initial Lewis structure to be sure we've included all the valence electrons. We can also check our calculation of formal charges by adding them up algebraically. Their sum must equal the net charge on the molecule, which it does. When rearranging electrons, we should move them from the atom with the negative formal charge toward the atom with the positive formal charge, and we've done that as well. Having done all these steps correctly gives us confidence that our answer is correct.

Self-Test

25. Phosphorous acid has the Lewis structure shown at the left below when constructed as described in the preceding section. Assign formal charges to the atoms in this structure and, if possible, draw a "better" Lewis structure for the molecule.

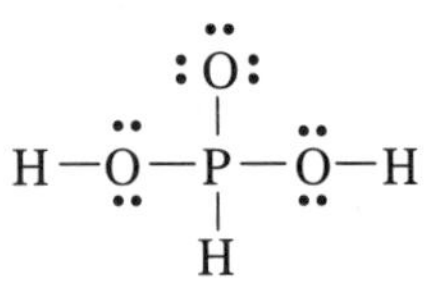

$$
\begin{array}{c}
:\ddot{O}: \\
| \\
H-\ddot{O}-P-\ddot{O}-H \\
| \\
H
\end{array}
$$

New Terms

Write the definitions of the following terms, which were introduced in this section. If necessary, refer to the Glossary at the end of the text.

bond length   bond energy   formal charge

bond order

9.9   Resonance applies when a single Lewis structure fails

Review

For some molecules and ions, a single Lewis structure cannot be drawn that adequately explains experimental bond lengths and bond energies. In these instances, we view the true structure of the particle as a sort of average of two or more Lewis structures, and we call this true structure a *resonance hybrid* of the contributing structures.

When you have a choice as to where to place a double bond in a Lewis structure, as when writing the Lewis structure for the HCO_2^- ion, the number of resonance structures you should draw is equal to the number of choices that you have. This rule works in most cases.

A resonance hybrid is more stable than any of its individual structures would be, if they were to actually exist. The extra stability is called the resonance energy.

Self-Test

26. The oxalate ion, $C_2O_4^{2-}$, has the skeletal structure shown below

$$
\begin{array}{cc}
O & O \\
C & C \\
O & O
\end{array}
$$

Draw the resonance structures for this ion in the space below.

27. How would the C—O bond lengths and bond energies in the $C_2O_4{}^{2-}$ ion (Question 26) compare to those in the following:

$$H-\overset{\overset{\displaystyle H}{|}}{\underset{\underset{\displaystyle H}{|}}{C}}-\overset{\cdot\cdot}{\underset{\cdot\cdot}{O}}-H \qquad \text{and} \qquad \overset{\cdot\cdot}{\underset{\cdot\cdot}{O}}=C=\overset{\cdot\cdot}{\underset{\cdot\cdot}{O}}$$

New Terms

Write the definitions of the following terms, which were introduced in this section. If necessary, refer to the Glossary at the end of the text.

resonance   resonance structure

contributing structure   resonance hybrid

9.10   Both electrons in a coordinate covalent bond come from the same atom

Review

When one atom donates a pair of electrons to another atom in order to form a single bond, we call the bond a *coordinate covalent bond*. Once formed, it is really no different than any other single bond because electrons can't "remember" where they came from. This is really just a bookkeeping device. When we want to note the origin of the electrons of a coordinate covalent bond, we use an arrow instead of a dash to represent the bond. The arrow points from the donor to the acceptor of the electrons.

Self-Test

28. Boron trifluoride, BF_3, can react with a fluoride ion, F^-, to form the tetrafluoroborate ion, $BF_4{}^-$. Diagram this reaction using Lewis symbols to show the formation of a coordinate covalent bond.

29. BF_3 reacts with organic chemicals called ethers to form addition compounds. Use Lewis formulas to diagram the reaction of BF_3 with dimethyl ether. The structure of dimethyl ether is

$$\overset{\overset{\displaystyle CH_3}{|}}{\underset{\underset{\displaystyle CH_3}{|}}{:\!O\!:}}$$

New Terms

Write the definition of the following term, which was introduced in this section. If necessary, refer to the Glossary at the end of the text.

coordinate covalent bond

Answers to Self-Test Questions

1. Ba ([Xe] $6s^2$) $\rightarrow$ Ba^{2+} ([Xe]) + $2e^-$; I ($5s^25p^5$) + e^- $\rightarrow$ I$^-$ ($5s^25p^6$)
   The compound is BaI$_2$.

2. (a) Mn ([Ar] $3d^54s^2$) $\rightarrow$ Mn^{2+} ([Ar] $3d^5$) + $2e^-$
   (b) Mn^{2+} ([Ar] $3d^5$) $\rightarrow$ Mn^{3+} ([Ar] $3d^4$)+ e^-

3. Removal of an electron from Rb$^+$ requires too much of a PE increase. The compound that does form contains Rb$^+$ and has the formula RbCl.

4. (a) Pb^{2+} [Xe] $4f^{14}5d^{10}6p^2$ (b) Pb^{4+} [Xe] $4f^{14}5d^{10}$

5. (a) $\ddot{\text{:}}\underset{\bullet}{\overset{\bullet}{\text{P}}}\bullet$ (b) $\text{:}\underset{}{\overset{\bullet}{\text{Te}}}\bullet$ (c) $\text{:}\underset{\bullet\bullet}{\overset{\bullet\bullet}{\text{Cl}}}\bullet$ (d) $\bullet\,\overset{\bullet}{\text{B}}\,\bullet$

6. Li $\bullet$ $\overset{\bullet\bullet}{\underset{\bullet\bullet}{\text{Se}}}\bullet$ $\bullet$ Li $\longrightarrow$ $2\text{Li}^+\left[\text{ : }\overset{\bullet\bullet}{\underset{\bullet\bullet}{\text{Se}}}\text{: }\right]^{2-}$

7. It shifts toward the region between the two atoms.

8. See Figure 8.4.

9. (a) GeH$_4$ (b) H$_2$Te (c) HI

10. (a) $\text{H}-\overset{\displaystyle \text{H}}{\underset{\displaystyle \text{H}}{\text{Ge}}}-\text{H}$ (b) $\text{H}-\overset{\bullet\bullet}{\text{Te}}-\text{H}$ (c) $\text{H}-\overset{\bullet\bullet}{\underset{\bullet\bullet}{\text{I}}}\text{:}$

11. $\underset{\displaystyle \text{H}}{\overset{\displaystyle \text{H}}{>}}\text{C}=\overset{\bullet\bullet}{\underset{\bullet\bullet}{\text{O}}}$

   Hydrogen can complete its valence shell by forming one bond, carbon by four bonds, and oxygen by forming two bonds. This is the only arrangement of atoms that satisfies this condition.

12. Two around each H, eight around each N, C, and O.

13. (a) C$_7$H$_{16}$, (b) C$_4$H$_8$, (c) C$_6$H$_{10}$

14. A

15. See page 368.

16. (a) B, (b) A, (c) D, (d) A, (e) E, (f) C, (g) C, D, and E

17. Charge = 0.12 e^-

18. N—O < As—Cl < Li—I < Ba—F

19. (a) K, (b) Mg, (c) Ti, (d) Cl, (e) O, (f) Ga

20. Two electrons have to be in each bond. Since $2 \times 4 = 8$, more than four bonds means more than an octet.

21. (a) eight (b) eighteen (c) A phosphorus atom has d orbitals in its empty $3d$ subshell.

22. (a) 32 (b) 36 (c) 16 (d) 42 (e) 24 (f) 10

23. (a)   (b)

   $\text{H}-\overset{\bullet\bullet}{\underset{\bullet\bullet}{\text{O}}}-\overset{\bullet\bullet}{\underset{\bullet\bullet}{\text{Cl}}}-\overset{\bullet\bullet}{\underset{\bullet\bullet}{\text{O}}}\text{:}$ $\text{H}-\overset{\bullet\bullet}{\underset{\bullet\bullet}{\text{O}}}-\overset{\displaystyle \overset{\text{:O:}}{\|}}{\text{C}}-\overset{\bullet\bullet}{\underset{\bullet\bullet}{\text{O}}}-\text{H}$

24. (a) $[PO_4]^{3-}$ (b) $[IF_4]^-$ (c) $[N \equiv O]^+$ (d) IF_5

(e) $[CO_3]^{2-}$ (f) $[C \equiv C]^{2-}$

25.

$$H-O-\overset{\ominus \ \ddot{O}}{\underset{H}{\overset{|}{P^{\oplus}}}}-O-H \qquad \text{Preferred structure is ...} \qquad H-O-\overset{O}{\underset{H}{\overset{||}{P}}}-O-H$$

26.

$$\left[\overset{O}{\underset{O}{\overset{||}{C}}}-\overset{O}{\underset{O}{\overset{|}{C}}} \right]^{2-} \leftrightarrow \left[\; \right]^{2-} \leftrightarrow \left[\; \right]^{2-} \leftrightarrow \left[\; \right]^{2-}$$

27. The C—O bond length decreases: $CH_3OH > C_2O_4^{2-} > CO_2$
    The C—O bond energy increases: $CH_3OH < C_2O_4^{2-} < CO_2$

28.

$$BF_3 + [F]^- \longrightarrow [BF_4]^-$$

29.

$$BF_3 + O(CH_3)_2 \longrightarrow F_3B-O(CH_3)_2$$

Tools you have learned

Consider removing this chart from the Study Guide so you can have it handy when tackling homework problems.

Tool	*How It Works*
Rules for electron configurations of ions	We use these rules when we need to write electron configurations for cations and anions.
Lewis symbols	These help us keep track of valence electrons in atoms and ions. Learn to use the periodic table to construct the Lewis symbol.
Periodic trends in electronegativity	Knowing these trends enables us to compare the relative polarities of chemical bonds and to identify the most (or least) electronegative element in a compound.
Method for drawing Lewis structures	We use this method when we need to have a Lewis structure for a molecule or polyatomic ion.
Correlation between bond properties and bond order	Knowing how bond order affects bond properties enables us to compare experimental quantities related to covalent bonds with those predicted by theory.
Formal charges	We use formal charges when we need to select the best Lewis structure for a molecule or polyatomic ion.
Method for determining resonance structures	Resonance structures provide a way of expressing the structures of molecules for which a single satisfactory Lewis structure cannot be drawn.

Summary of Useful Information

Rules for drawing Lewis structure (Also see Figure 9.8.)

1   Decide which atoms are bonded to each other and construct the skeleton structure. (Remember, H cannot be a central atom.)

2   Count valence electrons. (Remember: add an electron for each negative charge on an ion or subtract an electron for each positive charge.)

3   Place two electrons in each bond.

4   Complete octets of atoms (except H) that surround the central atom.

5   Place any remaining electrons on the central atom *in pairs*.

6   Form multiple bonds if the central atom has less than an octet.

    (Rule 6 is omitted for the elements Be and B.)

Chapter 10

Chemical Bonding and Molecular Structure

There are two principal goals in this chapter. One of them is to introduce you to the variety of three-dimensional shapes molecules can have. You will learn to sketch the shapes and also how to use Lewis structures to predict them. A knowledge of molecular shape is important because many of the physical properties of substances depend on the shapes of their molecules. As we point out in the introduction to the chapter, molecular shape is a prime factor in determining the behavior of biologically active molecules.

In this chapter we will also present more advanced views of covalent bonding that are based on the results of quantum theory. Our aim is to enable you to understand how the theories *explain* covalent bonding in terms of the way the orbitals of atoms interact.

Learning Objectives

Throughout your study of this chapter, keep in mind the following objectives:

1   To learn the basic molecular shapes found for various molecules and how to draw them. You should try very hard to visualize these molecular shapes as three-dimensional objects.

2   To learn a method for predicting the shapes of molecules by the use of Lewis structures.

3   To learn how to predict whether a molecule is polar based on its molecular shape.

4   To learn how a theory called Valence Bond Theory explains the formation of a covalent bond by the interactions of the orbitals of atoms forming the bond.

5   To learn how simple atomic orbitals are able to combine to produce hybrid orbitals which are better able to explain the shapes of molecules.

6   To learn how double and triple bonds are formed by the interaction of atomic orbitals.

7   To learn how a theory called Molecular Orbital Theory views electron energy levels and orbitals in molecules.

8   To learn how molecular orbital theory is able to avoid the concept of resonance.

9   To learn about the electronic structures of solids and how substances can be electrical conductors, semiconductors, and nonconductors (insulators).

10.1 Molecular shapes are built from five basic arrangements

Review

This section describes five basic geometric shapes that form the basis for most molecular structures you will encounter. The shapes are: *linear*, *planar triangular*, *tetrahedral*, *trigonal bipyramidal*, and *octahedral*. It is important that you develop the ability to visualize these shapes in three dimensions. It is also wise to learn to sketch them. Even if you don't consider yourself much of an artist, you should make the effort, because these three-dimensional concepts are likely to become important tools in other science courses you take.

The linear and planar triangular structures can be drawn on a two-dimensional surface, so you should have no difficulty with them. Instructions for drawing a tetrahedron are found in Figure 10.1, and simplified representations of the trigonal bipyramidal and octahedral structures are shown on page 403. Instructions for drawing the trigonal bipyramidal and octahedral structures are given in Study Guide Figure 10.1 below.

You should know the bond angles in the various structures. Notice that except for the trigonal bipyramid, all the other structures have equal angles between their bonds. Thus, in the tetrahedron, the bond angles are all 109.5°. In the octahedron, the bond angles are all 90°. In the trigonal bipyramid, the axial bonds make 90° angles with the equatorial bonds; the equatorial bonds are at 120° angles to each other.

Self-Test

1. On a separate sheet of paper, make sketches of the five basic molecular shapes described in this section. Indicate the bond angles on the drawings. (Continue to practice this until the structures you draw make three-dimensional sense to *you*.)

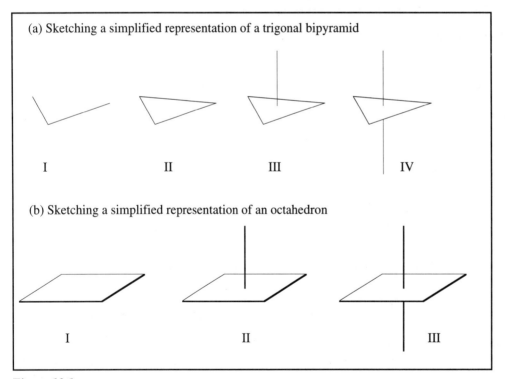

Figure 10.1

(a) To draw a trigonal bipyramid, begin by making a check mark (I), then draw a line across the top (II). Next, draw a line from the center of the triangle upward, representing an axial bond (III). Finally, draw a line downward (projecting from below the triangular plane), which represents the other axial bond (IV). (b) To draw an octahedron, begin with a parallelogram, representing a square viewed from the edge (I). Next, draw a line upward from the center (II), and finally, draw a line downward as if it comes out from behind the plane in the center (III).

New Terms

Write the definitions of the following terms, which were introduced in this section. If necessary, refer to the Glossary at the end of the text.

linear molecule

planar triangular molecule

tetrahedron

tetrahedral molecule

trigonal bipyramid

trigonal bipyramidal molecule

octahedron

octahedral molecule

bond angle

axial and equatorial bonds

10.2   Molecular shapes are predicted using the VSEPR model

Review

The basic postulate of the VSEPR theory is very simple—*electron groups (electron domains) in the valence shell of an atom stay as far apart as possible because they repel each other.* As we describe in this section, this simple concept allows us to predict how the electron domains will arrange themselves around the central atom in a molecule or polyatomic ion. Before we look at this, let's examine the two types of electron domains we encounter in molecules.

One type consists of *bonding domains*, which contain electron pairs involved in bonds between two atoms. A bonding domain can consist of one electron pair (a single bond), two electron pairs (a double bond), or three electron pairs (a triple bond). *Nonbonding domains* make up the other type. A nonbonding domain consist of one unshared electron pair (also called a lone pair). Study Example 10.1 to be sure you know how to identify these two types of electron domains. Here's another example.

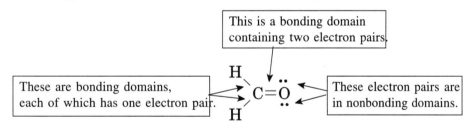

If all the electron pairs are in bonding domains, the arrangement that minimizes repulsions also defines the shape of the molecule or ion. When some electron pairs in the valence shell of a central atom are unshared (i.e., when they are *lone pairs*), the description we use for arrangement of the bonded atoms isn't the same as for the arrangement of electron domains. In ammonia, for example, the theory predicts that the electron domains are arranged tetrahedrally, but there are only three hydrogen atoms to attach to the nitrogen. When we look at how the nitrogen and three hydrogens are arranged, we see a pyramid with the nitrogen at the top. Therefore, even though the electron domains are distributed tetrahedrally, we describe the ammonia molecule as pyramid-shaped. (In fact, we say that NH_3 has a *trigonal* pyramidal shape because the pyramid has a three-sided base.)

The point of this is that when we apply the VSEPR theory to a molecule or ion, there are *two* shapes that concern us. One is the arrangement of the *electron domains* in the valence shell of the central atom. The other is the arrangement of the *bonded atoms* around the central atom. Remember, however, ***the shape of a*** *molecule is described according to the arrangement of its atoms, not the predicted orientations of the electron domains*. Although we use the theory to tell us where the electron pairs are, we use the orientations of the *atoms* when describing the molecular structure.

You should be able to predict the shape of the molecule or ion if you are able to draw its Lewis structure following the simple rules in Figure 9.8 on page 377. The shape will be one of those shown in Figures 10.2, 10.3, 10.4, and 10.5. The procedure is as follows:

1   Draw the Lewis structure for the molecule or ion.

2   Count the number of electron groups (domains) around the central atom. (Remember, a bonding domain can consist of a single, double, or triple bond.)

3   Determine the arrangement of electron groups around the central atom:

two groups	linear	five groups	trigonal pipyramidal
three groups	planar triangular	six groups	octahedral
four groups	tetrahedral		

4   Attach the necessary number of bonded atoms and then determine which shape from Figures 10.2 to 10.5 it corresponds to. (Try to draw the appropriate structure so you can visualize the shape.) Be particularly careful if the fundamental geometry is trigonal bipyramidal. Lone pairs always go in the equatorial plane.

Study Examples 10.2 through 10.6 in the text.

Self-Test

2.   Draw Lewis structures and, without referring to Figures 10.2 through 10.5, predict the geometry of each of the following:

(a) SO_4^{2-} (b) SF_2

(c) CS_2 (d) BrF_6^+

(e) PH_3 (f) SeF_4

(g) HCO_2^- (formate ion)   (h) BrF_2^-

New Terms

Write the definitions of the following terms, which were introduced in this section. If necessary, refer to the Glossary at the end of the text.

VSEPR theory   electron domain   electron group

lone pairs   nonlinear (bent) shape   trigonal pyramidal

square pyramidal   square planar

10.3   Polar molecules are asymmetric

Review

In this section we see that even when the bonds in a molecule are polar bonds, the effects of the individual bond dipoles can cancel if the molecule is symmetrical. If there are no lone pairs in the valence shell of an atom M in a molecule MX_n, and if all the atoms X that surround M are the same, then the molecule will have one of the symmetrical structures discussed in Section 10.2 and will be nonpolar.

Usually, molecules in which the central atom has lone pairs in its valence shell will be polar. The effects of their bond dipoles don't cancel. Two exceptions are linear molecules in which the central atom has three lone pairs, and square planar molecules in which the central atom has two lone pairs. These exceptions are illustrated in Figure 10.8 .

Self-Test

3.   Which of the following molecules will be polar?

(a) SCl_2 (b) SF_6 (c) ClF_3 (d) SO_3 _____

4.   The molecule CH_2Cl_2 has a tetrahedral shape, but is a polar molecule. Why?

_____

New Terms

None

10.4   Valence bond theory explains bonding as an overlap of orbitals

Review

The two main bonding theories based on the principles of wave mechanics are valence bond theory (VB theory) and molecular orbital theory (MO theory). The VB theory retains the image of individual atoms coming together to form bonds. The MO theory, in its simplest form, considers the molecule after the nuclei are in their proper positions and examines the electron energy levels that extend over *all* the nuclei. Both theories ultimately give the same results when they are refined by eliminating simplifying assumptions.

Valence Bond Theory

According to VB theory, a pair of electrons can be shared between two atoms when orbitals (one from each atom) overlap. By *overlap* we mean that two orbitals from different atoms simultaneously share some region in space. This overlap provides the means by which the electrons spread themselves over both nuclei.

Two other main principles in VB theory are:

(1)   Only two electrons, with their spins paired, can be shared between two overlapping orbitals.

(2)   The strength of a bond is proportional to the amount of overlap of the orbitals.

Atoms tend to form bonds that are as strong as possible because this lowers the energy of the atoms the most. This means that when several atoms surround some central atom, they tend to position themselves to give the maximum possible overlap of the orbitals that are used for bonding. According to VB theory, this is what determines molecular shapes and bond angles.

Self-Test   (Use separate sheets of paper to answer these questions.)

5.   Describe the formation of the chlorine-chlorine bond in Cl_2 using the principles of VB theory.

6.   Hydrogen telluride, H_2Te, has a H—Te—H bond angle of 89.5°. Explain the bonding in H_2Te in terms of the overlap of orbitals .

7.   Arsine, AsH_3, has H—AS—H bond angles equal to 92°. Explain the bonding in AsH_3.

New Terms

Write the definitions of the following terms, which were introduced in this section. If necessary, refer to the Glossary at the end of the text.

valence bond theory (VB theory)	overlap of orbitals
molecular orbital theory (MO theory)	bond angle

10.5   Hybrid orbitals are used to explain experimental molecular geometries

Review

Because many molecules have bond angles that can't be explained by the directional properties of simple atomic orbitals, it is necessary to consider how these simple orbitals can "mix" or blend when bonds are formed. Mixing simple atomic orbitals gives *hybrid orbitals* with directional properties that differ from the basic atomic orbitals. In general, hybrid orbitals form stronger bonds than simple atomic orbitals because they give better overlap with orbitals of other atoms. Be sure you learn the directional properties of the hybrids described in Figures 10.17 and 10.22. These are summarized in the following table.

Hybrid type	Orbitals Mixed	Orientations of Orbitals
sp	$s + p$	linear
sp^2	$s + p + p$	planar triangular
sp^3	$s + p + p + p$	tetrahedral
sp^3d	$s + p + p + p + d$	trigonal bipyramidal
sp^3d^2	$s + p + p + p + d + d$	octahedral

When unshared electron pairs exist in the valence shell of an atom, they are also found in hybrid orbitals. Study the explanations of the bonding for water and ammonia on page 429, as well as Example 10.13. Notice that the central atom needs one hybrid orbital for each attached atom *plus* one hybrid orbital for each lone pair.

Free rotation of groups of atoms around a single bond is possible because such rotation doesn't affect the overlap of the orbitals that form the bond, and therefore it does not affect appreciably the strength of the bond. Free rotation about bonds makes possible large numbers of different conformations. (In Section 9.6, you will see that quite a different situation exists with respect to rotation about a double bond.)

VSEPR theory helps us select the kind of hybrid orbitals an atom uses to form its bonds

We can use the VSEPR theory to figure out which kind of hybrid orbitals are used by an atom when it forms bonds. The VSEPR theory predicts the orientations of the electron groups around an atom, which we then use to deduce the kinds of hybrid orbitals that are used.

Example 10.1   Using VSEPR Theory to Deduce Hybridization

Which kind of hybrid orbitals are used by iodine in the ICl_2^- ion?

Analysis:
We can anticipate the kind of hybrid orbitals the central atom will use if we know the geometry of the molecule. To obtain the geometry, we will need a Lewis structure to which we can apply the VSEPR theory. Therefore, the first step is to construct the Lewis structure. Next, we apply the VSEPR theory. Once we know the geometry around the central atom, we can select the hybrid orbitals used.

Solution:
First, let's draw the Lewis structure for the ion.

$$\left[\ddot{\underset{\displaystyle\cdot\cdot}{Cl}} - \ddot{\underset{\displaystyle\cdot\cdot}{I}} - \ddot{\underset{\displaystyle\cdot\cdot}{Cl}} \right]^-$$

There are five electron pairs (domains) around iodine, which the VSEPR theory predicts should be arranged in a trigonal bipyramid. The kind of hybrids that have this orientation is sp^3d, so we conclude that in this ion the iodine uses two sp^3d hybrids to form the bonds and the other three to house the three lone pairs.

Is the Answer Reasonable?
Here are some things to check: Be sure you've accounted for all the valence electrons in the Lewis structure. If so, you've likely obtained the right Lewis structure. Next, count again the number of domains around the central atom. This number equals the number of hybrid orbitals needed. In this

ion, we have 5 domains around the iodine atom, so we will need 5 hybrid orbitals. The only set that consists of 5 orbitals is sp^3d. Our answer seems reasonable.

Coordinate covalent bonds and hybrid orbitals

According to valence bond theory, a coordinate covalent bond is formed by the overlap of a filled orbital of one atom with an empty orbital of another atom. In the discussion describing the formation of BF_4^- from BF_3 and F^-, notice that sufficient hybrid orbitals are made available to hold all of the electrons in the bonds.

Self-Test

8.   Why do atoms often use hybrid orbitals to form bonds instead of unhybridized atomic orbitals?

_____

9.   Use the VSEPR theory to predict which kinds of hybrid orbitals the central atom uses to form its bonds in each of the following molecules. To do this, use a separate sheet of paper to construct the Lewis structure for each.

(a) $AsCl_5$ _____

(b) $SeCl_2$ _____

(c) $AlCl_3$ _____

(d) NF_3 _____

10.   Use orbital diagrams to give explanations using hybrid orbitals of the bonding and geometry of the following:

(a) $GeCl_4$

(b) $SeCl_4$

(c) SCl_2

(d) IF_5

11. Explain, in terms of valence bond theory, the reaction for the formation of an ammonium ion from an ammonia molecule and a hydrogen ion.

12. Which type of hybrid orbitals are used by silicon in the $SiCl_6^{2-}$ ion? Give an orbital diagram for silicon that illustrates the bonding in this ion.

New Terms

Write the definitions of the following terms, which were introduced in this section. If necessary, refer to the Glossary at the end of the text.

hybrid orbitals	sp^3d hybrid orbitals	sp hybrid orbitals
sp^3d^2 hybrid orbitals	sp^2 hybrid orbitals	sp^3 hybrid orbitals
conformations	free rotation	

10.6   Hybrid orbitals can be used to describe multiple bonds

Review

In this section we see that two kinds of bonds can be formed by the overlap of orbitals:

σ bonds   in which the electron density is concentrated along an imaginary line joining the nuclei.

π bonds   in which the electron density is divided into two regions that lie on opposite sides of an imaginary line joining the nuclei.

In every molecule that you will study, a single bond is a σ bond. A double bond consists of *one* σ and *one* π bond; a triple bond consists of *one* σ bond and *two* π bonds. Thus, every bond between two atoms is composed of at least a σ bond. In complex molecules, the molecular geometry is determined by this σ bond framework, and the kind of hybrid orbitals that an atom uses is determined by the number of σ bonds it forms and the number of unshared pairs of electrons it holds. Review the brief summary on page 436. The following example illustrates how we apply these concepts.

Example 10.2   Determining the Kinds of Hybrids that Atoms Use in Molecules

What kinds of hybrid orbitals do the carbon and nitrogen atoms use in the molecule

$$\begin{array}{cccc} H & H & H & H \\ | & | & | & | \\ H-C & = C - C - N & -H \\ & & | & \cdot\cdot \\ & & H & \end{array}$$

Analysis:

For each atom, we can add up the number of atoms bonded to it and the number of lone pairs in its valence shell. The sum is equal to the number of hybrid orbitals required by the atom in the molecule. The number of hybrid orbitals tells us which kind of hybrid orbitals are involved. (For example, if an atom needs four hybrid orbitals, the hybrids would be sp^3.)

Solution:

The carbon on the left and the one in the center each form three σ bonds and neither has any unshared electrons. Therefore, each of these carbons needs three hybrid orbitals, so sp^2 hybrids are used. The carbon at the right forms four σ bonds and therefore uses sp^3 hybrids. The nitrogen forms three σ bonds, which requires three hybrid orbitals, and it needs a fourth hybrid to house the lone pair; the total is four, so sp^3 hybrids are used by the nitrogen.

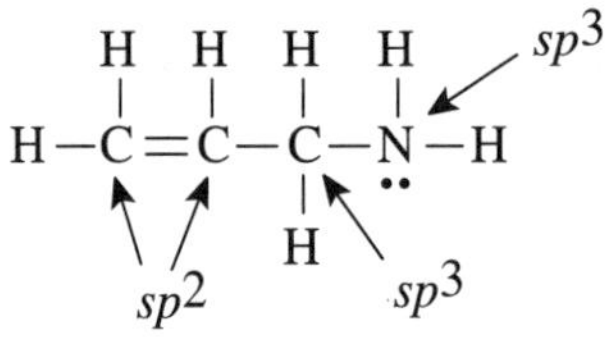

$$
\begin{array}{ccccc}
H & H & H & H & \\
| & | & | & | & \nearrow sp^3 \\
H-C & = C & - C & - N & -H \\
\nwarrow & \nearrow & | & | \nwarrow & \\
& & H & & \\
sp^2 & & & sp^3 &
\end{array}
$$

Is the Answer Reasonable?

All we can do here is recheck our work to be sure we haven't made any mistakes.

An important feature of double bonds is the restricted rotation around the axis of the bond. This is because such rotation misaligns the unhybridized p orbitals that form the π bond. As a result, rotation around a double bond axis involves bond breaking, which is very difficult.

Self-Test

13. Which kinds of hybrid orbitals are used by the carbon and oxygen atoms in acetic acid, which has the following structure?

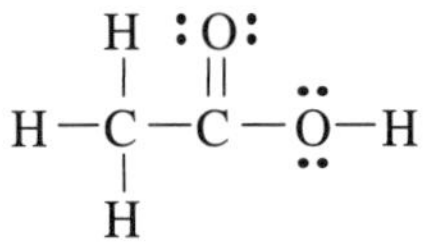

$$
\begin{array}{cc}
H & :O: \\
| & || \\
H-C - C & -\ddot{O}-H \\
| & \\
H &
\end{array}
$$

_____

_____

_____

_____

14. Explain why it should be possible to isolate three compounds with the formula $C_2H_2Cl_2$ and in which there are carbon-carbon double bonds.

15. Why doesn't it matter whether or not rotation about a triple bond is restricted?

_____

_____

New Terms

Write the definitions of the following terms, which were introduced in this section. If necessary, refer to the Glossary at the end of the text.

sigma bond (σ bond)

pi bond (π bond)

restricted rotation

10.7   Molecular orbital theory explains bonding as constructive interference of atomic orbitals

Review

According to molecular orbital theory (MO theory), when two orbitals overlap, their electron waves interact by constructive and destructive interference to give bonding and antibonding molecular orbitals. A *bonding orbital* is one that concentrates electron density between nuclei and leads to a lowering of the energy when occupied by electrons. An *antibonding orbital* concentrates electron density in regions that do not lie between nuclei. When occupied by electrons, an antibonding MO leads to a raising of the energy. Remember that compared to the atomic orbitals from which they are formed, bonding MOs are lower in energy and antibonding MOs are higher in energy.

The electronic structure of a molecule is obtained by feeding the appropriate number of electrons into the molecule's set of molecular orbitals. The rules that apply to the filling of molecular orbitals are the same as those for atomic orbitals.

1   An electron enters the lowest energy MO available.

2   No more than two electrons with spins paired can occupy any given MO.

3   When filling an energy level consisting of two or more MOs of equal energy, electrons are spread over the orbitals as much as possible with their spins in the same direction.

You should study the shapes of the σ and σ^* orbitals formed by the overlap of s orbitals as well as the shapes of the σ, σ^*, π, and π^* orbitals that arise from the overlap of p orbitals. Also study the order of filling of the MOs given by the energy level diagram in Figure 10.34 and Table 10.1.

Self-Test

16. On a separate piece of paper, construct the MO energy level diagrams and give the molecular orbital electron populations for (a) Li_2^- and (b) F_2^-.

17. Which should be more stable, C_2^+ or C_2^-? Explain.

_____

_____

18. Which has the more stable bond, NO or NO^+? (See Practice Exercise 14 on page 442.).

_____

New Terms

Write the definitions of the following terms, which were introduced in this section. If necessary, refer to the Glossary at the end of the text.

molecular orbitals   antibonding molecular orbital

bonding molecular orbital   bond order

10.8   Molecular orbital theory uses delocalized orbitals to describe molecules with resonance structures

Review

Molecular orbital theory avoids resonance by allowing for the simultaneous overlap of more than two orbitals in such a way as to produce a large π-type cloud that extends over three or more atoms. This kind of an extended molecular orbital is said to be *delocalized* because the bond is not localized between just two atoms. One of the most important molecules having delocalized MOs is benzene.

Molecules in which there are delocalized bonds are more stable than they would be if the bonds were localized. The extra stability produced by delocalization is called the delocalization energy.

Self-Test

19.   Draw resonance structures for the nitrate ion and show how the bonding would be represented using a delocalized MO.

20.   How is the benzene molecule represented to show the delocalized molecular orbitals of the ring?

New Terms

Write the definitions of the following terms, which were introduced in this section. If necessary, refer to the Glossary at the end of the text.

delocalized bond   delocalization energy

Answers to Self-Test Questions

1.   Check your answers by referring to the drawings of pages 401 through 403.

2.   (a)

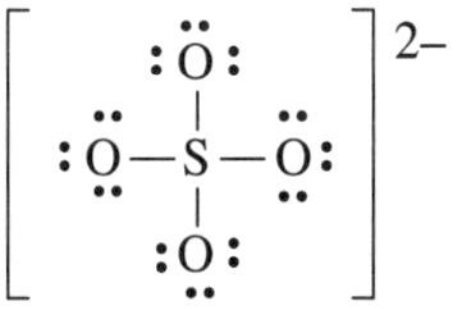

$$\begin{bmatrix} \ddot{:}\ddot{O}\ddot{:} \\ | \\ :\ddot{O}-S-\ddot{O}: \\ | \\ :\ddot{O}: \end{bmatrix}^{2-}$$ tetrahedral

(b) 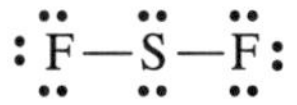 $:\ddot{F}-\ddot{S}-\ddot{F}:$ nonlinear

(c) 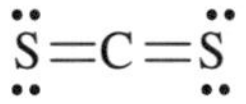 $\ddot{S}=C=\ddot{S}$ linear

(d)

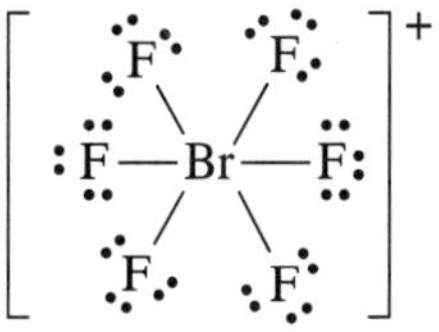

$$\begin{bmatrix} \ddot{F}\cdot \quad \cdot\ddot{F}\cdot \\ :\ddot{F}-Br-\ddot{F}: \\ \ddot{F}\cdot \quad \cdot\ddot{F}\cdot \end{bmatrix}^{+}$$ octahedral

(e) 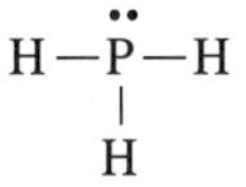 $H-\overset{\displaystyle\cdot\cdot}{P}-H$ with H below   trigonal pyramidal

(f) 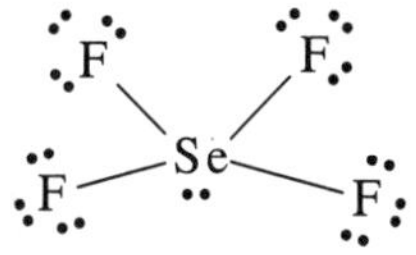 $\ddot{F}$, $\ddot{F}$ attached to Se, $\ddot{F}$, $\ddot{F}$ distorted tetrahedra

(g)

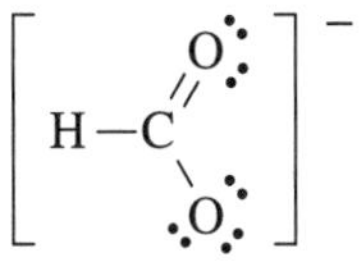

$$\begin{bmatrix} H-C \overset{\displaystyle/\!/ \ddot{O}:}{\underset{\displaystyle\backslash \cdot\ddot{O}:}{}} \end{bmatrix}^{-}$$ planar triangular

(h) 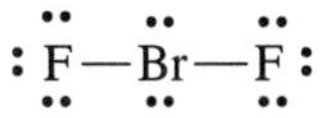 $:\ddot{F}-\ddot{Br}-\ddot{F}:$ linear

3. SCl_2 and ClF_3 are polar. (Draw Lewis structures; note lone pairs.)
4.   The C—Cl and C—H bonds differ in polarity, so the bond dipoles can't cancel to give a nonpolar molecule.
5.   Overlap of the half-filled p orbitals of the chlorine atoms produces the bond. See Figure 10.14 for F_2.
6.   Hydrogen $1s$ orbitals overlap two half-filled p orbitals of tellurium. Since p orbitals are at 90°, the bond angle is very close to 90°.

Te (in H_2Te)   (••) $5s$ (••)(•×)(•×) $5p$ • = H electron   × = Te electron

7. Hydrogen $1s$ orbitals overlap 3 half-filled p orbitals of arsenic.

As (in AsH_3)   (••) $4s$ (•×)(•×)(•×) $4p$ • = H electron   × = As electron

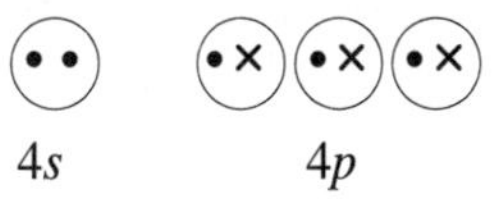

8.   Hybrid orbitals overlap better with orbitals on neighboring atoms and form stronger bonds.
9.   (a) sp^3d (b) sp^3 (c) sp^2 (d) sp^3
10.   (a)   Ge (in $GeCl_4$)   (•×)(•×)(•×)(•×) sp^3 × = Cl electron

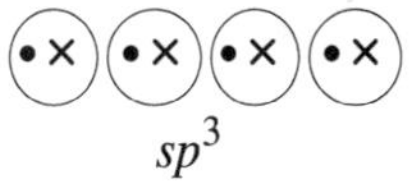

molecule is tetrahedral

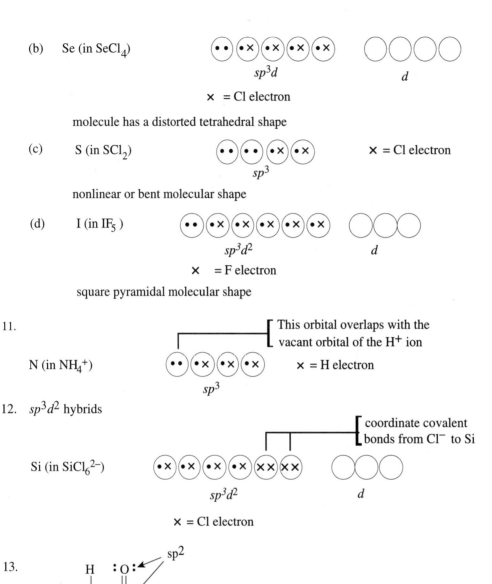

(b)   Se (in $SeCl_4$)

sp^3d

d

× = Cl electron

molecule has a distorted tetrahedral shape

(c)   S (in SCl_2)

sp^3

× = Cl electron

nonlinear or bent molecular shape

(d)   I (in IF_5)

sp^3d^2

d

× = F electron

square pyramidal molecular shape

11.

This orbital overlaps with the vacant orbital of the H^+ ion

N (in NH_4^+)

× = H electron

sp^3

12. sp^3d^2 hybrids

coordinate covalent bonds from Cl^- to Si

Si (in $SiCl_6^{2-}$)

sp^3d^2

d

× = Cl electron

13.

sp^2

sp^3

14. Restricted rotation around the double bond permits structures II and III (below) to be isolated as well as structure I.

I   II   III

15. The atomic arrangement is linear.

16.

	Li_2^-	F_2^-
$\sigma_{2p_z}^*$	◯	(↑)
$\pi_{2p_x}^*$, $\pi_{2p_y}^*$	◯◯	(↑↓)(↑↓)
σ_{2p_z}	◯	(↑↓)
π_{2p_x} , π_{2p_y}	◯◯	(↑↓)(↑↓)
σ_{2s}^*	(↑)	(↑↓)
σ_{2s}	(↑↓)	(↑↓)

17. C_2^+, bond order $= \dfrac{5-2}{2} = \dfrac{3}{2}$

C_2^-, bond order $= \dfrac{7-2}{2} = \dfrac{5}{2}$

Therefore C_2^- is more stable than C_2^+.

18. NO, bond order is $\dfrac{8-3}{2} = \dfrac{5}{2}$

NO^+, bond order is $\dfrac{8-2}{2} = \dfrac{6}{2}$

NO^+ has a more stable bond than NO.

19.

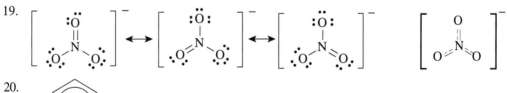

20.

⬡

Tools you have learned

Consider removing this chart from the Study Guide so you can have it handy when tackling homework problems.

Tool	*How It Works*
Basic molecular shapes linear, planar triangular, tetrahedral, trigonal bipyramidal, octahedral	Knowing how to draw them enables you to sketch the shapes of most molecules.
VSEPR Theory	This theory enables you to determine the shape of a molecule or polyatomic ion when its Lewis structure is known. It also enables you to determine the kind of hybrid orbitals used by the central atom in a molecule or polyatomic ion.
Molecular shapes	In this chapter you learned that knowing the shape of a molecule enables you to determine whether or not it is polar.
Molecular orbital energy diagram	We use the diagram to explain the bonding in period 2 diatomic molecules.

Summary of Useful Information

The various structures obtained by applying the VSEPR theory can be summarized as shown in the table below. In the formulas in the table, the symbol M stands for the central atom, X stands for an atom bonded to the central atom, and E stands for a lone pair of electrons in the valence shell of the central atom. Thus, MX_2E_3 represents a molecule in which the central atom is bonded to two other atoms and has three lone pairs in its valence shell. To use the information in the table, construct the Lewis structure for the substance. Then write its formula using the symbols as defined above. Find the formula in the table and read the name for the molecular shape alongside.

Formula	Structure	Formula	Structure
MX_2	linear	MX_5	trigonal bipyramidal
MX_3	planar triangular	MX_4E	distorted tetrahedral
MX_2E	nonlinear (bent)	MX_3E_2	T-shaped
MX_4	tetrahedral	MX_2E_3	linear
MX_3E	trigonal pyramidal	MX_6	octahedral
MX_2E_2	nonlinear (bent)	MX_5E	square pyramidal
		MX_4E_2	square planar

Correlation between VSEPR theory and orbital hybridization.

Number of sets of electrons in valence shell of central atom	Geometric arrangement of electron pairs	Hybrid orbitals that give this geometry
2	linear	sp
3	planar triangular	sp^2
4	tetrahedral	sp^3
5	trigonal bipyramidal	sp^3d
6	octahedral	sp^3d^2

Chapter 11
Properties of Gases

This chapter and the next examine the physical properties of the states of matter. Gases are the most easily understood, and many of the properties of gases are quite familiar to everyone. We will study the four variables that control gas behavior—pressure, volume, temperature, and amount (which we will measure in moles). These variables are interrelated, because one cannot be changed without changing one or more of the others. The relationships among them are expressed in the *gas laws*, which we describe in this chapter. We will also see how these laws can be explained in terms of a single theory, the kinetic theory of gases.

Learning Objectives

In this chapter, you should keep in mind the following goals.

1　To become aware of properties that characterize gases in general and how these properties can be understood in terms of a molecular model of a gas.

2　To learn the units of pressure and how barometers and manometers work to measure pressure.

3　To learn how the pressure and the volume of a fixed amount of gas are interrelated at constant temperature (Boyle's law).

4　To learn how the volume of a fixed amount of gas changes with temperature if the pressure is kept constant (Charles' law).

5　To learn how the pressure of a fixed amount of gas changes with temperature if the volume is kept constant (Gay-Lussac's law).

6　To be able to carry out gas law calculations using the combined gas law.

7　To know the ideal gas law equation, the value of the universal gas constant, R, and to be able to use this law to do gas law calculations, including molecular mass calculations.

8　To be able to carry out stoichiometric calculations when some or all of the substances are gases.

9　To learn what *partial pressure* means, what the relationship is between the total pressure of a gas mixture and the partial pressures of the mixture's components, and how to find what pressure a wet gas would have if made dry.

10　To learn the definition of mole fraction and how to relate partial pressures to total pressure using mole fraction.

11　To learn how the rate of effusion of a gas is related to its density and molecular mass.

12　To learn the postulates of the kinetic theory of gases and the model for an ideal gas that these postulates describe.

13　To be able to explain the gas laws in terms of the kinetic model of gases.

14　To learn why real gases do not behave as an ideal gas and how van der Waals corrected the ideal gas law to make it fit real gases better.

11.1   Familiar properties of gases can be explained at the molecular level

Review

The purpose of this section is to review familiar properties of gases and show that a very simple molecular model of this state of matter is able to explain gas behavior. Let's review the list.

Gases can be compressed

Gases exert a pressure that depends on the amount of gas in a container.

Gas pressure increases with increasing temperature.

Gases have low density, meaning a given volume doesn't contain much matter.

Gases expand to fill any container they're in.

Gases freely mix with each other.

All of these properties can be explained by a model that describes a gas as a collection of widely spaced molecules in rapid motion. Their collisions with the walls of the container produces the pressure, which increases with temperature because molecules go faster and hit harder as their temperatures increase.

Self-Test

1.   What four physical properties of gases are interrelated?

    _____

2.   If we compare gases at different temperatures, would the speed at which they mix with each other increase, decrease, or stay the same as the temperature in raised? Explain your answer in terms of the molecular model of a gas.

    _____

    _____

New Terms

None

11.2   Pressure is a measured property of gases

Review

Pressure, as you learned in Chapter 7, is the ratio of force to area. The earth exerts a gravitational force that pulls on the surrounding envelope of air—the atmosphere—and causes the air to be most concentrated near the earth's surface where it exerts a pressure that we call the *atmospheric pressure.* The simplest device to measure it is the Torricelli barometer.

At sea level, the atmospheric pressure holds a column of mercury in a barometer at an average height of about 760 mm. The pressure unit called the *standard atmosphere (atm)* was originally defined as the pressure exerted by a column of mercury 760 mm high at a temperature of 0 °C. With the introduction of the SI, the *atmosphere* was redefined in terms of the *pascal*, the SI unit of pressure (1 atm = 101,325 Pa = 101.325 kPa). The pressure unit *bar* is also defined in terms of the pascal (1 bar = 100 kPa). A smaller unit of pressure called the *torr* is defined as 760 torr = 1 atm. For most purposes, we can use the relationship 1 torr = 1 mm Hg, meaning a pressure of 1 torr will support a column of mercury 1 mm high. This makes measuring pressures in torr relatively simple by using mercury as the fluid in barometers and manometers (the topic to be discussed next).

You will frequently find it necessary to convert between torr and atm, so it is important that you remember the relationship

$$1 \text{ atm} = 760 \text{ torr} = 760 \text{ mm Hg}$$

To check yourself in calculations, remember that *the pressure expressed in torr is always numerically larger than if it is expressed in atm*. This will help you avoid errors. For example, if you are converting 0.700 atm to torr and use the conversion incorrectly, you would obtain an answer of 9.21×10^{-4} torr. But the number of torr should be much larger than the number of atm, so the answer must be wrong and we can check the calculation. (The correct answer is 532 torr.)

Your teacher may also want you to learn the relationships to convert between the units atmosphere and pascal, and between units of bar and pascal. If so, you will need to memorize the following:

$$1 \text{ atm} = 101,325 \text{ Pa}$$

$$1 \text{ bar} = 100,000 \text{ Pa}$$

Pressures of trapped gases are measured using manometers

Manometers, either the open-end or the closed-end type, are employed to measure the pressure of a confined gas. The difference in the heights of the two liquid levels in a U-shaped tube is proportional to the difference between the pressures exerted by the gases on opposite sides. If mercury is used in the manometer, the height difference is measured in mm Hg, which gives the pressure difference in torr. (Remember, 1 mm Hg = 1 torr.) Study the discussions of open and closed end manometers on pages 455 through 457.

Sometimes, a liquid other than mercury is used in a manometer. In these situations, the difference in the heights of the liquid levels is *larger* than if mercury were used. The explanation for this is given in Figure 11.6. Often it is necessary to convert this difference in heights to what the difference would be if mercury were in the manometer instead. If we call the liquid in the manometer A and its density d_A, the height of the equivalent column of mercury is given by the expression

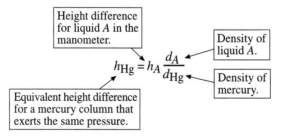

The value we calculate for h_{Hg}, expressed in millimeters, is the difference in torr between the pressures exerted by the gases connected to the two arms of the manometer. Study Example 11.2 on page 458.

Self-Test

3.   A pressure of 745 torr has what value in atm?   _____

4. On a TV weather report, the pressure was reported to be 29.4 in., meaning "inches of mercury." What does this correspond to in torr?

_____

In atm?

_____

5. A pressure of 0.980 atm has what value in torr?

_____

6. A pressure of 746 torr has what value in kilopascals (kPa)?

_____

7. The pressure in a storm center was reported to be 967 mb. Express this pressure in

(a) atm _____   (b) torr _____

8. In using an open-end manometer containing mercury, what measurements are required in determining the pressure of a confined gas?

_____

9. In using a closed-end manometer containing mercury, what measurements are required in determining the pressure of confined gas?

_____

10. On a day when the atmospheric pressure was 754 torr, the manometer reading on a gas sample was found to be 82 torr. What is the gas pressure in torr if the manometer is of the closed-end design?

_____ The open-end design?   _____

New Terms

Write the definitions of the following terms, which were introduced in this section. If necessary, refer to the Glossary at the end of the text.

barometer	pascal (Pa)
manometer	pressure
manometer, closed-end	standard atmosphere (atm)
manometer, open-end	torr
bar	

11.3 The gas laws summarize experimental observations

Review

In this section we examine some of the gas laws. We begin with the relationships among *P, V,* and *T* for a fixed (constant) amount of gas. These deal with situations in which we change one or two of the variables and observe (or calculate) how one of the others change. Historically, the scientists who studied gases, held one variable constant and observed the relationship between the other two. This led to three separate gas laws. We will also examine one of the laws that deal with situations in which the amount of gas changes.

Boyle's law (pressure–volume law). Gases generally obey the rule that their volumes are inversely proportional to their pressures, provided that the temperature and the amount of gas, *n*, are kept constant.

$$V \propto 1/P \quad \text{(constant } T \text{ and } n) \quad \text{or} \quad PV = \text{constant}$$

The hypothetical gas that would obey this relationship (as well as the others that we will discuss) exactly is called an *ideal gas.*

Charles's law (temperature–volume law). Provided we express the temperature of a gas in kelvins, the volume of a fixed amount of gas is directly proportional to the temperature if the pressure is kept constant.

$$V \propto T \quad \text{(constant } P \text{ and } n) \quad \text{or} \quad \frac{V}{T} = \text{constant}$$

Notice, once again, that for this relationship to hold, *temperatures must be expressed in kelvins*.

Gay-Lussac's law (pressure–temperature law). Provided that the volume is held constant, the pressure of a fixed quantity of gas is directly proportional to the Kelvin temperature.

$$P \propto T \quad \text{(constant } V \text{ and } n) \quad \text{or} \quad \frac{P}{T} = \text{constant}$$

As with Charles' law, calculations using this law require Kelvin temperatures.

The combined gas law

A single equation can be written that expresses all three of these laws.

$$\frac{PV}{T} = \text{constant}$$

Usually, we use these gas laws in calculations where one or two of the variables change and we wish to know the effect on another. If we use subscripts of 1 and 2 to identify the *P, V,* and *T* before and after the change, we can express the combined gas law as given by Equation 11.4 in the text.

$$\frac{P_1 V_1}{T_1} = \frac{P_2 V_2}{T_2} \qquad \text{(Equation 11.4)}$$

When $P_1 = P_2$, the equation reduces to that of Charles's law (the temperature–volume law). When $T_1 = T_2$, the combined gas law equation expresses Boyle's law (the pressure–volume law). When $V_1 = V_2$, the combined gas law equation reduces to Gay-Lussac's law (the pressure–temperature law).

Solving the combined gas law for one of its variables

Examples 11.3 and 11.4 illustrate how to apply the combined gas law in calculations. In using this equation, it is necessary to solve it for the unknown quantity. This is where some students have difficulty, especially if their algebra skills are rusty. Here is a tip that may help you avoid mistakes.

Suppose we wished to solve the equation for P_2. The first step is to rewrite the equation as shown below.

$$\boxed{\text{Ratio of volumes}} \qquad \boxed{\text{Ratio of temperatures}}$$

$$P_2 = P_1 \times \left(\frac{V}{V}\right) \times \left(\frac{T}{T}\right)$$

The volume and temperature ratios will ensure that volume and temperature units cancel and the answer will be in pressure units.

Next, we have to decide where to put the 1's and 2's. To follow the reasoning here, first look at the left side of Equation 11.4.

$$\frac{P_1 V_1}{T_1} \longleftarrow \boxed{\text{Fraction bar}}$$

Notice that the P_1 and V_1 are *both* above the fraction bar and T_1 is below the fraction bar. No matter how we rearrange things, if P_1 is on one side of the fraction bar (either in the numerator or denominator), then V_1 will be on the same side.[1] Furthermore, whichever side of the fraction bar we find P_1 and V_1 on, T_1 will be on the *opposite* side. Therefore, let's insert the subscripts 1 in the equation we're solving for P_2. In doing this, we have to realize that P_1 is considered to be in the numerator, as though it's in the fraction $P_1/1$.

$$\boxed{P_1 \text{ and } V_1 \text{ are both in the numerator.}}$$

$$P_2 = P_1 \times \left(\frac{V_1}{V} \right) \times \left(\frac{T}{T_1} \right)$$

$$\boxed{T_1 \text{ is in the denominator.}}$$

Once the 1's are in place, the 2's follow. Here's the equation solved for P_2.

$$P_2 = P_1 \times \left(\frac{V_1}{V_2} \right) \times \left(\frac{T_2}{T_1} \right)$$

Even if your algebra skills are good, the approach just described serves as a good check to see if you've solved the equation correctly.

Avogadro's principle deals with varying amounts of gas

Avogadro's principle describes how changing the amount of gas affects the other gas variables. If we keep T and P constant, the volume increases when we increase the amount of gas. (This is what you observe when you blow up a balloon; adding more air makes the balloon bigger.)

$$V \propto n \quad \text{(at constant } T \text{ and } P\text{)}$$

Similarly, if we keep T and V constant, the pressure increases when we increase the amount of gas. (This is what happens when you pump air into a tire. The tire doesn't get larger, but the pressure inside it increases.)

$$P \propto n \quad \text{(at constant } T \text{ and } V\text{)}$$

Avogadro's principle leads to the concept of a molar volume—the volume occupied by one mole of a gas. For comparisons, conditions of standard temperature and pressure (STP) are chosen. STP corresponds to 0 °C (273 K) and 1 atm. Under these conditions, 1 mol of an ideal gas occupies a volume of 22.4 L.

When gases react, their volumes are in the same ratio as their coefficients if the volumes are compared at the same T and P. (This is called the *law of combining volumes*.) Under these conditions, gas stoichiometry problems are very simple, as we will discuss in Section 11.6.

---

[1] This applies as long as P_1 and V_1 are on the same side of the equals sign.

Thinking It Through

1   Consider a sample of 2.15 g of krypton, one of the noble gases, occupying a volume of 640 mL at 25.0 °C under a pressure of 745 torr. It is desired to make the conditions of the final state of this amount of sample be a temperature of 35.0 °C and a pressure of 740 torr.

(a)   Are enough data provided to enable another scientist to duplicate the sample in its initial state? If not, what additional data would be needed?

(b)   For this specific quantity of sample to be in its final state, what other variable must change? To what value?

Self-Test

11.   The pressure on 750 mL of a gas is changed at constant temperature from 720 torr to 760 torr. What is the new volume?

_____

12.   To change the volume of a gas at 755 torr from 400 mL to 350 mL at constant temperature, what new pressure must be provided?

_____

13.   What do we call the hypothetical gas that would obey the pressure–volume law exactly?

_____

14.   At constant temperature and amount, the volume of an ideal gas is inversely proportional to

(a) $\frac{1}{P}$ (b) P (c) P^2 (d) $\sqrt{P}$ _____

15.   What will be the new volume if 250 mL of oxygen is cooled from 25 °C to 15 °C at constant pressure?

_____

16.   A sample of 750 mL of nitrogen gas is heated from 20 °C to 100 °C. In order to keep its pressure constant, to what new volume should the gas be allowed to change?

_____

17.   A sample of 656 mL of oxygen gas at 740 torr is heated from 25.0°C to 55.0 °C in a sealed container. What gas variable changes and to what new value?

_____

18.   What will be the final volume of a 500-mL sample of helium if its pressure is changed from 740 torr to 780 torr and its temperature is changed from 30 °C to 50 °C?

_____

19.   In order for the pressure of a fixed mass of gas at 10 °C to change from 1 atm to 2 atm and its volume to go from 3 L to 4 L, what must the final temperature become?

_____

20. What is the final pressure in torr exerted by 375 mL of argon at 755 torr and 25 °C if its temperature goes to 50 °C and its volume changes to 425 mL?

_____

21. What is the standard molar volume (including its units)?

_____

22. Use the combined gas law to calculate the molar volume of a gas at 25 °C and 2 atm.

_____

23. What is Avogadro's principle?   _____

_____

24. What do we call the reference conditions for gas temperature and pressure, and what are their values?

_____

25. A sample of 11.2 L of a gas was trapped at STP. How many moles of the gas was this?

_____

New Terms

Write the definitions of the following terms, which were introduced in this section. If necessary, refer to the Glossary at the end of the text.

absolute zero	pressure–temperature law (Boyle's law)
combined gas law	pressure–volume law (Gay-Lussac's law)
ideal gas	temperature–volume law (Charles's law)
standard temperature and pressure (STP)	Avogadro's principle
law of combining volumes	standard molar volume

11.4   The ideal gas law relates *P*, *V*, *T*, and the number of moles of gas, *n*

Review

The ratio *PV/T* is proportional to the number of moles of gas, *n*. The proportionality constant is given the symbol *R*. Inserting *R* into the proportionality and clearing fractions gives the usual form of the *ideal gas law*,

$$PV = nRT \qquad\qquad \text{(Equation 11.5)}$$

The ideal gas law gives us the relationship among all four physical variables of a gas—pressure, volume, temperature, and number of moles. The *universal gas constant*, *R*, has different numerical values, depending on the units used for *P* and *V*. The value we will use in calculations is 0.0821 L atm/mol K. Remember that when you use this value of *R* in the ideal gas law equation, the unit of pressure must be atm and the unit of volume must be liter (the temperature, of course, must be in kelvins). The ideal gas law equation can be used in the following kinds of calculations.

1   To find the *volume* that a given amount of gas will occupy at some value of pressure and temperature.

2   To find the *pressure* that a given amount of gas will exert if it is kept at a given temperature in some particular volume.

3   To find the *temperature* a given amount of a gas must have if it is confined in a specified volume and is to exert a certain pressure.

4   To use the measured values of *P*, *V*, and *T* to find the number of moles (*n*) of gas in the sample. If we then divide the mass of the gas sample in grams by the number of moles, we get the *molar mass*, which is numerically the same as the molecular mass.

5   To calculate the density of a gas at a particular temperature and pressure, and to use gas density to calculate the molar mass of a gas.

Be sure to study the worked examples in this section and answer the practice exercises. Then work on the exercises below.

Thinking It Through

2   A gaseous compound of sulfur and fluorine with an empirical formula of SF has a density of 4.102 g L^{-1} at 25.0 °C and 750 torr. What is the molecular mass and the molecular formula of this compound?

Self-Test

26.   Write the equation for the ideal gas law.   _____

27.   Write the value of *R*, including the units.   _____

28.   What volume will 0.982 g of O_2 occupy if its temperature is 21 °C and its pressure is 748 torr?

_____

29.   If 0.156 g of N_2 is confined at 18.0 °C in a volume of 135 mL, what pressure will it have?

_____

30.   If a 0.138-gram sample of CO_2 is to have a pressure of 4.50 atm when kept in a container with a volume of 30.0 mL, what temperature in degrees Celsius must it have?

_____

31.   A 0.390-gram sample of a gas at 25 °C occupied a volume of 350 mL at a pressure of 740 torr. Calculate its molecular mass.

_____

32.   What is the density of argon gas to 25 °C and 785 torr?

_____

33. Arrange the following gases in order of increasing density (lowest to highest) if the densities are compared at the same T and P.

$$N_2, \ CH_4, \ C_4H_{10}, \ H_2, \ CO_2$$

New Terms

Write the definitions of the following terms, which were introduced in this section. If necessary, refer to the Glossary at the end of the text.

    ideal gas law (equation of state of an ideal gas)   universal gas constant (R)

11.5   In a mixture each gas exerts its own partial pressure

Review

Gases in a mixture exert their own pressures, called *partial pressures,* independently of the other gases present. The total pressure of the mixture is a simple function of the partial pressures, P_a, P_b, P_c and so forth.

$$P_{total} = P_a + P_b + P_c + \ldots$$

Called *Dalton's law of partial pressures,* this law is useful in doing calculations involving gases collected over water. When collected in this way, the gas is a really a mixture. It contains the desired gas mixed with water vapor. Often we want to calculate the pressure of the collected gas without the water vapor ($P_{dry\ gas}$). All we need to do is subtract the partial pressure of the water vapor from the total pressure of the gas mixture. (The total pressure is all we can actually measure; partial pressures are calculated.)

$$P_{dry\ gas} = P_{total} - P_{water}$$

The partial pressure of water vapor, when it is in the presence of liquid water, is called the vapor pressure of the liquid water. All liquids, including water, have their own *vapor pressures* that depend only on the temperature of the liquid that's in contact with the vapor. For any given liquid the value of its vapor pressure increases with increasing temperature of the liquid. The vapor pressure of water at various temperatures is found in Table 11.2 on page 478.

    The partial pressure, P_A, of a gas A in a mixture is a fraction of the total pressure, P_{total} For a given V and T, pressure is proportional to n. Therefore,

$$\frac{P_A}{P_{total}} = \frac{n_A}{n_{total}} = X_A$$

where X_A is called the mole fraction of A. The mole fraction is the ratio of the number of moles of a substance (such as A) to the total number of moles of all the substances present.

$$X_A = \frac{n_A}{n_A + n_B + n_C + \ldots} = \frac{n_A}{n_{total}}$$

Multiplying a mole fraction by 100 gives the *mole percent.*

$$\text{mol } \% A = X_A \times 100\%$$

The relationship between the partial pressure of a gas and the total pressure is often written as

$$P_A = X_A P_{total}$$

Thinking It Through

3   A 500-mL sample of nitrogen is collected over water at a temperature of 20 °C. The atmospheric pressure at the time of the experiment is 746 torr. What would be the volume of the nitrogen at this same temperature if it were free of water and at a pressure of 760 torr?

Self-Test

34.   Suppose that the total pressure on a mixture of helium and neon is 745 torr and the partial pressure of the helium is 640 torr.

   (a)   What is the partial pressure of the argon?   _____

   (b)   What is the mole fraction of argon in the mixture?   _____

35.   Suppose that all of the argon were removed from the sample in the previous question but the volume of the container was left the same and the temperature was not changed. What would a pressure gauge read for the remaining gas?

   _____

36.   The average composition of the air we exhale in terms of the partial pressures of the gases is as follows: N_2, 569 torr; O_2, 116 torr; CO_2, 28 torr; and $H_2O(g)$, 47 torr. Calculate the composition of the air in mole percents.

New Terms

Write the definitions of the following terms, which were introduced in this section. If necessary, refer to the Glossary at the end of the text.

   Dalton's law of partial pressures   partial pressure

   mole fraction   vapor pressure

   mole percent (mol %)

11.6   Gas volumes are used in solving stoichiometry problems

Review

For reactions involving gases, Avogadro's principle provides a new way to state stoichiometric equivalencies. For example, from the coefficients in the equation

$$N_2(g) + 3H_2(g) \rightarrow 2NH_3(g)$$

we know that in terms of *moles* 1 mol $N_2 \Leftrightarrow$ 3 mol H_2. We can now rephrase this in terms of gas *volumes* provided that the volumes are compared at identical pressures and temperatures.

$$1 \text{ vol. } N_2 \Leftrightarrow 3 \text{ vol. } H_2$$

When we measure gaseous reactants and products at the same temperature and pressure, gas stoichiometry problems become very simple. Study Examples 11.13 and 11.14. Notice how we deal with problems in which the T and P are not the same for all the gases.

Thinking It Through

4   Sodium hydroxide granules can remove carbon dioxide from an air stream by its reaction according to the following equation.

$$NaOH(s) + CO_2(g) \rightarrow NaHCO_3(s)$$

The air of a room measuring 6.00 m × 6.00 m × 3.00 m containing CO_2 at a concentration of 1.00×10^{-6} mol/L is passed through a bed of NaOH granules. What is the minimum number of grams of NaOH needed to remove all of the CO_2 from this quantity of air?

Self-Test

37.   In the following reaction, all substances are gases.

$$2NO + O_2 \rightarrow 2NO_2$$

Assuming that the gases are measured at the same values of P and T, how many liters each of NO and O_2 are needed to prepare 10 L of NO_2?

_____

38.   Nitrogen and hydrogen can be made to combine to give ammonia according to the following equation:
$$N_2 + 3H_2 \rightarrow 2NH_3$$

All the substances are gases. If one liter of nitrogen is used, what volume of hydrogen is needed and what volume of ammonia is made (assuming all volumes are measured under the same conditions of temperature and pressure)?

_____

39.   One step in the industrial synthesis of sulfuric acid is the following reaction:

$$2SO_2(g) + O_2(g) \rightarrow 2SO_3(g)$$

If 640 L of O_2, initially at 740 torr and 20.0 °C, is used to make SO_3, how many moles and how many grams of SO_2 are needed?

_____

New Terms

None

11.7   Effusion and diffusion in gases leads to Graham's Law

Review

Diffusion is not the same as *effusion*. Effusion is the movement of a gas into a vacuum through an extremely small hole in the wall of the gas container, the intent being to minimize collisions of the effusing gas molecules with other gas molecules or with themselves. Effusion is an uncommon phenomenon because it has to be set up experimentally. Diffusion, a common phenomenon, is simply the spreading out and intermingling of the molecules of one gas with another.

According to *Graham's law of effusion*, the rate of effusion of a gas is inversely proportional to the square root of the density of a gas, or to the square root of the molecular mass of the gas, provided that the pressure and temperature of the gas are kept constant. Equation 11.10 is the easiest equation to use in calculations.

Self-Test

40.   State the law of gas effusion.

_____

_____

41.   At room temperature and 760 torr the density of helium is 0.000160 g/mL. Under these conditions the density of hydrogen is 0.0000818 g/mL. How much more rapidly will hydrogen effuse than helium through the same pinhole?

(a)   1.95 times as rapidly   (c)   1.22 times as rapidly

(b)   1.11 times as rapidly   (d)   1.40 times as rapidly   _____

42.   Arrange the following gases in order of increasing (smallest to largest) rates of effusion.

$$N_2, \ CH_4, \ C_4H_{10}, \ H_2, \ CO_2$$

_____

43.   Look back at your answer to Self-Test Question 33. What relationship is there between the density of a gas and its rate of effusion?

_____

New Terms

Write the definitions of the following terms, which were introduced in this section. If necessary, refer to the Glossary at the end of the text.

diffusion   Graham's law (law of gas effusion)

effusion

11.8   The kinetic-molecular theory explains the gas laws

Review

The gas laws describe *observations*; they *explain* nothing. They would be true even if atoms and molecules were yet to be discovered. The kinetic theory of gases, on the other hand, answers the question, "What *must* be true about a gas at the submicroscopic level to explain why the gas laws exist?" This section takes each of the gas laws in turn and in a qualitative way, uses the kinetic theory and its model of an ideal gas to explain them.

The *kinetic theory of gases* proposes a model for the structure of a gas in the form of a set of postulates. They are given on page 487. They are simple and should be learned. In summary, the theory describes a gas as a collection of extremely tiny particles in random motion. Collisions of the particles explains *how* gases have pressures and how the average kinetic energy of the particles is proportional to the temperature. The higher the average kinetic energy of gaseous particles, the higher the pressure and the higher the temperature.

The theory accounts for the gas laws without much difficulty:

Pressure–temperature law ($P \propto T$, at constant n and V). If raising the temperature makes the average kinetic energy of the gas particles increase, they hit the walls harder, giving a higher pressure.

Temperature–volume law ($V \propto T$, at constant n and P). If gas particles are given higher average kinetic energies and velocities by giving the gas a higher temperature, the pressure tends to increase. To keep the pressure constant, the collisions with the walls have to be spread out over a larger area, which is accomplished by letting the particles take up a larger space or volume.

Pressure–volume law ($P \propto V$, at constant n and T). If the volume is reduced, the container's walls are hit more frequently by gas particles, thereby increasing the pressure.

Effusion law (effusion rate $\propto 1/\sqrt{\text{gas density}}$, at constant pressure and temperature). A gas having a low density has less massive particles than a gas with a high density. At the same temperature, the lighter particles move with a greater average root mean square speed than the heavier particles. This means that particles of the less dense gas get from one place to another more quickly than those of the more dense gas, so the gas with the lower density must effuse more rapidly.

Law of partial pressures ($P_{\text{total}} = P_a + P_b + P_c + \ldots$). According to the model of an ideal gas, its particles do not repel or attract neighboring particles. This causes the particles to act independently, so those of each gas contribute separately toward the total pressure. Therefore, their partial pressures add up to the total pressure, not some value less than or greater than the total.

Avogadro's principle ($V \propto n$, at constant P and T; or $P \propto n$, at constant V and T). At the same T, regardless of the gas, gas molecules have the same average kinetic energy and they exert the same average force per hit when they strike a unit area of the walls. The frequency of these hits per unit area is proportional to the molar concentration of the gas (moles per unit of volume), so the total force per unit area—the pressure—is proportional to the moles of the gas and not the molecular mass of the gas. Hence $P \propto n$, which is one way of stating the law.

Gas compressibility. Gases, but not liquids and solids, are relatively easily compressed because the volume occupied by a gas is mostly empty space.

Absolute zero. When all of the particles in a gas stop moving, the gas particles must have zero average kinetic energy, regardless of their mass. Since gas temperature is proportional to this average kinetic energy, a state of zero kinetic energy must be the coldest condition possible—absolute zero (0 K or −273.15 °C).

Self-Test

44. What are the postulates of the kinetic theory of gases?

    _____

    _____

    _____

45. The Kelvin temperature of a gas is proportional to what property of the gas particles?

    _____

46. What is the fundamental difference between the gas laws—taken as a group— and the postulates of the kinetic theory?

    _____

    _____

47. In your own words explain each of the gas laws in terms of the model of the ideal gas.

    _____

    _____

48. Which of the following statements are false?

    (a) Compressing the gas molecules into a smaller volume causes them to become smaller.

    (b) Raising the temperature makes the gas molecules strike the walls of a container with greater average force.

    (c) The theory predicts that at absolute zero, molecules should be motionless.

    (d) Increasing the volume of a gas causes the molecules to slow down, so they don't hit the walls as hard and the pressure is less.

    (e) The pressure of a gas is related to the number of collisions per second with a given area of the walls as well as to the force with which the molecules collide.

    (f) Raising the temperature causes a gas to expand because raising the temperature makes the gas molecules larger.

    _____

New Terms

Write the definitions of the following terms, which were introduced in this section. If necessary, refer to the Glossary at the end of the text.

kinetic theory of gases   root mean square speed

11.9   Real gases don't obey the ideal gas law perfectly

Review

Real gases do not obey the ideal gas law perfectly. There are two reasons for this that can be traced to inaccuracies in the original kinetic theory explanation for ideal gas behavior. First, the molecules of a real gas have finite volumes (the original kinetic theory described gas particles as points without volume). Second, real molecules have weak attractions toward each other (a fact ignored by the theory).

The finite volumes of the gas molecules cause the ideal gas law to be disobeyed when gases are squeezed into very small volumes at high pressure. The attractions between the molecules cause paths of the molecules to swerve as the molecules pass near each other. As a result, real molecules travel longer distances between collisions, so they don't collide with the walls as often and produce a lower pressure. This phenomenon becomes most important when the molecules are slowed down by cooling them to low temperatures. Therefore, high P and low T are conditions under which there are the greatest deviations from ideal gas behavior.

To adapt the ideal gas law to fit real gases, van der Waals applied corrections to the real P and V. He subtracted a term, nb (where n = moles and b is a factor that depends on the size of the gas molecules), from the measured volume of the gas. Because the actual pressure is smaller that the pressure an ideal gas would exert, a correction was added to the real pressure. He added the term n^2a/V^2 (where n = moles, a is a correction factor that is proportional to the strengths of the attractive forces between the molecules) to the measured pressure of the gas. The final result is called the van der Waals equation of state.

$$\left(P + \frac{n^2a}{V^2}\right)(V - nb) = nRT$$

Be sure you know the significance of the van der Waals factors a and b.

Self-Test

49. What are the two principal reasons why real gases deviate from ideal gas behavior?

     _____

50. Why is the measured volume of a real gas larger than the volume that would be occupied by an equivalent number of moles of an ideal gas?

     _____

     _____

51. Why is the measured pressure of a real gas smaller than the pressure that would be exerted by an equivalent number of moles of an ideal gas?

     _____

     _____

52. The gas Kr has a larger van der Waals a constant than Ne. Which of the two gases has the greater attractive forces between their molecules?

     _____

53. Which gas would be expected to have the larger value of the van der Waals constant *b*, HF or HCl? Why?

New Terms

None

Solutions to Thinking It Through

1   (a)   No additional data are needed. (b) The volume must change. The combined gas law equation applies; $V_2 = 648$ mL.

2   Whenever you are stuck about what to do, go back to the basic definitions and let them lead you. In this problem, remember that a molecular mass is always the ratio of the grams of a substance to the moles of the substance—the "grams per mole." We're told that 1 L of the gas has a mass of 4.102 g, so what we have to do next is find out how many moles this mass equals while being able to occupy 1 L at 750 torr and 25.0 °C. The universal gas law, $PV = nRT$, will help us because we know R, P (750 torr), T (25.0 °C), and V (for which we can assume at least 3 significant figures since "per liter" in the density's units can be taken as an exact number). Before we calculate n, we have to convert 750 torr into atm (0.987 atm) and 25.0 °C into kelvins (298 K) in order to use $R = 0.0821$ L atm mol^{-1} K^{-1}. Using these figures, the number of moles n is found to be

$$n = \frac{PV}{RT} = \frac{0.987 \text{ atm} \times 1.00 \text{ L}}{0.0821 \text{ L atm mol}^{-1}\text{K}^{-1} \times 298 \text{ K}} = 4.03 \times 10^{-2} \text{ mol}$$

The ratio of grams to moles is

$$\frac{4.102 \text{ g}}{4.03 \times 10^{-2} \text{ mol}} = 102 \text{ g mol}^{-1}$$

The molecular mass, therefore, is 102. The formula mass for the empirical formula, SF, calculates to be 51.07. This is half the molecular mass (102/51.07), so each subscript in SF has to be multiplied by 2. The molecular formula of the compound is S_2F_2.

3   The initial state of the gas is that it is wet with a total pressure of 746 torr in a volume of 500 mL at a temperature of 20 °C. The final state is at the same temperature, so we have ultimately a combined gas law problem (where $T_1 = T_2$, only we need the initial pressure of the dry nitrogen, which Dalton's law lets us calculate). At 20 °C, the vapor pressure of water is 17.54 torr (from Table 11.2 of the text).

$$P_{\text{total}} = 746 \text{ torr} = P_{\text{nitrogen}} + 17.54 \text{ torr}$$

$$P_{\text{nitrogen}} = 728 \text{ torr}$$

Canceling T_1 and T_2 from the combined gas law equation, we have

$$P_1V_2 = P_2V_2$$

$$(728 \text{ torr})(500 \text{ mL}) = (760 \text{ torr})(V_2)$$

$$V_2 = 479 \text{ mL}$$

4   Don't let all the business about the room's dimensions throw you. With a 1:1 mole ratio of NaOH to CO_2, the stoichiometry is simple; it's getting the number of moles of CO_2 that requires work. Once we find the number of moles of CO_2 we automatically know the number of moles of NaOH. Then it's a moles to grams conversion to find the number of grams of NaOH. We're told that the concentration of CO_2 in the room is 1.00×10^{-6} mol L^{-1}. So we need the volume of the room *in liters,* because

$$\text{(vol, in L)} \times 1.00 \times 10^{-6}\,\frac{\text{mol } CO_2}{\text{L}} = \text{mol } CO_2$$

The room's volume in cubic meters is easy:

$$6.00 \text{ m} \times 6.00 \text{ m} \times 3.00 \text{ m} = 108 \text{ m}^3$$

To convert 108 m^3 to liters is just a unit-conversion problem like those in Chapter 3.

$$1 \text{ m}^3 = 1 \times 10^3 \text{ L}$$

So 108 $m^3 = 108 \times 10^3$ L. Now we can calculate the number of moles of CO_2 using our first equation above.

$$(108 \times 10^3 \text{ L}) \times 1.00 \times 10^{-6}\,\frac{\text{mol } CO_2}{\text{L}} = 1.08 \times 10^{-1} \text{ mol NaOH}$$

Because the formula mass of NaOH is 40.00,

$$1.08 \times 10^{-1} \text{ mol NaOH} \times \frac{40.00 \text{ g NaOH}}{1 \text{ mol NaOH}} = 4.32 \text{ g NaOH}$$

The CO_2 in the room air requires a minimum of 4.32 g NaOH to be removed by the chemical reaction given.

Answers to Self-Test Questions

1. Pressure, volume, temperature and mass (or moles)
2. Increase. Molecules move faster and can mix faster at the higher temperature.
3. 0.980 atm
4. 747 torr 0.983 atm
5. 745 torr
6. 99.5 kPa
7. (a) 0.954 atm, (b) 725 torr
8. Atmospheric pressure and the difference in heights of the liquid level in the manometer.
9. The difference in heights of the liquid level in the manometer.
10. 82 torr, 836 torr
11. 711 mL
12. 863 torr
13. An ideal gas
14. b
15. 242 mL
16. 955 mL
17. Pressure changes to 814 torr.

18. 506 mL
19. 755 K (482 °C)
20. 722 torr
21. 22.4 L/mol at STP
22. 12.2 L/mol
23. Equal volumes of gases have equal numbers of moles when compared under identical conditions of pressure and temperature.
24. Standard temperature and pressure, 273 K and 760 torr.
25. 0.500 mol
26. $PV = nRT$
27. $R = 0.0821$ L atm/mol K
28. 0.753 L or 753 mL
29. $P = 0.986$ atm
30. 251 °C
31. 28.0 g/mol
32. 1.69 g L^{-1}
33. H_2, CH_4, N_2, CO_2, C_4H_{10}
34. (a) 105 torr (b) 0.141
35. 640 torr (This question gets at the fundamental meaning of partial pressure.)
36. 74.9 mol % N_2, 15.3 mol % O_2, 3.7 mol % CO_2, 6.2 mol % H_2O
37. 10 L of NO and 5 L of O_2
38. 3 L of H_2 and 2 L of NH_3
39. 51.8 mol of SO_2; 3.32×10^3 g of SO_2
40. The rate of effusion of a gas is inversely proportional to the square root of the gas density.
41. d
42. C_4H_{10}, CO_2, N_2, CH_4, H_2
43. At a given T and P, rate of effusion is inversely proportional to gas density.
44. Compare your answer with the three postulates of the kinetic theory given on page 450 of the text.
45. To the average kinetic energy of the gas particles.
46. The gas laws describe observations. The kinetic theory theorizes about the structure of a gas that is consistent with the observations.
47. Compare what you write with the explanations given in Section 11.8.
48. (a), (d), (f)
49. Individual gas particles have real volumes; individual particles in a gas sample do have small attractions for each other.
50. Because the molecules of a real gas have finite volumes and therefore take up some space.
51. Molecular attractions cause real gas molecules to travel farther between collisions with the walls, thereby reducing the number of collisions per second.
52. Kr
53. HCl, because HCl is a larger molecule than HF.

Tools you have learned

Consider removing this chart from the Study Guide so you can have it handy when tackling homework problems.

Tool	*How It Works*
Combined gas law	Use this law when you need to calculate a particular value of *P*, *V*, or *T*, given other values and the quantity of gas as fixed. Use this law also in applications of Boyle's, Charles' or Gay-Lussac's law. Just drop out the variable that doesn't change.
Ideal gas law	Use this law when any three of the four variables of the physical state of a gas, *P*, *V*, *T*, or *n*, are known and you want to calculate the value of the fourth.
Dalton's law of partial pressures	Use this law to calculate the partial pressure of one gas in a mixture of gases from the total pressure and either the partial pressures of the other gases or their mole fractions. Use this law when you're given the volume and temperature of a gas collected over water and the atmospheric pressure, and you need to calculate the pressure the gas would have when dry.
Mole fractions	Use the defining equation to calculate the mole fraction or mole percent of one component of a mixture from other data about the mixture. Use mole fraction and the total pressure to calculate the partial pressure of one gas in a mixture.
Graham's law of effusion	Use this law to calculate relative effusion rates of gases. Use this law to calculate molecular masses of gases from relative rates of effusion.

Summary of Important Equations

Combined gas law equation

$$\frac{P_1 V_1}{T_1} = \frac{P_2 V_2}{T_2}$$

Ideal gas law equation

$$PV = nRT$$

Dalton's law of partial pressures

$$P_{total} = P_a + P_b + P_c + \ldots$$

Mole fraction, X_A, of component A in a mixture

$$X_A = \frac{n_A}{\text{total number moles of all components}}$$

Mole fraction of a gas, X_A, and its partial pressure, P_A,

$$P_A = X_A P_{total}$$

Graham's Law equation

$$\frac{\text{effusion rate } (A)}{\text{effusion rate } (B)} = \sqrt{\frac{d_B}{d_A}} = \sqrt{\frac{M_B}{M_A}}$$

Chapter *12*

Intermolecular Attractions and the Properties of Liquids and Solids

In Chapter 11 you learned about the physical properties of gases, and you saw that gas behavior is essentially independent of the chemical composition of the gas molecules. In this chapter we turn our attention to the other two states of matter, where attractions between molecules play a dominant role, and where physical properties are strongly affected by the chemical makeup of the particles.

Learning Objectives

Throughout your study of this chapter, keep in mind the following objectives:

1   To learn why the physical properties of liquids and solids depend so heavily on their chemical composition while the properties of gases do not.

2   To learn the nature and relative strengths of the principal kinds of intermolecular attractions. You should also learn the factors that influence the strengths of intermolecular attractions.

3   To be able to use a Lewis structure and a predicted molecular geometry to anticipate the nature of the intermolecular attractions that exist in liquid and solid states of a substance

4   To learn some general properties of liquids and solids, and how these properties are related to the closeness of packing of molecules and to intermolecular attractive forces.

5   To learn about the kinds of changes of state and to learn about the concept of dynamic equilibrium.

6   To learn about the factors that control the pressure that a vapor exerts when it is in equilibrium with a liquid or a solid.

7   To learn how boiling point is defined in scientific terms and the factors that influence the boiling point.

8   To be able to use the structures of molecules to compare physical properties such as vapor pressure and boiling point.

9   To examine the kinds of energy changes that accompany changes of state.

10   To be able to use energy changes associated with vaporization to compare the strengths of the intermolecular attractive forces in liquids.

11   To learn how we can predict the way that the composition of an equilibrium system is affected by outside influences that are able to upset the equilibrium.

12   To learn how the pressure–temperature relationships between the states of substances can be represented graphically.

12.1 Gases, liquids, and solids differ because intermolecular forces depend on the distances between molecules

Review

Intermolecular attractions are attractions that exist between molecules. They are most significant when molecules are close together and are practically insignificant when molecules are far apart. In gases, the molecules hardly feel the intermolecular attractions at all because the particles are so widely spaced. Differences in the strengths of these attractions caused by differences in chemical makeup are so small that all gases behave in nearly the same way. This is why we are able to have gas laws. But in liquids and solids, where the molecules are practically touching, the intermolecular attractions are very strong and differences in their strengths are significant and are reflected in differences in physical properties.

Self-Test

1. Under what conditions do the properties of real gases deviate *most* from the predicted properties of an ideal gas?

_____

New Term

Write the definition of the following term, which was introduced in this section. If necessary, refer to the Glossary at the end of the text.

intermolecular forces

12.2 Intermolecular attractions involve electrical charges

Review

The attractions between neighboring molecules (*intermolecular attractions*) are always much weaker than the chemical bonds (*intramolecular attractions*) within molecules. Although the strengths of bonds determine the chemical properties of substances, it is the strengths of intermolecular attractions that determine many of the physical properties. The principal kinds of intermolecular attractions are *dipole–dipole attractions, hydrogen bonds, London forces, ion-dipole, and ion–induced attractions.*

Dipole–dipole attractions occur between polar molecules and are about 1% as strong as normal covalent bonds.

Hydrogen bonding is an especially strong type of dipole-dipole attraction (about 5 times the strength of the usual dipole-dipole attraction). It occurs between molecules in which hydrogen is covalently attached to nitrogen, oxygen, or fluorine. Therefore, molecules that contain O–H and N–H bonds experience hydrogen bonding. (This is something to look for in the structure of a molecule when you wish to know if hydrogen bonding will be a factor.)

London forces (also called *dispersion forces*) result from attractions between *instantaneous dipoles* and *induced dipoles* in neighboring molecules. London forces flicker on and off and their average strength is usually less than most dipole-dipole attractions. Remember that London forces are present in *all* substances.

Large molecules and atoms have larger, more easily distorted electron clouds than small molecules and atoms. As a result, the strengths of London forces increase as the sizes of the atoms in a molecule become larger. When comparing molecules that contain atoms of about the same size, the longer the chain of atoms, the greater is the *total* strength of the London forces of attraction. Molecular shape also affects the strengths of London forces; even with the same number of atoms, small compact shapes yield weaker London forces than long chain-like shapes.

When there are many atoms in a molecule, the total attractions produced by London forces often outweigh other forces, such as dipole-dipole attractions.

Ion–dipole attractions are simply the attractions between ions and polar molecules, like the attraction of a sodium ion for the partially negative end of a permanently polarized water molecule.

Ion-induced dipole attractions result when an ion induces a dipole in a nearby molecule and thus sets up an attraction between the two.

Self-Test

2. How do the strengths of London forces, dipole–dipole attractions and hydrogen bonds compare?

    _____

3. What kinds of attractive forces (both intermolecular forces and chemical bonds) are present between the particles in each of the following substances?

    (a) helium   _____

    (b) sodium nitrate, $NaNO_3$ _____

    (c) ethyl alcohol, CH_3CH_2OH _____

    (d) sulfur dioxide   _____

    (e) butane, $CH_3CH_2CH_2CH_3$ _____

4. For each pair of substances below, choose the one in which there will be the stronger intermolecular attractions.

    (a) Cl_2 or Br_2 _____   (d) CH_4 or SiH_4 _____

    (b) HF or HCl   _____   (e) PF_3 or PCl_3 _____

    (c) C_2H_6 or C_6H_{14} _____   (f) CH_4 or CH_3Cl _____

5. Arrange the following in order of increasing (smallest to largest) strengths of intermolecular attractions.

    (A)   (B)   (C)

    $$CH_3-\underset{\underset{\displaystyle OH}{|}}{CH}-CH_3 \qquad CH_3-CH_2-CH_3 \qquad CH_3-\underset{\overset{\displaystyle O}{||}}{C}-CH_3$$

    _____

6.   In the preceding question:

(a)   Which experience London forces?   _____

(b)   Which experience hydrogen bonding?   _____

(c)   Which experience moderate dipole-dipole forces?   _____

New Terms

Write the definitions of the following terms, which were introduced in this section. If necessary, refer to the Glossary at the end of the text.

intermolecular attraction	intramolecular attraction
dipole–dipole attractions	ion–dipole attractions
hydrogen bond	ion–induced dipole attractions
induced dipole	London forces
instantaneous dipole	polarizability

12.3   Intermolecular forces and tightness of packing affect the properties of liquids and solids

Review

The physical properties of liquids and solids depend on both the tightness of packing of their particles as well as on the strengths of the intermolecular attractions. However, some properties depend more on one of these factors than the other.

Properties that depend mostly on tightness of packing

The closeness of packing is the reason that liquids and solids resist compression when pressure is applied to them, and it is also primarily responsible for the slow rates of diffusion in these states of matter.

Liquids and solids are nearly *incompressible* because there is almost no empty space into which to squeeze the molecules when pressure is applied. *Diffusion* is slow in liquids because to move from one place to another the particles must work their way around and past so many other near neighbors. In human terms, it's like attempting to cross a crowded room; you must squeeze past so many other people that movement is slow. Diffusion in solids is nearly nonexistent; the particles are not free to move about the way they can in liquids.

Properties that depend mostly on intermolecular attractions

Properties mentioned in this section that are controlled mostly by the strengths of intermolecular attractions are retention of shape and volume, surface tension, wetting of a surface by a liquid, viscosity, and evaporation and sublimation.

Shape and volume. Attractive forces hold the particles close together in liquids and solids, so their volumes don't change when transferred from one container to another. In solids, the attractive forces are even greater and the particles cannot easily move away from their equilibrium positions. This rigidity allows solids to keep their shapes when transferred from one container to another.

Surface tension. To understand *surface tension,* as well as many of the other physical properties of liquids and solids, it is necessary to understand the relationship between attractive forces and potential energy, so let's review these important concepts. First, *whenever something feels an attractive force, it has potential energy.* A large reservoir of water high in the mountains has potential energy because the water feels the earth's gravitational attraction. Allowing the water to flow to a lower altitude releases some of this potential energy, which we can use to generate electricity or for some other energy-consuming activity. If there were no attractive forces between the earth and the water, it would not flow downhill and it would not release energy that we could use. Second, *the greater the sum total of the attractive forces felt by something, the lower will be its potential energy.* Consider the water, for example. You probably know that the earth's gravitational attraction decreases with increasing distance from the earth's center. When the water flows to a lower altitude, its comes closer to the center of the earth, so it feels a *greater* gravitational attraction at the same time that its potential energy decreases.

Now let's consider the phenomenon known as *surface tension.* This property of liquids arises because molecules at the surface feel fewer attractions than the molecules within the liquid. This is because the molecules at the surface are only partially surrounded by other molecules, whereas those inside the liquid are completely surrounded. This causes the molecules within the liquid to experience greater *total* attractive forces than molecules at the surface, and as a result, the molecules in the interior of the liquid have lower potential energies than those at the surface. The *surface tension* is a quantity that's related to the energy difference between a molecule at the surface and a molecule inside the liquid. We can also say that this is the amount of energy needed to bring a molecule from the interior to the surface. When the intermolecular attractions are large, the surface tension is large. The tendency of a liquid to achieve a minimum potential energy causes the liquid to minimize its surface area and thereby minimize the number of higher-energy surface molecules.

Wetting of a surface. For a liquid to engage in the *wetting* of a surface, the attractive forces between the liquid and the surface must be about as strong as between molecules of the liquid. Substances with low surface tensions easily wet solid surfaces because the attractions in the liquid are weak. Water has a large surface tension and can wet a glass surface because the surface contains oxygen atoms to which water molecules can hydrogen bond. Water doesn't wet waxy or greasy surfaces because the water molecules are only weakly attracted to hydrocarbon molecules by London forces. The London forces are much weaker than the hydrogen bonding between water molecules in the liquid, so molecules of water tend to cling together and don't spread out over the greasy surface.

Viscosity. *Viscosity* is the internal "friction" within a liquid and makes it difficult for a liquid to flow. Viscosity increases with increasing intermolecular attractions and increasing molecular size. Viscosity decreases with temperature.

Evaporation and sublimation. *Evaporation* of a liquid or *sublimation* of a solid occurs when molecules leave their surfaces and enter the vapor space that surrounds them. Remember that it is always the molecules with very large kinetic energies that escape, so the average kinetic energy of molecules left behind is lowered, and this means that the temperature is also lowered. (In other words, evaporation leads to a cooling effect, and so is endothermic.)

Factors affecting the rate of evaporation

Two factors control the rate of evaporation. One is the temperature and the other is the strengths of the intermolecular attractions. Increasing the temperature increases the rate of evaporation. Study Figure 12.17 to be sure you understand why this is so. Substances with weak intermolecular attractive forces evaporate more rapidly than those that have strong intermolecular attractive forces. Study Figure 12.18.

Thinking It Through

1   For the following pairs of substances, choose the one that will have the greater rate of evaporation at a given temperature.

(a) $CH_3CH_2CH_3$ or $CH_3CH_2CH_2CH_2CH_3$

(b)
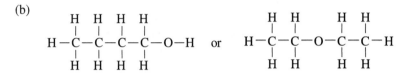

Self-Test

7.   Which of the properties discussed in this section are primarily the result of the tightness of packing of molecules in liquids and solids?

_____

8.   In ethylene glycol (antifreeze), the intermolecular attractive forces are much greater than between molecules in gasoline. Molecules that evaporate from ethylene glycol, therefore, carry with them much more kinetic energy than most of the molecules that evaporate from gasoline. Yet, if you spill these two liquids on yourself, the gasoline gives a much greater cooling effect than the ethylene glycol. Why? (Think!)

_____

_____

9.   Arrange the molecules in Question 5 in order of increasing rate of evaporation (slowest to fastest).

_____

10.   Which of the molecules below is expected to have the greater viscosity? Why?

$$HO-CH_2-CH_2-OH \qquad\qquad CH_3-CH_2-OH$$
ethylene glycol   ethanol

_____

_____

New Terms

Write the definitions of the following terms, which were introduced in this section. If necessary, refer to the Glossary at the end of the text.

compressibility	incompressibility	surface tension
wetting	surfactant	viscosity
evaporation	sublimation	

12.4   Changes of state lead to dynamic equilibria

Review

Changes of state include:

solid → liquid	liquid → vapor	vapor → solid
solid → vapor	liquid → solid	vapor → liquid

Changes of state can take place under conditions of dynamic equilibrium. Equilibrium is one of the most important concepts for you to grasp in this chapter. In a *dynamic equilibrium,* two opposing events are taking place at equal rates, and the dynamic events associated with a *phase change*, or change of state, such as those involving evaporation and *condensation,* provide simple illustrations. For example, when a liquid evaporates into an enclosed space, the concentration of molecules in the vapor rises until the rate at which molecules condense becomes equal to the rate at which they evaporate. Once this happens, the *number* of molecules in the vapor remains constant with time; in the time it takes for a hundred molecules to evaporate into the vapor, another hundred molecules condense and leave the vapor. Thus a dynamic equilibrium is a condition in which an unending sequence of balanced events leads to a constant composition within the equilibrium system. Melting and freezing also involve an equilibrium, and the *melting point* of a solid is simply the temperature at which dynamic equilibrium exists between the solid and liquid phases. Sublimation also involves an equilibrium, one between the solid and vapor phases of a substance.

Self-Test

11. In some high-altitude cities, such as Denver, Colorado, snow sometimes disappears gradually without ever melting. What happens to it?

_____

12. What is another term that has the same meaning as "change of state?"

_____

13. What changes are constantly taking place during the equilibrium involved in sublimation?

_____

14. Dynamic equilibrium was discussed earlier in the text when we dealt with a chemical process in aqueous solution. What was that process?

_____

New Terms

Write the definitions of the following terms, which were introduced in this section. If necessary, refer to the Glossary at the end of the text.

change of state

melting point

condensation

12.5   Vapor pressures of liquids and solids are controlled by temperature and intermolecular attractions.

Review

When a liquid evaporates into an enclosed space, the vapor exerts a pressure called the *vapor pressure.* When the rates of evaporation and condensation become equal, the pressure exerted by the vapor is called the *equilibrium vapor pressure of the liquid.* Usually, when we use the term *vapor pressure,* we mean equilibrium vapor pressure. Its value is independent of the size of the container, the surface area of the liquid, or the amount of liquid present, as long as at least some liquid remains at equilibrium.

A similar situation exists for solids. The vapor that's in equilibrium with a solid also exerts a pressure called the *equilibrium vapor pressure of the solid,* which is also independent of the size of the container, the surface area of the solid, or the amount of solid that remains at equilibrium.

Vapor pressures differ for different substances, and when measured at the same temperature, can be used to compare the strengths of intermolecular attractions. When the intermolecular attractions are strong, vapor pressures are low.

Vapor pressure depends on temperature, as we saw in Chapter 11 when we discussed collecting gases over water. For any given liquid, the vapor pressure increases as the temperature rises. This is because at the higher temperature, a larger fraction of the molecules have sufficient energy to escape from the liquid's surface. The higher rate of evaporation at the higher temperature requires a greater concentration of molecules in the vapor to establish equilibrium, so the vapor pressure is higher. For the same reason, the equilibrium vapor pressure of a solid also rises with increasing temperature.

Self-Test

15.   The following are the vapor pressures of some common solvents at a temperature of 20 °C. Arrange these substances in order of increasing strengths of their intermolecular attractive forces.

carbon disulfide, CS_2	294 torr
ethyl alcohol, C_2H_5OH	44 torr
methyl alcohol, CH_3OH	96 torr
acetone, $(CH_3)_2CO$	186 torr

_____

16.   On a separate sheet of paper, sketch a graph that shows how vapor pressure varies with temperature for a typical liquid.

New Terms

Write the definitions of the following terms, which were introduced in this section. If necessary, refer to the Glossary at the end of the text.

equilibrium vapor pressure of a liquid   vapor pressure

equilibrium vapor pressure of a solid

12.6   Boiling occurs when a liquid's vapor pressure equals atmospheric pressure

Review

Boiling occurs when bubbles of vapor form below the surface of a liquid. The *boiling point* is the *temperature* at which a liquid boils, and it is the temperature at which the liquid's vapor pressure becomes equal to the atmospheric pressure exerted on the liquid's surface. The temperature at which the vapor pressure equals 1 atm is the liquid's *normal boiling point*. Substances with strong intermolecular attractions generally have high boiling points.

Thinking It Through

2   Arrange the following in order of increasing normal boiling point.

$(CH_3)_2CHNH_2$, $(CH_3)_2CH_2$, $(CH_3)_2CO$

Self-Test

17.   Why does a pressure cooker cook foods more rapidly than when they are boiled in an open pot?

_____

_____

_____

18.   Arrange the substances in Question 15 in increasing order according to their expected boiling points.

_____

19.   What happens to the temperature if heat is added more rapidly to an already boiling liquid? Explain.

_____

_____

20.   Why is boiling point a useful physical property for the purposes of identifying liquids?

_____

21.   Which would be expected to have a higher boiling point, argon or krypton?   _____

22.   Which would be expected to have a higher boiling point, hydrogen sulfide or hydrogen chloride?

_____

23.   Which would be expected to have the higher vapor pressure at 20 °C, C_3H_8 or C_5H_{12}?

_____

24. Suppose a vacuum pump is attached to two separate containers, one holding ethanol and the other ethylene glycol. When the pump is turned on, the pressure above each liquid will drop. As the pressure is reduced, which liquid will boil first? Why?

$$HO-CH_2-CH_2-OH \qquad\qquad CH_3-CH_2-OH$$
$$\text{ethylene glycol} \qquad\qquad\qquad \text{ethanol}$$

_____

_____

New Terms

Write the definitions of the following terms, which were introduced in this section. If necessary, refer to the Glossary at the end of the text.

boiling point   normal boiling point

12.7   Energy changes occur during changes of state

Review

Figure 12.27 describes the shapes of heating and cooling curves, which show how the temperature of a substance changes as heat is added or removed. The flat portions of the graphs correspond to phase changes that represent potential energy changes associated with solid ↔ liquid and liquid ↔ vapor phase changes. Learn the general shapes of such graphs and what kinds of energy changes take place along the various line segments.

Every change of state is accompanied by an energy change. They are *enthalpy* changes because phase changes take place under conditions of constant pressure. Study the definitions of *molar heat of fusion, molar heat of vaporization,* and *molar heat of sublimation* given in this section. Melting of a solid, evaporation of a liquid, and sublimation are changes that absorb energy (they are endothermic), so their values of ΔH are positive. Phase changes in the opposite direction—freezing of a liquid and condensation of a vapor to either a liquid or to a solid—release energy and are exothermic. Their values of ΔH are equal in magnitude to, but opposite in sign to the heats of fusion, vaporization, and sublimation.

The size of the enthalpy change associated with a change of state depends on the strengths of the intermolecular attractions in the substance and also by how much these forces are altered during the change of state. The $\Delta H_{vaporization}$ is larger than ΔH_{fusion}, and the $\Delta H_{sublimation}$ is larger than $\Delta H_{vaporization}$.

Measuring heats of vaporization

Facets of Chemistry 12.2 discusses how heats of vaporization can be measured by observing how the vapor pressure changes with temperature. The *Clausius–Clapeyron equation* (Equation 1, and its alternative form, Equation 2) relates the heat of vaporization to the variation of vapor pressure with temperature.

Thinking It Through

3   (Facets of Chemistry 12.2) Butane, C_4H_{10}, is the liquid fuel in disposable cigarette lighters. The normal boiling point of butane is –0.5 °C and its heat of vaporization is 24.27 kJ/mol. What is the

approximate pressure of butane gas over liquid butane in a cigarette lighter at room temperature (25 °C)? Express your answer in units of atm.

Self-Test

25. If you know the values of the molar heat of fusion and the molar heat of vaporization, you can calculate the value of the molar heat of sublimation. Why can this be done, and how can it be done?

_____

26. The molar heat of vaporization of benzene (C_6H_6) is 30.8 kJ/mol and water has a molar heat of vaporization of 40.7 kJ/mol. Which of these substances has the larger intermolecular attractive forces?

_____

27. Referring to the data in Question 26, which substance, water or benzene, should have the larger molar heat of sublimation?

_____

28. How many kilojoules are required to vaporize 10.0 g of liquid benzene? (See Question 26.)

_____

29. Arrange the following in increasing order of their expected molar heats of vaporization: HCl, NH_3, CH_4, HBr, He.

_____

30. According to Figure 12.26 (page 528), what would the boiling point of water be if it were not for hydrogen bonding? Could life as we know it exist without hydrogen bonding?

_____

31. Which of the following compounds should have the higher heat of vaporization?

    (a) $CH_3CH_2CH_3$ or $CH_3CH_2CH_2CH_2CH_3$ _____

    (b)

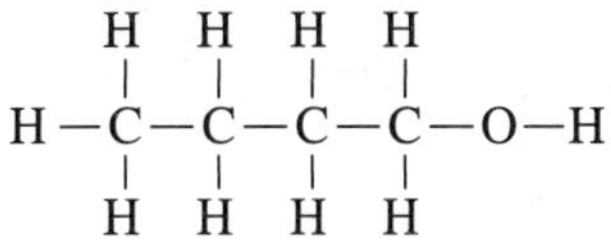

    ```
        H   H   H   H                    H   H       H   H
        |   |   |   |                    |   |       |   |
    H — C — C — C — C — O — H    or   H — C — C — O — C — C — H
        |   |   |   |                    |   |       |   |
        H   H   H   H                    H   H       H   H
    ```

    _____

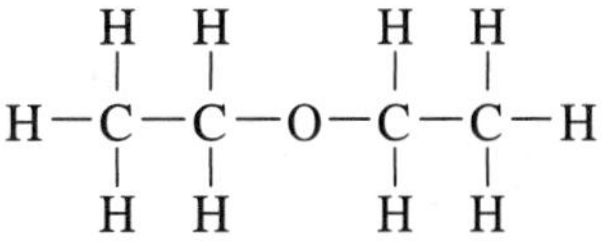

New Terms

Write the definitions of the following terms, which were introduced in this section. If necessary, refer to the Glossary at the end of the text.

cooling curve	heating curve	fusion
molar heat of fusion	molar heat of sublimation	molar heat of vaporization
supercooling		

12.8   Changes in a dynamic equilibrium can be analyzed using Le Châtelier's principle

Review

Le Châtelier's principle must be learned. We will use it frequently when studying both physical and chemical equilibria. The principle, for example, lets us generalize about physical changes as follows.

1   An increase in temperature shifts an equilibrium in the direction of an endothermic change.

2   An increase in pressure shifts an equilibrium in a direction that favors a decrease in volume.

Self-Test

32.   State Le Châtelier's principle.   _____

_____

33.   The equilibrium between a liquid and a vapor can be written

$$\text{liquid} \;\rightleftharpoons\; \text{vapor}$$

(a)   The change is endothermic in one direction and exothermic in the other. Rewrite the equation, placing heat as a reactant or product, whichever is appropriate.

_____

(b)   If the temperature of a container in which this equilibrium is established is decreased while the volume is kept the same, what happens to the position of equilibrium and the relative amounts of liquid and vapor in the container?

_____

_____

New Terms

Write the definitions of the following terms, which were introduced in this section. If necessary, refer to the Glossary at the end of the text.

Le Châtelier's principle   position of equilibrium

12.9   Phase diagrams graphically represent pressure-temperature relationships

Review

A typical *phase diagram* has three lines that serve as temperature-pressure boundaries for the three physical states of the substance—solid, liquid, and gas. Any point on one of these lines represents a temperature and pressure in which there is an equilibrium between *two* phases. The three equilibrium lines intersect at a common point called the *triple point.* Each substance has a unique triple point corresponding to the temperature and pressure at which solid, liquid, and gas are simultaneously in equilibrium with each other.

The liquid-vapor line terminates at the *critical point*, corresponding to the *critical temperature* and *critical pressure.* Below the critical temperature, a substance can be liquefied by the application of pressure. Above the critical temperature, no amount of pressure can create a distinct liquid phase. Above the critical temperature the substance is said to be a *supercritical fluid.*

Self-Test

34. According to the phase diagram for water (Figure 12.28), what phase will exist at

    (a)   –5 °C and 330 torr   _____

    (b)   15 °C and 3.00 torr   _____

    (c)   15 °C and 330 torr   _____

    (d)   0 °C and 850 torr   _____

    (e)   100 °C and 330 torr   _____

35. What changes occur if water at –15 °C and 20 torr is gradually warmed at a constant pressure?

    _____

36. What changes will be observed if water is heated from –15 °C to 100 °C at a pressure of 4.58 torr?

    _____

37. (a) At what temperature does liquid and gaseous water exist in equilibrium at 1 atm? (b) What do we call this temperature?

    (a)   _____°C   (b)   _____

New Terms

Write the definitions of the following terms, which were introduced in this section. If necessary, refer to the Glossary at the end of the text.

critical point	phase diagram
critical pressure	supercritical fluid
critical temperature	triple point

Solutions to Thinking It Through

1   The "tool" we use to answer this question is the relationship between the rate of evaporation and the strengths of intermolecular attractions: the weaker the intermolecular attractions, the faster the rate of evaporation. This then raises the question, which of these has the weaker intermolecular attractions? Three kinds of attractions were discussed: dipole-dipole, hydrogen bonding, and London forces.

In (a), we have hydrocarbon molecules, which are not particularly polar molecules. They don't have N–H or O–H bonds, so hydrogen bonding isn't important either. The principal attractions are therefore London forces. Then we ask, what factors affect the strengths of London forces. For molecules formed by the same elements, the longer the molecular chain, the stronger the attractions, so $CH_3CH_2CH_3$ has the weaker attractions and the higher vapor pressure.

In (b) we have one molecule with an O–H bond, so this molecule experiences hydrogen bonding, which is so much stronger than other intermolecular attractions we can safely assume the molecule with the O–H bond will have the stronger intermolecular attractions. Therefore, it is the second molecule, C_2H_5–O–C_2H_5, that will have the higher vapor pressure.

2   Boiling point increases with increasing strengths of the intermolecular attractive forces. The compound $(CH_3)_2CHNH_2$ has N–H bonds and so experiences hydrogen bonding, a very strong kind of intermolecular attraction. The compound $(CH_3)_2CH_2$ is a hydrocarbon, so we only expect to find London forces here. The compound $(CH_3)_2CO$ has a polar carbon-oxygen bond, so it will have dipole-dipole attractions in addition to London forces. All the molecules are about the same size, so the London forces in each liquid should be about the same strength. Therefore, the order of overall intermolecular attractions is $(CH_3)_2CH_2 < (CH_3)_2CO < (CH_3)_2CHNH_2$. The boiling points should therefore increase from $(CH_3)_2CH_2$ to $(CH_3)_2CHNH_2$.

3   To calculate the vapor pressure of the butane at room temperature, our tool is the Clausius–Clapeyron equation in Facets of Chemistry 12.2. Equation 2 on page 532 is appropriate if we have both an initial pressure and an initial temperature. By definition, the normal boiling point is the boiling point at a pressure of 1 atm, so the initial conditions are P = 1 atm and t = –0.5 °C. The final temperature is 25 °C. We take R = 8.314 J mol^{-1} K^{-1}, convert temperatures to kelvins, change ΔH_{vap} to joules, and then substitute into the Equation 2 to solve for P at the higher temperature. You should obtain a vapor pressure of 2.5 atm.

Answers to Self-Test Questions

1. High pressure and low temperature
2. Hydrogen bonding is stronger than ordinary dipole-dipole attractions, which are usually stronger than London forces.
3. (a) London forces, (b) ionic bonding, covalent bonding, London forces,
   (c) covalent bonding, hydrogen bonding, London forces, (d) dipole-dipole forces, London forces,
   (e) London forces
4. (a) Br_2, (b) HF, (c) C_6H_{14}, (d) SiH_4, (e) PF_3, (f) $CHCl_3$
5. (B) < (C) < (A)
6. (a) All of them, (b) Only (A), (c) Only (C)
7. Retention of volume, incompressibility, rate of diffusion
8. At room temperature gasoline has many molecules that have enough energy to escape the liquid and it evaporates very quickly. This removes heat very quickly.
9. (A) < (C) < (B)
10. Ethylene glycol. Because of the two –OH groups, it will have stronger intermolecular attractions than ethanol.
11. It sublimes.
12. Phase change
13. Evaporation from the solid; condensation.
14. Ionization of weak acids and weak bases.

15. $CS_2 < (CH_3)_2CO < CH_3OH < C_2H_5OH$
16. See Figure 12.23 on page 524.
17. The higher pressure causes the water to boil at a higher temperature.
18. $CS_2 < (CH_3)_2CO < CH_3OH < C_2H_5OH$
19. Temperature remains constant. Adding heat just makes the liquid boil more rapidly.
20. It is easily measured.
21. Krypton
22. HCl (more polar)
23. C_3H_8 (weaker total London forces)
24. Ethanol will boil first. It has a higher vapor pressure than ethylene glycol, so as the pressure is reduced, the external pressure will match the vapor pressure of ethanol before that of ethylene glycol.
25. Enthalpy is a state function, so $\Delta H_{sublimation} \cong \Delta H_{fusion} + \Delta H_{vaporization}$
26. Water
27. Water
28. 3.95 kJ
29. $He < CH_4 < HBr < HCl < NH_3$
30. Water would boil at about $-80\ °C$. Life as we know it couldn't exist because it would be too cold.
31. (a) $CH_3CH_2CH_2CH_2CH_3$, (b) $CH_3CH_2CH_2CH_2OH$
32. Check your answer on page 533.
33. (a) heat + liquid $\rightleftharpoons$ vapor, (b) The position of equilibrium shifts to the left. There will be more liquid and less vapor.
34. (a) solid, (b) gas, (c) liquid, (d) liquid, (e) gas
35. solid $\rightarrow$ liquid $\rightarrow$ gas
36. It would begin as a solid. At $0.01\ °C$, all three phases would appear. Above $0.01\ °C$ it would all be gas.
37. (a) $100\ °C$, (b) the boiling point.

Summary of Important Information

Intermolecular attractions and physical properties

As the strengths of intermolecular attractions *increase*, there is an *increase* in

surface tension

viscosity

molar heat of vaporization

molar heat of sublimation

normal boiling point

critical temperature

and a *decrease* in

rate of evaporation

vapor pressure

Tools you have learned

Consider removing this chart from the Study Guide so you can have it handy when tackling homework problems.

Tool	*How it Works*
Relationship between intermolecular forces and molecular structure	From molecular structure, we can determine whether a molecule is polar or not and whether it has N–H or O–H bonds. This lets us predict and compare the strengths of intermolecular attractions. You should be able to identify when dipole-dipole, London, and hydrogen bonding occurs.
Factors that affect rates of evaporation and vapor pressure	They allow us to compare the relative rates of evaporation based on temperature and the strengths of intermolecular forces in substances. They also allow us to compare the strengths of intermolecular forces based on the relative magnitudes of vapor pressures at a given temperature.
Boiling points of substances	They allow us to compare the strengths of intermolecular forces in substances based on their boiling points.
Enthalpy changes during phase changes	They allow us to the strengths of intermolecular forces in substances based on relative values of $\Delta H_{\text{vaporization}}$ and $\Delta H_{\text{sublimation}}$.
Le Châtelier's principle	Le Châtelier's principle enables us to predict the direction in which the position of equilibrium is shifted when a dynamic equilibrium is upset by a disturbance. You should be able to predict how the position of equilibrium between phases is affected by temperature and pressure changes.
Phase diagram	We use a phase diagram to identify temperatures and pressures at which equilibrium can exist between phases of a substance, and to identify conditions under which only a single phase can exist.

Chapter *13*

Structures, Properties, and Applications of Solids

Crystalline solids have structures that are highly ordered, which has made their study much easier than liquids. We begin this chapter with discussions of the nature of this internal order, how it is described in simple ways, and how the properties of crystalline materials reflect the kinds of particles that make up a solid. The applications of solid materials, however, extend beyond those that are crystalline. Our goal in this chapter is to also introduce you to a variety of such materials and their uses, ranging from semiconductors to polymers to high tech ceramics found in modern electronic devices. We conclude by introducing you to a field of cutting-edge science called nanotechnology, which deals with structures having dimensions ranging from a few to several hundred atoms in size.

Learning Objectives

Throughout your study of this chapter, keep in mind the following objectives:

1. To learn how atoms, molecules, or ions are arranged in crystalline solids and how we are able to describe their structures in simple ways.

2. To understand the concept of a lattice and how structures of crystalline solids can be described by giving the properties of a repeating unit called a unit cell.

3. To learn the properties of the three kinds of cubic unit cells.

4. To be able to count the number of atoms per unit cell given the arrangement of atoms in the cell.

5. To learn what distinguishes cubic and hexagonal closest packed structures.

6. To learn how the properties of amorphous solids differ from those of crystalline solids.

7. To see how data obtained by X-ray diffraction experiments are used to obtain structural information about crystals.

8. To learn how the physical properties of crystalline solids can be related to the kinds of particles at lattice positions and the kinds of attractive forces between the particles.

9. To learn how some substances solidify without forming a crystalline solid.

10. To learn how energy levels of atoms in a crystal merge to form energy bands. You should learn the distinction between valence and conduction bands, and how n- and p-type semiconductors differ.

11. To learn how a polymer is formed by linking many monomer units.

12. To learn how to write the formula for a polymer.

13. To learn how addition polymers are formed through free radical polymerization.

14. To learn the structures of polyethylene and polypropylene.

15. To learn how condensation polymers are formed by elimination of water or other small molecule.

16   To learn the basic structures of polyesters and nylons.

17   To learn what crosslinking means and how it affects the properties of a polymer.

18   To learn how crystallinity affects the properties of a polymer.

19   To learn the nature of the different kinds of liquid crystalline phases.

20   To learn the structural properties possessed by molecules that form liquid crystals.

21   To learn how cholesteric liquid crystals affect polarized light and how that property is used in liquid crystal displays.

22   To learn the desirable properties of ceramics and the kinds of substances that form useful ceramic materials.

23   To learn the essential features of the sol-gel process.

24   To learn what nanotechnology hopes to achieve and the tools used to explore structures at the atomic level of detail.

25   To learn the basic structures of graphite, diamond, fullerenes, and carbon nanotubes.

13.1   Crystalline solids have an ordered internal structure

Review

A crystalline solid is characterized by a highly organized, regular, repeating pattern of particles. To describe such structures, we use the concept of a lattice, which is a pattern of points that have the same repeat distances arranged at the same angles as the particles in the crystal. Study the discussion of two-dimensional lattices to be sure you understand the concept.

When describing crystals, we often use the term crystal lattice for the three-dimensional lattice that describes their structures. Especially important is the concept that a single kind of lattice can be used to characterize many different structures (just as many different wallpaper patterns can be created using the same basic two-dimensional lattice.)

In a lattice, the smallest repeating unit is called the *unit cell*. The number of kinds of lattices (or unit cells) is very small, and all substances can be described in terms of this limited set. The most symmetrical lattices (and unit cells) are cubic. There are three kinds of cubic unit cells—*simple cubic, face-centered cubic,* and *body-centered cubic.* Figure 13.6 illustrates how different substances can have the same kind of unit cell but with different cell dimensions.

Counting atoms in a unit cell

You should learn how to calculate the number of atoms in a unit cell. This requires that you realize that we have to assemble the parts actually enclosed within the cell into whole atoms, so we can count their number. Remember the following:

An atom entirely with a unit cell counts as 1 atom.

An atom at a corner contributes 1/8 of an atom to the unit cell.

An atom in a face contributes 1/2 of an atom to the unit cell.

An atom along an edge contributes 1/4 of an atom to the unit cell.

The rock salt structure, which is the structure of NaCl, is quite common among the alkali halides. You should learn how the atoms are arranged in the unit cell. Study Figures 13.8 and 13.11.

Closest-packed structures

There are two kinds of closest-packed structures—cubic closest packed (ccp) and hexagonal closest packed (hcp). Using letters A, B, and C to designate the relative orientations of layers of atoms, the ccp structure has the layer stacking arrangement A-B-C-A-B C ..., whereas the hcp structure has the layer stacking arrangement A-B-A-B-A-B .... Study Figures 13.12 through 13.14.

Amorphous solids

Unlike crystals, *amorphous solids* lack long-range order and resemble liquids in their arrangements of particles. They are often called *supercooled liquids*. Examples include glass and many plastics.

Self-Test

1.   Why do crystals have such regular surface features?

2.   What is the difference between a lattice and a structure based on the lattice?

3.   Below is an illustration of atoms arranged in a two-dimensional lattice. On the drawing, sketch the outline of a unit cell for the lattice.

4.   How many atoms are there per unit cell in the structure in the preceding problem?

5.   The compound ZnS forms a face centered cubic structure as shown in Figure 13.10. All four Zn^{2+} ions are located entirely within the unit cell. How many S^{2-} ions are there per unit cell?

6.   The unit cell for ZnS is shown in Figure 13.10. Could the compound $AlCl_3$ crystallize with this same arrangement of ions in a cubic unit cell? Explain your answer.

7.   On a separate sheet of paper, make sketches of simple cubic, face-centered cubic and body-centered cubic unit cells. Compare your drawings to Figures 13.4, 13.5, and 13.7.

8.   How do the unit cells of KCl and NaCl compare?   _____

_____

9.   Which kind of unit cell is depicted in each of the following?

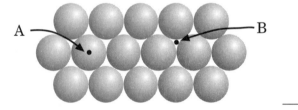

(a)   (b)   (c)

(a) _____   (b) _____   (c) _____

10.   The radius of a K⁺ ion is 133 pm and the radius of a Cl⁻ ion is 181 pm. The salt KCl has the same kind of unit cell that NaCl has (Figure 13.8). Calculate the length of an edge of this unit cell in units of picometers.

_____

11.   Below is a sketch of part of the first layer in a closest-packed structure. If the second layer is started by placing an atom over point B and the third layer is started by placing an atom over point A, is the result a ccp or hcp structure?

A ——   —— B

_____

12.   Why aren't the structures of amorphous solids described by lattices?

_____

New Terms

Write the definitions of the following terms, which were introduced in this section. If necessary, refer to the Glossary at the end of the text.

simple (primitive) cubic unit cell	lattice	crystal lattice
body-centered cubic (bcc) unit cell	unit cell	rock salt structure
face-centered cubic (fcc) unit cell	closest-packed structure	cubic closest -packed (ccp)
hexagonal closest-packed (hcp)	amorphous solid	supercooled liquid

13.2   X-ray diffraction is used to study crystal structures

Review

When a crystal is bathed in X rays, it produces a *diffraction pattern* that can be recorded on film or by other detection devices. From the angles at which the diffracted beams emerge from the crystal and the wavelength of the X rays, the distances between planes of atoms can be computed using the *Bragg equation,*

$$n\lambda = 2d\sin\theta$$

where n is an integer, λ is the wavelength of the X rays, d is the distance between planes of atoms, and θ is the angle at which the X rays are diffracted. By a complex procedure, scientists can use these interplanar distances to figure out the structure of the crystal.

     X-ray diffraction data yields information about distances between atoms within a crystal, from which the sizes of unit cells and even atomic sizes can be measured. Study Example 13.2.

Self-Test

13. Why do diffracted X-ray beams occur in only certain directions when a crystal is bathed in X rays?

    _____

    _____

14. Draw three additional sets of parallel lines through the points in Figure 13.19. Do all the sets of lines have the same spacing?

    _____

15. In the Bragg equation (Equation 13.1):

    (a)   What does the symbol θ stand for?   _____

    (b)   What does the symbol λ stand for?   _____

    (c)   What does the symbol d stand for?   _____

New Terms

Write the definitions of the following terms, which were introduced in this section. If necessary, refer to the Glossary at the end of the text.

     diffraction pattern   Bragg equation

13.3   Physical properties are related to crystal types

Review

In this section we assign crystalline solids to four crystalline types: *ionic crystals, molecular crystals, covalent crystals,* and *metallic crystals.* For each type you should learn the kinds of particles found at the lattice sites and the kinds of attractive forces that exist between them. You should be able to relate these forces to the physical properties of the solid. Study Table 13.1 and Example 13.3.

Self-Test

16. What is the electron-sea model of a metallic crystal?

     _____

17. One of the elements was once named columbium, and that name is still used by some people. It is shiny, soft, ductile, and melts at 2468 °C. It also conducts electricity. What type of crystal does columbium form?

     _____

18. Antimony pentachloride forms white crystals that do not conduct electricity and that melt at 2.8 °C to give a nonconducting liquid. What kind of solid are $SbCl_5$ crystals?

     _____

19. Strontium fluoride forms brittle, nonconducting crystals that melt at 1450 °C and give an electrically conducting liquid. The solid itself is not an electrical conductor. What type of solid is SrF_2?

     _____

New Terms

Write the definitions of the following terms, which were introduced in this section. If necessary, refer to the Glossary at the end of the text.

covalent crystal   metallic crystal

ionic crystal   molecular crystal

13.4   Band theory explains the electronic structures of solids

Review

In a solid, the energy levels of the atoms merge to form sets of *energy bands*, each of which contains enormous numbers of closely spaced energy levels. The core electrons (electrons below the valence shell) of the atoms in a crystal are localized on the atoms. The valence shell energy levels combine to form a delocalized band called the *valence band*. This energy band contains the valence electrons of the atoms in the crystal. An energy band that is either filled or empty but which extends uninterrupted throughout the crystal is called a *conduction band*. (A conduction band is a delocalized band of energy levels.)

In a metal, the valence band is either a partially filled band or overlaps an empty conduction band, so metals are good electrical conductors. In a nonmetal, the valence band is filled and there is a large energy gap (called the *band gap*) between the filled band and the lowest-energy conduction band. At normal temperatures, the conduction band is empty and there is no means for the substance to conduct electricity, so it is an insulator.

In a semiconductor, the band gap between the filled valence band and the conduction band is small. Thermal kinetic energy is able to raise some electrons into the conduction band, so semiconductors are weak electrical conductors. Raising the temperature increases the number of electrons populating the conduction band, so semiconductors become better electrical conductors as their temperatures are raised.

p-type and n-type semiconductors

If an element from Group IIIA is added as an impurity to a semiconductor from Group IVA, the valence band contains one electron vacancy for each atom of the Group IIIA element added. Electrical conduction begins when an electron from a neighboring atom fills this hole, leaving a "positive hole" elsewhere. Migration of such positive holes through the solid is the mode of electrical conduction in this p-type semiconductor.

If an element from Group VA is added as an impurity to a semiconductor from Group IVA, the valence band contains one extra electron for each atom of the Group VA element added. Electrical conduction takes place by the migration of these extra electrons through the conduction band in this n-type semiconductor.

Self-Test

20. Which kind of semiconductor is formed when

   (a)   antimony is added to silicon?   _____

   (b)   boron is added to germanium?   _____

New Terms

Write the definitions of the following terms, which were introduced in this section. If necessary, refer to the Glossary at the end of the text.

band theory	conduction band	n-type semiconductor
energy band	band gap	solar battery
valence band	p-type semiconductor	

13.5   Polymers are composed of many repeating molecular units

Review

Polymers are *macromolecules*, which are large molecules containing hundreds or even thousands of atoms. They are formed by linking together many smaller molecules called monomers. The result is long chainlike molecules composed of large numbers of identical repeating units. Not all the molecules of a given polymer are identical in size, but they do contain the same repeating unit. Study the example of polypropylene shown on pages 564 and 565.

Addition polymers are formed by just adding monomer units to form the chain. This is usually accomplished by a process involving *free radicals* (very reactive molecular fragments containing one or more unpaired electrons). Branching or linear chains are possible, depending on how the polymer is formed. Study the formation of polyethylene on page 566, as well the structure of polystyrene on page 567. Also, study Example 13.4 to be sure you know how to write the formula for an addition polymer.

Condensation polymers are formed by the elimination of a small molecule and the joining of monomer units. The monomer units being joined do not have to be identical; when they are different, the polymer formed is called a copolymer. Nylon is an example; learn the formula for the repeating unit of nylon 6,6.

Crosslinking occurs by the formation of bridges between adjacent polymer strands and increases the rigidity of the polymer. Rubber is formed by crosslinking the latex polymer called polyisoprene by heating it with sulfur. Polymer strands are linked by sulfur–sulfur bonds.

Crystallinity affects the properties of polymers, which become stronger as the polymer becomes more crystalline. Learn the differences between the different polyethylene polymers (LDPE, HDPE, UHMWPE).

Inorganic polymers

Most polymers have a carbon backbone and are called organic polymers. Polymers in which the backbone chain consists of atoms other than carbon are called inorganic polymers.

Silicone polymers are formed by reaction of water with compounds in which silicon is bonded to organic groups (e.g., —CH_3) and chlorine. Study the formation of the methylsilicone polymer described on page 574.

Self-Test

21. What is the difference between an addition polymer and a condensation polymer?

_____

22. What is the function of an initiator in the creation of an addition polymer?

_____

23. Write the structural formula for the addition polymer formed from $CH_3CH_2CH{=}CH_2$.

24. Write the structural formula for the condensation polymer formed by elimination of CH_3OH from the following two monomer units.

$$CH_3-O-\overset{\overset{\textstyle O}{\|}}{C}-CH_2-CH_2-CH_2-\overset{\overset{\textstyle O}{\|}}{C}-O-CH_3 \quad \text{and} \quad HO-CH_2-CH_2-OH$$

25. What would be the repeating unit in nylon 4,4? (Sketch its structure.)

26. What is the difference between LDPE and HDPE?

_____

27. How does crosslinking affect the physical properties of a polymer?

_____

28. Write the structural formula of the compound formed by the reaction of water with $Si(CH_3)_3Cl$.

New Terms

Write the definitions of the following terms, which were introduced in this section. If necessary, refer to the Glossary at the end of the text.

addition polymers	amide bond	backbone of a polymer
branching	condensation polymer	condensation polymerization
copolymer	crosslinks	initiator
macromolecules	monomer	nylon 6,6
polymer	polymerization	polystyrene
vulcanized rubber		

13.6   Liquid crystals have properties of both liquids and crystals

Review

Liquid crystals are substances with properties of *both* crystals and liquids. For example, they possess a relatively high degree of internal order, like a crystal, but they also have the ability to flow like a liquid. Different kinds of liquid crystals have different degrees of order within them. Here's a summary.

- *Nematic phase*: Molecules are aligned parallel to each other but are able to move in any direction.
- *Smectic phase*: Molecules are aligned parallel to each other in layers that can slide over each other.
- *Smectic C phase*: Like the smectic phase, but the axes of the molecules tilt relative to a line perpendicular to the plane of a layer.
- *Cholesteric phase*: Molecules are aligned in thin layers. The axes of the molecule rotate from one layer to the next, imparting a twist going down through a stack of layers.

The cholesteric phase is able to rotate polarized light and is the kind of liquid crystal used in LCD displays. Study Figure 13.36 to be sure you understand how polarized light differs from unpolarized light.

Self-Test

29. (a)   How is a liquid crystal like a liquid?

_____

(b)   How is a liquid crystal like a crystal?

_____

30. What property characterizes molecules that are able to form a liquid crystalline phase?

_____

31. Which molecule below would be more likely to form a liquid crystalline phase? (Note: At each intersection of bonds there is a carbon atom plus enough hydrogen atoms to give a total of four bonds to carbon.) Justify your choice.

(a)

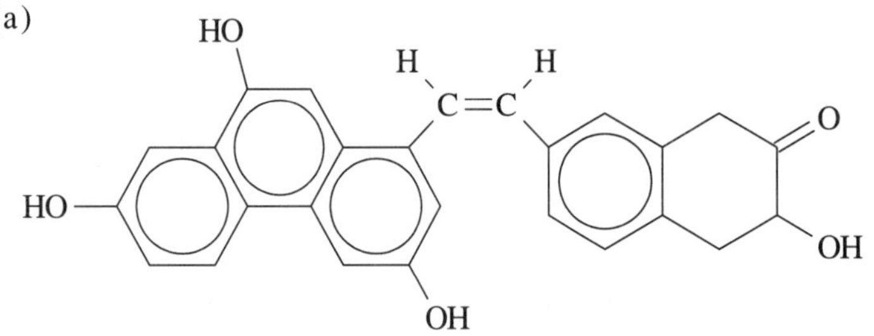

(b)

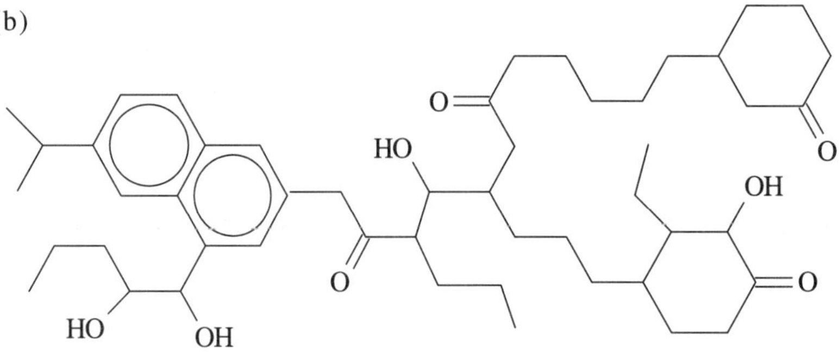

_____

New Terms

Write the definitions of the following terms, which were introduced in this section. If necessary, refer to the Glossary at the end of the text.

liquid crystal	nematic phase	smectic phase
cholesteric phase	smectic C phase	plane polarized light

13.7   Modern ceramics have applications far beyond porcelain and pottery

Review

This section deals mainly with *advanced ceramics*, which are specialized ceramic materials that have exceptional properties. All ceramics have high melting points and are very hard. Some general uses include:

- *Abrasives*: Materials such as silicon carbide (SiC) and alumina (Al_2O_3) are used in sandpaper and grinding wheels.

- *Refractories*: Because of their high melting points, ceramics are used in special bricks that line high temperature furnaces. The surface of the Space Shuttle is fitted with special light-weight ceramic tiles to protect the Shuttle from the high temperatures generated when the space vehicle reenters Earth's atmosphere. (Failure of the tiles has been blamed for the tragic loss of the Columbia, February 1, 2003.)

- *Electrical insulators*: Most ceramics have very low electrical conductivities and are used as insulators in high-voltage electrical systems.

Hardness and high melting points reflect the large lattice energies of the ceramic materials, which arise because the metal ions are small and have large positive charges.

Ceramics can be made by mixing the raw materials and then heating the mixture to the point where it almost melts. At the high temperature, the fine particles begin to join, producing a solid. The process is called *sintering*.

The sol-gel process enables the synthesis of some ceramics that can't be made by other methods. The process involves *hydrolysis* (reaction with water) of metal alkoxides. These are metal compounds formed with alcohol molecules that have lost a proton to yield an anion. Using ethanol as an example,

$$CH_3-CH_2-OH \quad \xrightarrow{\ -H^+\ } \quad CH_3-CH_2-O^-$$

ethanol   ethoxide
(an alkoxide ion)

The hydrolysis takes place stepwise, with polymerization occurring during the hydrolysis. For example, the first step for the hydrolysis of a metal ethoxide and subsequent polymerization is

C₂H₅O—M—OC₂H₅ + H₂O ⟶ C₂H₅O—M—OH

$$\begin{array}{ccc} & C_2H_5O & \\ & | & \\ C_2H_5O-\!\!\!\!&M&\!\!\!\!-OC_2H_5 \\ & | & \\ & C_2H_5O & \end{array} + H_2O \longrightarrow \begin{array}{c} C_2H_5O \\ | \\ C_2H_5O-M-OH \\ | \\ C_2H_5O \end{array}$$

$$\begin{array}{c} C_2H_5O \qquad\qquad OC_2H_5 \\ | \qquad\qquad\quad | \\ C_2H_5O-M-O-\boxed{H \qquad H-O}-M-OC_2H_5 \\ | \qquad\qquad\quad | \\ C_2H_5O \qquad\qquad OC_2H_5 \end{array} \longrightarrow \begin{array}{c} C_2H_5O \qquad\quad OC_2H_5 \\ | \qquad\qquad | \\ C_2H_5O-M-O-M-OC_2H_5 \\ | \qquad\qquad | \\ C_2H_5O \qquad\quad OC_2H_5 \end{array}$$

H₂O

Continued hydrolysis yields further polymerization. The ultimate product is a gel-like substance that can be treated in a variety of ways, as described in Figure 13.39. Study the applications described on page 581.

Silicon carbide fibers can be made from poly(methylsilane). See the structure on page 582.

Superconducting ceramics lose resistance to the flow of electricity at significantly higher temperatures than metals. The temperature at and below which superconductivity occurs is given the symbol T_c. Among the most well characterized materials are mixed oxides of yttrium, barium, and copper. An example is

$YBa_2Cu_3O_7$, which has T_c = 93 K. This means inexpensive liquid nitrogen, with a boiling point of 77K, is cold enough to bring the solid into a superconducting condition.

Superconducting solids are repelled by a magnetic field. If problems related to transforming superconducting materials into wires can be overcome, it may someday be possible to have *mag-lev trains* suspended in air and traveling at high speed with no more resistance than experienced by an airplane.

Self-Test

32. Which oxide is more likely to be used to form a ceramic, SnO or ZrO_2? Why?

_____

33. Which oxide has the larger lattice energy, Na_2O or Al_2O_3? _____

34. What is the formula for the methoxide ion? _____

35. What is a xerogel? _____

36. What is an aerogel? _____

37. What electrical property does a *piezoelectric ceramic* possess?

_____

38. Write an equation for the thermal decomposition of poly(methylsilane), $(CH_3SiH)_n$.

_____

39. Which ceramic is used to coat the surfaces of tools to make them ore wear resistant? _____

40. Why is boron nitride used in cosmetics? _____

_____

New Terms

Write the definitions of the following terms, which were introduced in this section. If necessary, refer to the Glossary at the end of the text.

ceramic	sintering	sol-gel process
alkoxide	hydrolysis	xerogel
aerogel	superconductor	

13.8   Nanotechnology deals with controlling structure at the molecular level

Review

Nanotechnology (also called molecular nanotechnology) deals with extremely small-scale structures that have dimensions on the order of tens to hundreds of atoms. The goal is ultimately to be able to build structures atom by atom so that properties can be controlled in precise ways. At present, scientists are not capable of doing this, but the tools are being developed that permit manipulation of very small structures and sometimes, individual atoms themselves.

Like the scanning tunneling microscope (STM) discussed in Chapter 1, the atomic force microscope (AFM) enables the imaging of surface features down to the atomic scale. A description of how the instrument works is given in Figure 13.46. Unlike the STM, the AFM is able to study surface features of nonconducting samples.

This section describes different forms of the element carbon. In diamond, carbon atoms are joined by single bonds only, so each carbon is at the center of a tetrahedron of other carbon atoms. The most stable form is graphite, which consists of layers of fused (joined) benzene-like rings, as illustrated on page 587 and in Figure 13.48. A more recently discovered form of carbon consists of tiny balls of carbon atoms arranged in five- and six-member rings. The simplest is C_{60}, called buckminsterfullerene (buckyball for short). Much interest by scientists working in the field of nanotechnology has recently been focused on still another form of carbon referred to as carbon nanotubes. These consist of long tubes of carbon atoms that can be viewed as rolled up sheets of graphite, capped at each end by part of a buckyball. The computer-generated image on the cover of the textbook shows the structure of a carbon nanotube looking down its axis. See also Figure 13.48*d*. Study some of the properties of carbon nanotubes described on page 588 that make them unique.

Self-Test

41. Why does the STM require that the sample be electrically conducting?

_____

42. In what major way is the structure of diamond different from the other forms of carbon?

_____

43. Why was the C_{60} molecule named buckminsterfullerene?

_____

44. How is the structure of a carbon nanotube related to the structure of graphite? How is it related to the structure of a buckyball?

_____

_____

45. Which other substances besides carbon are known to form nanotubes?

_____

46. How does the strength of a carbon nanotube compare with that of stainless steel?

_____

47. Why are nanosized rods of $BaTiO_3$ and $SrTiO_3$ of interest to makers of computer storage devices?

_____

48. The textbook describes potential uses for nanosized fibers of alumina. What is alumina? (Hint: you may have to look elsewhere in this chapter to find the answer.)

_____

New Terms

Write the definitions of the following terms, which were introduced in this section. If necessary, refer to the Glossary at the end of the text.

nanotechnology	molecular nanotechnology
atomic force microscope	graphite
fullerene	buckminsterfullerene
buckyball	carbon nanotube

Answers to Self-Test Questions

1. Because they have such ordered internal structures.
2. A lattice is a symmetrical array of points; a structure based on a lattice has chemical units associated with the lattice points.
3.
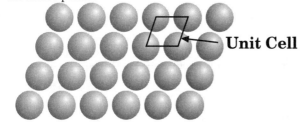

4. One.
5. Four.
6. No. $AlCl_3$ would require three times as many anion sites as cation sites, but the ratio on the ZnS structure is one to one.
7. See Figures 13.4, 13.5, and 13.7.
8. Both are fcc, but the edge length is greater for KCl than for NaCl because K^+ is larger than Na^+.
9. (a) face centered cubic, (b) body centered cubic, (c) simple cubic
10. Edge length = 2(133 pm) + 2(181 pm) = 628 pm
11. hcp
12. Because amorphous solids lack long range order.
13. Constructive and destructive interference
14. More sets can be drawn. Their spacings differ.
15. (a) The angle at which X rays are reflected from a set of planes of atoms in a crystal. (b) The wavelength of the X rays. (c) The distance between planes of atoms that are giving the reflection.
16. Positive ions of the metal at lattice sites surrounded by a sea of mobile electrons.
17. Metallic (the element is niobium, Nb)
18. Molecular
19. Ionic
20. (a) n-type, (b) p-type
21. In an addition polymer, monomer units are simply joined end to end. In a condensation polymer, a small molecule is eliminated when the monomer units become joined.
22. To start a free radical polymerization of monomer units. (See pages 565 and 566).

23.

$$\left(\begin{array}{cc} \overset{\displaystyle H}{\underset{\displaystyle CH_2}{\overset{\displaystyle |}{\underset{\displaystyle |}{C}}}} & \overset{\displaystyle H}{\underset{\displaystyle H}{\overset{\displaystyle |}{\underset{\displaystyle |}{C}}}} \\ | \\ CH_3 \end{array}\right)_n$$

24.

$$\left(O-\overset{O}{\overset{||}{C}}-CH_2-CH_2-CH_2-\overset{O}{\overset{||}{C}}-O-CH_2-CH_2\right)_n$$

25.

$$\left(\overset{O}{\overset{||}{C}}-CH_2-CH_2-\overset{O}{\overset{||}{C}}-\overset{H}{\overset{|}{N}}-CH_2-CH_2-CH_2-CH_2-\overset{H}{\overset{|}{N}}\right)_n$$

26. LDPE has shorter chains and more branching than HDPE.

27. Crosslinking makes a polymer stiffer, stronger, and raises its melting point.

28.

$$CH_3-\underset{\underset{\displaystyle CH_3}{|}}{\overset{\overset{\displaystyle CH_3}{|}}{Si}}-O-\underset{\underset{\displaystyle CH_3}{|}}{\overset{\overset{\displaystyle CH_3}{|}}{Si}}-CH_3$$

29. (a) It is able to flow like a liquid. (b) It has some internal order similar to a crystal.

30. The molecules tend to be long, thin, and have a rigid central portion.

31. Molecule (a), because molecule (b) has many branches and single bonds, which makes it more flexible and more difficult to align with other molecules of the same type.

32. ZrO_2, because the more highly charged Zr^{4+} ion would give the oxide a high lattice energy.

33. Al_2O_3 (because of the large charge on the Al^{3+} cation.)

34. CH_3O^-.

35. A very porous solid formed by removal of solvent from the product of the sol-gel process.

36. An extremely low density ceramic solid formed by rapid removal of solvent at a temperature above which the solvent cannot exist as a liquid.

37. They produce an electric potential when their shapes are deformed, as when made to bend or twist.

38. $(CH_3SiH)_n(s) \overset{\text{heat}}{\longrightarrow} (SiC)_n(s) + 2nH_2(g)$

39. Titanium nitride, TiN.

40. It consists of flat platelike crystals that slide over each other easily and thereby give the cosmetic a smooth and silky texture.

41. It relies on a flow of electricity.

42. In diamond, carbon uses only single bonds. In the other forms, a degree of multiple bonding is involved between carbon atoms.

43. The arrangement of carbon rings is similar to the structural members of the geodesic dome, a type of architectural structure designed by R. Buckminster Fuller.

44. The tube is structured like a rolled-up sheet of graphite. The tubes are capped by parts of buckyballs.

45. Molybdenum disulfide, MoS_2, and boron nitride, BN.

46. A carbon nanotube is about 100 times stronger than the same weight of stainless steel.

47. The rods can be electrically polarized in either of two directions.

48. Al_2O_3

Tools you have learned

Consider removing this chart from the Study Guide so you can have it handy when tackling homework problems.

Tool	How it Works
Bragg equation	Using the wavelength of X rays and angles at which X rays are diffracted from a crystal, the distances between planes of atoms can be calculated.
Unit cell structures for simple cubic, face-centered cubic, and body-centered cubic lattices	By knowing the arrangements of atoms in these unit cells, we can use the dimensions of the unit cell to calculate atomic radii and other properties.
Properties of crystal types	By examining certain physical properties of a solid (hardness, melting point, electrical conductivity in the solid and liquid state), we can often predict the nature of the particles that occupy lattice sites in the solid and the kinds of attractive forces between them.
Polymerization reactions forming addition polymers	From the structure of a monomer, we can predict the structure of the polymer.
Polymerization reactions forming condensation polymers	From the structures of the monomers, we can predict the structure of the polymer.
Hydrolysis reactions in the sol-gel process	We can use structural formulas to write reactions for the formation of oxygen bridges between metals and nonmetals as part of the creation of ceramic materials.

Chapter **14**

Solutions

The emphasis in this chapter is on the *physical* properties common to all solutions, not on their chemical properties. The latter depend on the chemicals themselves.

Learning Objectives

In this chapter, you should keep in mind the following goals.

1   To learn how the random motions of molecules tend to lead naturally to the spontaneous mixing of a solute with a solvent.

2   To learn how intermolecular attractive forces, such as dipole-dipole attractions, ion-dipole attractions and hydrogen bonds, can affect to the ability of a solute to dissolve in a solvent.

3   To learn how the like-dissolves-like rule helps us qualitatively predict whether a solution can be made.

4   To learn how the heat of solution for a solid dissolving in a liquid is the net effect of lattice energies and heats of solvation.

5   To learn how the relative strengths of intermolecular attractions in two liquids affect their heat of solution.

6   To learn how the effect of temperature on solubility correlates with the sign of the heat of solution.

7   To learn how to apply Henry's law in calculations.

8   To learn the general response to pressure of any equilibrium between a saturated solution of a gas in a liquid.

9   To learn how to work problems involving weight percentage concentration and molal concentration.

10 To learn how to convert a concentration in one set of units to another set.

11 To learn how solutes affect the vapor pressure of a solution and to use the Raoult's law equation.

12 To learn how a nonvolatile solute elevates the boiling point and depresses the freezing point of a solution and how the data involved in these changes can be used to calculate a molecular mass.

13 To learn how osmotic pressure data can be used to determine molecular masses.

14 To learn that the colligative properties of solutions of ionic compounds depend not just on the molality in terms of the formula units of the compound but on the number of ions into which the compound breaks up in solution.

15 To learn the chief physical characteristics of colloidal dispersions and to see how particle size is the key factor in understanding the differences among solutions, colloidal dispersions, and suspensions.

14.1   Substances mix spontaneously when there is no energy barrier to mixing

Review

When two substances are in contact, the random motions of their molecules tend to cause them to mix. In the absence of forces to prevent this, mixing occurs and the substances dissolve in one another. We can also view the mixed and unmixed states in terms of their relative probabilities once the two are in contact, as we do here for two gases. The mixed state has a vastly higher probability than the unmixed one, so the system changes spontaneously from a state of low probability to one of high probability. This illustrates a general principle of nature: *a system, left to itself, will tend toward the most probable state.*

For the mixing of gases, there are virtually no attractive forces to hinder the natural tendency toward mixing, so gases always mix freely. However, for the dissolving of liquids or solids in liquid solvents, even though probability is still a significant factor, intermolecular attractions also play a very important role. Analysis of the solution process for various solute-solvent combinations leads to the *like-dissolves-like rule*, where "like" refers to similarities in polarity. This rule tells us that *substances with similar strengths of inter-molecular attractions tend to be soluble in each other.* On the other hand, when the intermolecular attractions are much different in the solute and solvent, the substances are mutually insoluble.

For solutions of molecular substances, the intermolecular forces to be considered are London forces, dipole-dipole attractions, and hydrogen bonding. For an ionic compound dissolving in a liquid, ion-dipole attractions are important.

The ability of water to *solvate* (*hydrate*) ions largely explains why ionic compounds dissolve better in water than in any other solvent.

Self-Test

1. What are the driving forces behind the formation of the following solutions at room temperature and pressure? (Go beyond the like-dissolves-like rule and examine the solution process in detail.)

    (a) Ethyl alcohol in water.   *hydrogen bonding*

    _____

    (b) Potassium bromide in water.   *Ionization (Br⁻, K⁺)   causes   Ion-Induced dipole attractions.*

    (c) The fragrance of a flower bouquet spreading in a room's air.
        tendency towards more probable state.

2. Wax consists of nonpolar molecules. In terms of intermolecular attractions, explain why wax does not dissolve in water.
    does not form any attraction b/t H₂O b/c H₂O are attracted to eachother much more strongly.

3. Why should the hydration of the ions of NaCl help this compound to dissolve in water?
    They gain polar charges that are easily surrounded by water

4. Liquids *X* and *Y* consist of nonpolar molecules. Will *X* tend to dissolve in *Y*? ___Yes___

   Explain. ___There's nothing preventing the more probable state___

   _____

5. $NaNO_3$, an ionic solid, does not dissolve in gasoline. Explain.

   ___gasolines attractions are much weaker___

6. Which of the following is probably more soluble in benzene, C_6H_6? Why?

   (a) $CH_3-CH_2-CH_2-O-CH_3$ (b) $CH_2-CH-CH_2$

   $\qquad\qquad\qquad\qquad\qquad\qquad\qquad$ | $\quad$ | $\quad$ |

   $\qquad\qquad\qquad\qquad\qquad\qquad\qquad$ OH $\quad$ OH $\quad$ OH

   ___a) because its bonds are relatively the same strength___

   ___as those of benzene___

New Terms

Write the definitions of the following terms, which were introduced in this section. If necessary, refer to the Glossary at the end of the text.

like-dissolves-like rule $\qquad$ H_2O hydration H_2O $\qquad$ solvation $Na^+{(aq)} + Cl^-{(aq)}$

$\qquad\qquad\qquad\qquad\qquad\qquad\qquad\qquad$ H_2O $\qquad$ H_2O

$\qquad\qquad\qquad\qquad\qquad\qquad\qquad\qquad\qquad\quad$ H_2O

14.2 Enthalpy of solution comes from unbalanced intermolecular forces

Review

It costs energy to separate solute particles from each other and to separate solvent molecules from each other, because forces of attraction have to be overcome. But when new forces of attraction become established between solute and solvent molecules, energy is liberated. The *heat of solution* is the net enthalpy change, and enthalpy diagrams are helpful to show these relationships.

For salts, we have information about *lattice energies* and *hydration energies* from independent sources, and the net values of these energies are roughly of the same magnitude as measured heats of solution.

Two liquids form an *ideal solution* when the net heat of solution is zero, indicating that there is no change in forces of attraction between molecules in the separated components when compared to the forces between them in the solution.

For gases, no energy is needed to separate the molecules. Energy must be expended to open pockets in the solvent to accept the gas molecules; an exception is water, which already has empty spaces in its structure. Energy is released when the gas molecules enter the pockets. When gases dissolve in organic solvents, the process is often endothermic because more energy must be expended opening pockets that is released by gas molecules entering the solution. When gases dissolve in water, the process is usually exothermic because pockets already exist in water to accept the solute, so only the exothermic contribution occurs.

Self-Test

7.   What is another technical name for "heat of solution?"

_____Enthalpy_____

8.   How does the lattice energy affect the heat of solution of a solid solute?

_____

9.   Consider a sample of NaCl and a separate beaker of water. The potential energy of this system

_____*increases*_____when the NaCl is converted to its gaseous ions. Why?

(increases or decreases)

_____The molcules are Seperating requires outside energy_____

The separated ions now enter the water. Each becomes surrounded by water molecules, which is a

phenomenon called ___hydration_____.

The potential energy of the system _____*decreases*_____ as a result of this.

(increases or decreases)

Why? _____the bonds are reformed and some out of energy is lost to the surroundings____

10.   For a particular salt, the true lattice energy is –400 kJ/mol and the hydration energy is –500 kJ/mol. If the heat of solution for the salt were a function just of these two energies, what is its value?

_____–100 KJ/mol____

The formation of the solution is therefore exothermic or endothermic?

_____(X)_____

11.   When one liquid is dissolved in another, we can envision a 3-step process. Which step or steps is endothermic and why?

_____1, 2 because they involve Seperation of both Solute/Solvent_____

_____

Which step or steps is exothermic and why? ____intermingling of the liquids____

_____

12.   What is true about two liquids and their solution if the solution is properly described as ideal?

_____ΔH = 0° i/e forces required to break = forces to bond_____

_____

13.   Briefly explain why gases dissolve in water exothermically.

_____b/c H$_2$O already has pockets for gas and therefore_____

_____Less energy required to Spoone H$_2$O_____

New Terms

Write definitions of the following terms, which were introduced in this section. If necessary, refer to the Glossary at the end of the text.

molar enthalpy of solution

heat of solution

ideal solution   *heat Sol = 0*

hydration energy

solvation energy

LD Seperating molecules

LD read to form bonds

14.3   A substance's solubility changes with temperature.

Review

We can include the heat of solution as a product (when it is negative) or as a reactant (when it is positive) in the equilibrium expression for a saturated solution. When we do this, it is easy to apply Le Châtelier's principle to predict how the equilibrium will shift if the solution is heated or cooled. As a rule, gases become less soluble in water as the temperature increases.

Self-Test

14. Suppose a saturated solution involves the following equilibrium.

$$\text{solute}_{\text{undissolved}} + \text{heat} \rightleftharpoons \text{solute}_{\text{dissolved}}$$

(a) Is the formation of the solution endothermic or exothermic?

endo

(b) If the saturated solution is cooled, will more solute come *out* of solution or go *into* solution?

into

15. Write the equilibrium expression for a saturated solution of a gas in water, and include the heat of solution either as a product or as a reactant.

Solute un ⇌ Solute dis + heat

16. Write the equilibrium expression for a saturated solution of a typical gas in an organic solvent, and include the heat of solution as a product or reactant.

heat + Solute und ⇌ Solute dis

New Terms

None

14.4   Gases become more soluble at higher pressures

Review

The relevant equilibrium is

$$\text{gas + solvent} \rightleftharpoons \text{solution}$$

As the gas dissolves, a large decrease in the volume of this system occurs, so an increase in pressure—a volume-reducing action—will shift the equilibrium to the right.

   Henry's law (pressure–solubility law) describes the direct proportionality between gas solubility and pressure that applies for many gases.

$$C_{gas} = k_H P_{gas}$$

where C_{gas} is the concentration of gas in the solution, P_{gas} is the partial pressure of the gas over the solution, and k_H is the proportionality constant. Equation 14.2 gives a useful alternate expression of the law. Study Example 14.1.

   Some important gases, such as CO_2, SO_2, and NH_3, are helped into aqueous solution by strong inter-molecular forces or by reacting to some extent with the solvent.

$$\frac{43.0 \text{ mg/L}}{760 \text{ torr}} = \frac{X}{540 \text{ torr}}$$

Self-Test

17.   At 20 °C, the solubility of oxygen in water is 43.0 mg/L at a pressure of 760 torr. Calculate its solubility in these units at a pressure of 540 torr.

$$\underline{\qquad\qquad 30.55 \text{ mg/L} \qquad\qquad}$$

18.   Perform the following calculations and answer the question regarding gas solubility.

   (a)   Concentrated hydrochloric acid is a 12 *M* solution of HCl in water. How many liters of HCl gas (at 20 °C and 1 atm) dissolve in 1.0 L of water to give this solution?

   $$\underline{\qquad\qquad\qquad\qquad\qquad\qquad}$$

   (b)   At 20 °C and 1 atm, the solubility of N_2 in water is 0.0197 g L^{-1}. How many liters of N_2 at this temperature and pressure dissolve in 1.0 L of water?

   $$\underline{\qquad\qquad\qquad\qquad\qquad\qquad}$$

   (c)   What chemical fact accounts for the difference in the amounts of these two gases that dissolve in 1.0 L of water?

   $$\underline{\qquad\qquad\qquad\qquad\qquad\qquad\qquad\qquad\qquad}$$

   $$\underline{\qquad\qquad\qquad\qquad\qquad\qquad\qquad\qquad\qquad}$$

New Term

Write the definition of the following term, which was introduced in this section. If necessary, refer to the Glossary at the end of the text.

   Henry's law (pressure-solubility law)

14.5   Molarity changes with temperature; molality, mass percentages, and mole fraction do not

Review

The molarity of a give solution varies slightly with temperature because liquids expand when heated. As the temperature rises, the solute is distributed in a larger volume, and the molarity decreases. Although molarity is very useful for problems dealing with stoichiometry, when we are concerned with the physical properties of a solution, we need to have concentration expressed in a way that is invariant with temperature changes.

Mole fractions (Section 11.5) as well as the new concentration expressions introduced in this section, mass fractions (or mass percents) and molalities, do not change as the temperature of a solution changes. This is because they give us, directly or indirectly, information about the ratio of particles of solute to particles of solvent, and that ratio doesn't change when a solution's temperature varies.

All expressions of concentration are ratios, with quantities specified in the numerator and denominator. For example, in Chapter 5 you learned that molarity is a ratio of moles of solute to liters of solution. To be able to work with the various ways of expressing concentration, it is essential that you know precisely how they are defined. This means you must know the units of the numerator and those of the denominator. Those who do not learn these units invariably flounder when trying to work problems involving concentrations. Let's take a look at each of the concentration expressions presented in this chapter to see how we can extract from them information about the solution composition.

Mole Fraction

In Chapter 11 you learned that the units implied in a *mole fraction* are

$$\frac{\text{number of moles of one component}}{\text{total number of moles of all components}}$$

It is also useful to remember that the sum of the mole fractions of all the components equals one. Therefore, if a solution of sugar in water is described as having 0.10 mole fraction of sugar, the mole fraction of water must be 0.90, because 0.10 + 0.90 = 1.00. If we wished to use this information to construct conversion factors, the following two are available.

$$\frac{0.10 \text{ mol sugar}}{0.90 \text{ mol water}} \qquad \frac{0.90 \text{ mol water}}{0.10 \text{ mol sugar}}$$

Notice that we've used the mole fraction to derive the relative amounts of the two components.

A mole percent, of course, is simply a mole fraction multiplied by 100%.

Percentage Concentration

The most commonly used *percentage concentration* is *mass percent* (also called weight percent or weight/weight percent). In the lab, if mass percentage is not specified, you can assume that this is what is meant. Mass percentage is defined as follows

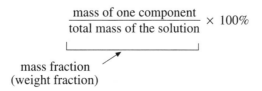

$$\frac{\text{mass of one component}}{\text{total mass of the solution}} \times 100\%$$

mass fraction
(weight fraction)

It's the mass fraction (also called weight fraction) multiplied by 100%. Mass is usually expressed in grams, but it can be expressed in whatever mass units you wish. The only condition is that the units must be the same in both numerator and denominator.

Quantities similar to mass percentage are parts per million (ppm) and parts per billion (ppb).

$$ppm = mass\ fraction \times 10^6 \qquad ppb = mass\ fraction \times 10^9$$

When we are given a mass percentage, it is not difficult to translate it to a ratio of masses. We just imaging having 100 g of solution. For example, suppose a solution were labeled 12.0 % (w/w) C_2H_5OH in water. If we had 100 g of this solution, it would contain 12.0 g of C_2H_5OH and we could express the concentration as

$$\frac{12.0\ g\ C_2H_5OH}{100\ g\ solution}$$

We can use this to make two conversion factors that could be used in calculations.

$$\frac{12.0\ g\ C_2H_5OH}{100\ g\ of\ solution} \qquad \frac{100\ g\ of\ solution}{12.0\ g\ C_2H_5OH}$$

Other percentage concentrations are mass/volume (or weight/volume, w/v) and volume/volume (v/v). Review the discussion on page 615.

Molality

The units for *molality* are moles of solute per kilogram of solvent. Thus, if an aqueous solution were labeled 0.25 *m* CH_3OH, we can represent the concentration as follows

$$0.25\ m\ CH_3OH = \frac{0.25\ mol\ CH_3OH}{1.00\ kg\ H_2O}$$

Often, it's desirable to substitute 1000 g of solvent for 1 kilogram, so we can express the molality as

$$0.25\ m\ CH_3OH = \frac{0.25\ mol\ CH_3OH}{1000\ g\ H_2O}$$

When we use molality in calculations, we can derive two conversion factors. For example,

$$\frac{0.25\ mol\ CH_3OH}{1000\ g\ of\ solvent} \qquad \frac{1000\ g\ of\ solvent}{mol\ CH_3OH}$$

Be sure you know the difference between *molarity* and *molality*. They sound a lot alike but mean quite different things (see page 616). Notice that the numerators are the same for both units, but the denominators are quite different.

Concentration Conversions

In converting among concentration units, remember that when you want to convert from a molar concentration to another unit, or from another unit to a molarity, you also have to know the density of the solution. Density provides the link between a solution's mass and its volume, because density is mass per unit volume.

If conversions such as the one studied in Examples 14.4 and 14.5 of the text give you a headache, you're probably a member of a large club. You will find that organizing facts, both given and sought, into tables provides the best help. Let's see how this would work by redoing Example 14.4. The problem asked, "What is the molality of 10.0% (w/w) aqueous NaCl?" The *Analysis* of this problem noted that it asks us to go from one set of units to another, that is, from

$$\frac{10.0 \; grams \; of \; NaCl}{100.0 \; grams \; of \; NaCl \; soln} \quad to \quad \frac{? \; mole \; of \; NaCl}{1 \; kg \; of \; water}$$

A way to organize the approach to any of these kinds of problems is to set up a small table like the one below. First, we identify the information that's needed to compute the target concentration. We'll put such data in boxes. Next, we will extract information from the numerator and denominator of the given concentration and place it into the table. Here's the empty table where we've identified the locations of both the given data and the data required to find the answer.

Substance	Mass	Moles
NaCl	_____ g	_____ mol
H$_2$O	_____ g = _____ kg	
Total	_____ g	

The answer will be found by applying the defining equation for molality to our specific situation.

$$\text{Molality} = \frac{\boxed{\text{mole NaCl}}}{\boxed{\text{kg H}_2\text{O}}}$$

Now, let's begin to fill in the table. First, we take apart the given concentration to give us two table entries, mass of NaCl and mass of H$_2$O (shaded entries in the table). Then we calculate the other table entries needed to obtain the remaining units for the desired concentration. The table can now be completed. The mass of water is the difference between the total mass and the mass of NaCl; a grams-to-moles calculation is performed to calculate the number of moles of NaCl. (Notice that we haven't bothered to fill in places in the table that are not relevant to the problem.)

Substance	Mass	Moles
NaCl	10.0 g	0.171 mol
H$_2$O	90.0 g = 0.0900 kg	
Total	100.0 g	

The molality, *m*, is now found by taking the ratio

$$\text{molality} = \frac{0.171 \text{ mol NaCl}}{0.0900 \text{ kg H}_2\text{O}} = 1.90 \; m$$

Let's next work an example that shows how the use of a table can help us carry out another conversion from one concentration expression to another.

Example 14.1   Percents by Mass Starting from Mole Fractions

So-called 100 proof alcohol is a solution of ethyl alcohol in water. It has the a mole fraction of C_2H_5OH of 0.235. Calculate the mass percent of ethyl alcohol in 100 proof ethyl alcohol. The molecular masses are: H_2O, 18.02; ethyl alcohol, C_2H_5OH, 46.07.

Analysis:
We are asked to carry out the following conversion from the given mole fraction of C_2H_5OH to its mass percent:

$$\frac{0.235 \ mol \ C_2H_5OH}{total \ of \ 1 \ mol} \quad to \quad \frac{? \ grams \ C_2H_5OH}{total \ grams \ soln} \times 100\% \ \text{(the mass percent of } C_2H_5OH)$$

Notice that we begin by assuming a total amount of solution equal to 1 mol. Then the mole fraction becomes equal to the number of moles of C_2H_5OH.

We will enter the data from the mole fraction ratio into the table (shaded areas). The data needed to calculate the answer is identified by boxes. By difference, we can calculate the moles of water in the solution. For the corresponding mass-percent concentration, we need to convert moles of C_2H_5OH to grams. For the total mass of the solution, we must also compute the mass of water that corresponds to 0.765 mol of water. Then we can add the masses of alcohol and water to find the total mass of the solution.

Solution:
The number of grams of the components are found as follows.
For C_2H_5OH:

$$0.235 \ \text{mol} \ C_2H_5OH \ \times \ \frac{46.07 \ \text{g} \ C_2H_5OH}{1 \ \text{mol} \ C_2H_5OH} \ = \ 10.8 \ \text{g} \ C_2H_5OH$$

For H_2O:

$$0.765 \ \text{mol} \ H_2O \ \times \ \frac{18.02 \ \text{g} \ H_2O}{1 \ \text{mol} \ H_2O} \ = \ 13.8 \ \text{g} \ H_2O$$

The completed table is as follows, with the given data shaded and the data we need to calculate the answer in boxes.

Substance	Mass	Moles
C_2H_5OH	10.8 g	0.235 mol
H_2O	13.8 g	0.765 mol
Total mass	24.6 g	1.000 mol

The desired mass percent is found by:

$$\text{Mass percent of } C_2H_5OH = \frac{\text{grams } C_2H_5OH}{\text{Total grams}} \times 100\%$$

$$= \frac{10.8 \text{ g}}{24.6 \text{ g}} \times 100\%$$

$$= 43.9\% \text{ C}_2\text{H}_5\text{OH}$$

Is the Answer Reasonable?

We can check to be sure we have the right units for numerator and denominator for both the given and desired concentrations. If these are correct, the rest should be okay as well. You could also do some approximate arithmetic for the moles to grams calculations and the mass percent calculation. Do this on your own, and you should be satisfied we've obtained the correct answer.

Example 14.2   Finding Mole Fractions Starting from Molality

Ammonium nitrate is sometimes a component in nitrogen fertilizers. Suppose that an aqueous solution of NH_4NO_3 has a molal concentration of 2.480 *m*. What are the mole fractions of NH_4NO_3 and water in this solution?

Analysis:

The requested conversion is:

$$\frac{2.480 \text{ mol NH}_4\text{NO}_3}{1000 \text{ g H}_2\text{O}} \quad \text{to} \quad \frac{? \text{ mol NH}_4\text{NO}_3}{\text{total moles}}$$

All we have to calculate is the *total number of moles*. We're given 2.480 mol of NH_4NO_3. So we calculate the number of moles of H_2O in 1000 g and add the two quantities together.

Solution:

We need to convert 1000 g of water to moles of water.

$$1000 \text{ g H}_2\text{O} \times \frac{18.02 \text{ g H}_2\text{O}}{1 \text{ mol H}_2\text{O}} = 55.49 \text{ mol H}_2\text{O}$$

The completed table follows.

Substance	Mass	Moles
NH_4NO_3		2.480 mol
H_2O	1000 g	55.49 mol
Total		57.97 mol

The mole fraction of water is simply

$$X_{H_2O} = \frac{55.49}{57.97} = 0.9572 \text{ or } 95.72 \text{ mol}\%$$

The mole fraction of NH_4NO_3 is

$$X_{NH_4NO_3} = \frac{2.480}{57.97} = 0.04278 \text{ or } 4.278 \text{ mol}\%$$

Is the Answer Reasonable?
Check the units for the given and desired concentration units and that we've transferred the information correctly into the table.

Thinking It Through

1   What are the molarity and the molality of a solution of NaBr with a concentration of 5.00%? The density of the solution is 1.04 g/mL.

Self-Test

19.   A solution for disinfecting clinical thermometers was made by dissolving together 160 g of ethyl alcohol and 40.0 g of water. Calculate the percentage by mass of each component.

_____

20.   A sample with a mass of 84.5 g of 12.5% sucrose solution was taken for an experiment. How many grams of sucrose were present?

_____

21.   A bottle bears the label: "0.250 molal sodium chloride." What are the two conversion factors made possible by this information?

_____

22.   If the bottle in the preceding question actually holds only 0.100 mol NaCl, what mass of *water* is present as the solvent?

_____

23.   What mass of solute in grams is needed to make a 0.100 molal glucose solution if you intended to use 250 g of water as the solvent? Use 180 as the molecular mass of glucose.

_____

24.   How would you prepare 500 g of 0.500% sugar?

_____

_____

25.   For an experiment you need 8.42 g of H_2SO_4, and it is available as 10.0% (w/w) H_2SO_4. How many grams of this solution should you take?

_____

26.   In a two-component solution of benzene (C_6H_6) in carbon tetrachloride (CCl_4), the mole fraction of benzene is 0.450. What is the mass percent of benzene in this solution?

_____

27. A solution of $CaCl_2$ has a concentration of 4.57 m. What are the mole fractions and mole percents of $CaCl_2$ and H_2O in this solution?

28. What is the molality of a 15.0% (w/w) LiCl solution?

29. What are the mole fractions and the mole percents of the components in 15.0% (w/w) LiCl (the same solution in Question 28)?

30. In a two-component solution of sugar in water, the mole fraction of sugar is 0.0150. Calculate the percentage by mass of the sugar. Use 342 as the molecular mass of sugar.

31. A solution of NaCl has a concentration of 1.15 m. What are the mole fractions and mole percents of NaCl in this solution?

32. What is the percentage concentration of a sodium carbonate solution with a concentration of 1.04 mol/L? The density of the solution is 1.105 g/mL? (Be sure to remember that both units in 1.105 g/mL refer to the *solution*. The value of the density tells us that there are 1.105 g Na_2CO_3 *solution* per milliliter of Na_2CO_3 *solution,* or 1 mL Na_2CO_3 solution per 1.105 g Na_2CO_3 solution.)

New Terms

Write the definitions of the following terms, which were introduced in this section. If necessary, refer to the Glossary at the end of the text.

mass percentage	weight percentage
mass/volume percentage	parts per million
parts per billion	molal concentration
molality	weight fraction

14.6 Substances have lower vapor pressures in solution

Review

The vapor pressure of a liquid is lowered when a nonvolatile solute (a solute that cannot evaporate) is dissolved in it because the solute particles interfere with the escape of molecules of the liquid into the vapor state but not their return to the liquid solution. For a molecular solute, a simple relationship exists between the vapor pressure of the solution, $P_{solution}$, the vapor pressure of the pure solvent, $P^\circ_{solvent}$ and the mole fraction of the *solvent* (not of the solute). This is the vapor-pressure concentration law or *Raoult's law*.

$$P_{\text{solution}} = X_{\text{solvent}} \times P^{\circ}{}_{\text{solvent}}$$

(If the solute breaks up into ions, we have to modify this, but that is a subject for Section 14.9.)

In a solution of two volatile liquids, each undergoes evaporation and thereby contributes a partial pressure toward the total vapor pressure. Each component also has its evaporation hindered by the other component, and so is subject the Raoult's law. As a result, we use a variation of the Raoult's law equation to calculate the *partial* pressure of the vapor of each component in the solution. Thus, for component A we have

$$P_A = P_A{}^{\circ} \times X_A$$

where P_A is the partial vapor pressure of component A; $P_A{}^{\circ}$ is the vapor pressure of A when it is a pure liquid; and X_A is the mole fraction of A in the solution.

The total vapor pressure exerted by a mixture of volatile substances is the sum of the individual partial vapor pressures. This is simply an application of Dalton's law of partial pressures.

Only *ideal solutions* obey Raoult's law exactly. In many real two-component solutions of volatile liquids, the plot of the total vapor pressure bows upward—a positive deviation from Raoult's law. This happens when intermolecular forces in the solution are less than they are in the separated components. Such solutions form endothermically because it costs more energy to overcome forces of intermolecular attraction in the separate components than is recovered when new forces of attraction operate in the solution.

When the plot bows downward—a negative deviation—intermolecular forces in the solution are greater than they are in the separated components. Such solutions form exothermically because less energy is needed to overcome attractive forces in the separate components than is recovered when attractive forces start to operate in the newly forming solution.

Vapor pressure lowering by a solute is only one example of a *colligative property*, a property relating to the relative *numbers* of particles of the mixture's components, not on their chemical identities.

Self-Test

33. What is the vapor pressure of a 1.00 molal sugar solution at 25 °C? Sugar, $C_{12}H_{22}O_{11}$, is a nonvolatile, non-ionizing solute and the vapor pressure of water at 25 °C is 23.8 torr.

34. At 20 °C the vapor pressure of pure toluene is 21.1 torr and of pure cyclohexane is 66.9 torr. What is the vapor pressure of a solution made of 25.0 g of cyclohexane and 25.0 g of toluene? (Cyclohexane is C_6H_{12} and toluene is C_7H_8.)

New Terms

Write the definitions of the following terms, which were introduced in this section. If necessary, refer to the Glossary at the end of the text.

colligative properties

Raoult's law (vapor pressure-concentration law)

14.7   Effects of Solutes on Freezing and Boiling Points of Solutions

Review

Two other colligative properties of solutions, *boiling point elevation* and *freezing point depression*, are chiefly exploited to obtain molecular masses.

To overcome the lowering of the vapor pressure by a nonvolatile solute, we have to raise the solution's temperature a small amount to get it to boil. To make the solution freeze, we must decrease the solution's temperature below the solvent's normal freezing point. The value of the change in temperature, Δt, is directly proportional to the solution's molal concentration. The proportionality constant is different for each solvent and is different for boiling and freezing. Depending on the actual physical change, the constant is called the *molal boiling point elevation constant, k_b* or the *molal freezing point depression constant, k_f*. If we measure Δt, and can find k_b or k_f for the solvent in a table, then we can calculate m, our symbol here for molal concentration:

$$\Delta t = k_b\, m$$

$$\Delta t = k_f\, m$$

When we know m by this procedure, we can calculate the moles of solute present in the mass of solute used for the solution. When we know both mass (in grams) and moles, it's simply a matter of taking the ratio of grams to moles to find the molar mass. To review the procedure:

Step 1   Select a solvent, and dissolve a measured mass of solute in a measured mass of the solvent.

Step 2   Measure the boiling point of this solution or its freezing point. (Unusually, precise thermometers have to be used because Δt is nearly always small.)

Step 3   Use Δt and k_b or k_f to calculate m.

Step 4   Multiply m by the number of kilograms of solvent actually used to find the moles of solute. Notice how the units work out to give "mol solute":

$$\underbrace{\frac{\text{mol solute}}{\text{kg solvent}}}_{\text{molality}} \times \text{ kg solvent} = \text{mol solute}$$

Step 5   Divide the number of grams of solute by the number of moles of solute to find the grams per mole, the molar mass. This numerically equals the molecular mass.

Restudy Example 14.9 in the text to see how these steps were used.

Thinking It Through

2   Vapona is the ingredient in flea collars and pesticide "strips." A sample of 0.347 g of vapona was melted together with 35.0 g of camphor, and this mixture was cooled to give a solid. The solid was pulverized and its melting point was found to be 35.99 °C. Using the identical thermometer, pure camphor was found to melt at 37.68 °C. For camphor,

$$k_f = 37.7 \ \frac{°C}{m} \ .$$

What is the molecular mass of vapona?

Self-Test

35.   A solution of 8.32 g of PABA, once widely used as a sunscreen agent, in 150 g of chloroform boiled at 62.62 °C. At the same pressure and with the same thermometer, pure chloroform boiled at 61.15 °C. What is the molecular mass of PABA? For chloroform,

$$k_b = 3.63 \ \frac{°C}{m}$$

_____

New Terms

Write the definitions of the following terms, which were introduced in this section. If necessary, refer to the Glossary at the end of the text.

boiling point elevation   freezing point depression constant

boiling point elevation constant   freezing point depression

14.8   Osmosis is flow of material through a semipermeable membrane due to unequal concentrations

Review

Whether the movement of a fluid through a semipermeable membrane is *osmosis* or *dialysis* depends on the kind of membrane separating the solutions or dispersions of unequal concentration. As this membrane becomes less and less permeable, we approach osmosis as the limiting case of dialysis in general.

With the help of Equation 14.6 in the text ($\Pi = MRT$) we can use data on temperature (T) and *osmotic pressure* (Π) to find moles per liter (M). From the molarity thus obtained and the volume of solution prepared for the experiment, we can obtain the number of moles of solute that were used. Then we take the ratio of grams of solute to moles of solute to obtain the molar mass, just as in the preceding section.

$$\frac{\text{grams solute}}{\text{moles solute}} = \text{molar mass}$$

The technique of using osmotic pressure to determine a molecular mass is particularly useful with high-molecular-mass substances, because even though a given mass gives a very low concentration in units of molarity, the osmotic pressure is still sizable enough for an accurate measurement.

Self-Test

36.   In some sciences, a body fluid might be described as having a "high osmotic pressure." How should we translate this—as meaning a high or a low concentration?

_____

37. When raisins or prunes are placed in warm water they soon swell up and the water becomes slightly sweet to the taste. What phenomenon is occurring, dialysis or osmosis?

_____

38. If a steak on the grill is heavily salted as it grills, barbecue experts say that the "salt draws the juices." Explain how that happens.

_____

_____

39. An aqueous solution of a protein with a concentration of 1.30 g/L at 25 °C had an osmotic pressure of 0.0160 torr. What was its molecular mass?

_____

New Terms

Write the definitions of the following terms, which were introduced in this section. If necessary, refer to the Glossary at the end of the text.

dialysis   osmotic pressure

osmosis

14.9   Ionic solutes affect colligative properties differently than nonionic solutes

Review

If you know that a solute breaks up into ions when it dissolves, then, as a rough estimate of the effect of this breakup on colligative properties, multiply the molality (or molarity) by the number of ions each formula unit gives as it dissociates. This works best for very dilute solutions.

As solutions of electrolytes become increasingly concentrated, their ions behave less and less well as fully independent particles. To compare the degrees of dissociation at different concentrations, the *van't Hoff factor* is determined. It is the ratio of the degrees of freezing point depression actually observed for the solution to the freezing point calculated by assuming that the electrolyte does not dissociate at all.

One way to become comfortable about this concept is to study Table 14.4 in the text. Look at the first row of data for NaCl. The last column tells us that if NaCl were 100% dissociated in solution, its van't Hoff factor would be 2.00. This is because two ions form from each NaCl unit. And in a very dilute solution (0.001 *m*), the van't Hoff factor is 1.97 or very close to 2.00. In this quite dilute solution, the electrolyte behaves as if it were almost 100% dissociated. In the more concentrated solution of 0.1 *m*, the van't Hoff factor is less, only 1.87. But even this is not far from 2.00, so the solute behaves as if it were still mostly dissociated. Notice in Table 14.4 that the salts that give the largest deviations are those made of ions with more than one charge. These are able to attract each other strongly, and so they are less able to act independently in a solution.

This section also describes how a *percent dissociation* can be estimated from freezing point depression data, but the method is not highly accurate.

The section also alerts you to the existence of some solutes that give weaker colligative properties than expected because of the *association* of solute particles, one with another. In particular, notice how molecules of benzoic acid associate by hydrogen bonding to give dimers (particles formed from two identical small molecules).

Self-Test

40. Calculate the freezing point of aqueous 0.250 *m* NaCl on the assumption that it is 100% dissociated.

   _____

   Now do this calculation on the assumption that it is not dissociated at all.

   _____

41. The van't Hoff factor for a salt at a concentration of 0.01 *m* is 2.70. At a concentration of 0.05 *m*, its van't Hoff factor would be greater or less?

   _____

   Explain. _____

   _____

New Terms

Write the definitions of the following terms, which were introduced in this section. If necessary, refer to the Glossary at the end of the text.

van't Hoff factor   association

14.10   Colloidal Dispersions

Review

In solutions, the solute particles are the smallest chemical species we know—atoms, ions or molecules of ordinary size (0.1 to 1 nm in average diameter). These are too small to scatter light and too easily buffeted about by solvent molecules to settle under the influence of gravity.

Particles dispersed in a *colloid* are said to be in a *colloidal dispersion.* They are next larger in size, ranging from 1 to 1000 nm (1 μm) in average diameter. Whether colloidally dispersed particles settle under the influence of gravity depends much on the presence of stabilizing agents (e.g., *emulsifying agents*). Colloidally dispersed particles are large enough to scatter light, a phenomenon called the *Tyndall effect.* They can also be observed (with the aid of a microscope) to be buffeted about, a phenomenon called *Brownian movement. Sols,* gels, smokes, *emulsions* and foams are among the common types of colloidal dispersions. If colloidal particles collect and merge to sizes larger than 1000 nm, the system becomes a *suspension* that must be continuously agitated to be kept somewhat (but imperfectly) homogeneous.

Soaps and detergents are good emulsifying agents for making an emulsion of oily material in water. The molecules (or ions) of these agents have water-attracting or *hydrophilic* portions and water-avoiding or *hydrophobic* sections. In the absence of oily or greasy material with which to form an emulsion, detergent molecules (or ions) in water spontaneously form colloidal sized particles called *micelles* in which the hydrophobic portions of the molecules mingle and arrange themselves so the hydrophilic portions are exposed to the aqueous medium.

Self-Test

42. What law of chemical combination do compounds obey that mixtures do not obey?

     law of definite proportions

43. The three chief kinds of mixtures are named

     s Solutions, Suspension, Colloids

44. Suspensions are mixtures in which one kind of intermixed particles have what dimensions?

     < 1000nm (10m)

45. Finely divided starch, a white solid, dispersed in water easily passes through a filter paper. Is this mixture a suspension?

     yes   no

46. A headlight set on high beam cutting into an early morning fog makes a strong glare. What is this effect called?

     Tyndall

47. Ions and molecules of ordinary size have mass and therefore are influenced by gravity. However, they do not settle under that influence when they are in a solution in a liquid or a gas. Why not?

     buffered,

48. What are the important differences between solutions and colloidal dispersions that are *observed* (as opposed to the explanations for the observations)?

     Tyndal / Gravity

49. What natural force is at work when a colloidal dispersion spontaneously separates?

     Gravity

50. Classify these substances as a foam, aerosol, smoke, emulsion, sol, or gel.

    (a) marshmallow   *foam*

    (b) milk   *emulsion*

    (c) grape jelly   *gel*

    (d) soap suds   *Sol*

    (e) cheese   *emulsion*

    (f) clouds   *aerosol*

    (g) cream   *emulsion*

(h)   mayonnaise   _____*emulsion*_____

(i)   the black discharge from a power plant stack   _____*smoke*_____

51.   What one fact about the particles in a colloidal dispersion gives rise to the Tyndall effect?

_____*size*_____

52.   A typical synthetic detergent consists of the following ions.

$$CH_3CH_2CH_2CH_2CH_2CH_2CH_2CH_2CH_2CH_2OSO_2^- \quad \text{and} \quad Na^+$$

phil

phob

Separately underline and label its hydrophilic and its hydrophobic parts. Explain in your own words how this substance can emulsify oily materials.

_____

_____

New Terms

Write the definitions of the following terms, which were introduced in this section. If necessary, refer to the Glossary at the end of the text.

Brownian movement	emulsifying agent	micelle
colloid	hydrophilic	sol
colloidal dispersion	hydrophobic	suspension
emulsion	Tyndall effect	

Solutions to Thinking It Through

1   The ability to solve a problem like this develops as you remember the basic definitions of the concentration expressions. Let's express them with all of their units displayed. To go from 5.00% NaBr to the molarity of this solution, we have to think of converting the quantities in the given mass percentage into the equivalent amounts of substances expressed in the units needed to calculate the molarity. In other words, we are changing

$$\frac{5.00 \text{ g NaBr}}{100 \text{ g NaBr soln}} \quad \text{to} \quad \frac{? \text{ mol NaBr}}{1 \text{ L NaBr soln}}$$

For the numerator, we have to do a grams-to-moles conversion on 5.00 g NaBr, so we'll need the formula mass of NaBr (which calculates to be 102.89).

$$5.00 \text{ g NaBr} \times \frac{1 \text{ mol NaBr}}{102.89 \text{ g NaBr}} = 4.86 \times 10^{-2} \text{ mol NaBr}$$

We also need to do a mass$_{soln}$ to volume$_{soln}$ calculation.

$$100 \text{ g NaBr soln} \Leftrightarrow ? \text{ L NaBr soln}$$

The mass-to-volume tool is provided by the solution's density, which is in units of g/mL.

$$100 \text{ g NaBr soln} \times \frac{1 \text{ mL NaBr soln}}{1.04 \text{ g NaBr soln}} = 96.2 \text{ mL NaBr soln}$$

The unit of "mL" is not what we want; we need the unit of "L," but we know that $1 \text{ mL} = 10^{-3} \text{ L}$, so we simply make the substitution:

$$96.2 \text{ mL NaBr soln} = 96.2 \times 10^{-3} \text{ L NaBr soln}$$

We now have "mol NaBr" and "L NaBr solution," so we take the ratio to calculate the molarity:

$$\text{molarity} = \frac{4.86 \times 10^{-2} \text{ mol NaBr}}{96.2 \times 10^{-3} \text{ L NaBr soln}} = 0.505 \text{ } M$$

To calculate the molality, we also begin with the basic definitions and proceed methodically to convert quantities from one set of units into the units we want. As above, to go from 5.00% NaBr to the molality of this solution, we have to think of converting the quantities in the given weight percentage:

$$\frac{5.00 \text{ g NaBr}}{100 \text{ g NaBr soln}}$$

into the following units:

$$\frac{? \text{ mole NaBr}}{1 \text{ kg water}} = \text{molality}$$

We already know the number of moles of NaBr—4.86×10^{-2} mol NaBr. So what's left is to convert 100 g NaBr solution into kilograms of just the solvent, water.

$$100 \text{ g NaBr solution} \Leftrightarrow ? \text{ kg } H_2O$$

But the 100 g of the solution contains 5.00 g of solute, so the net mass of the solvent, H_2O, is 95.0 g. Because $1 \text{ g} = 10^{-3}$ kg, we now have the mass of the solvent in kilograms: 95.0×10^{-3} kg. All we have left to do is take the ratio of the number of moles of NaBr to the number of kilograms of water.

$$\text{molality} = \frac{4.86 \times 10^{-2} \text{ mol NaBr}}{95.0 \times 10^{-3} \text{ kg water}} = 0.512 \text{ } m$$

2   Remember that the molecular mass is the ratio of grams to moles. We've been given the number of grams: 0.347 g vapona. We have to use the freezing point depression data to calculate how many moles this corresponds to. The equation that relates Δt to the molality of the solution is

$$\Delta t = k_f m.$$

The value of Δt is $(37.68 - 35.99) \text{ °C} = 1.69 \text{ °C}$. Knowing that $k_f = 37.7 \text{ °C/m}$, we can write:

$$1.69 \text{ °C} = 37.7 \frac{\text{°C}}{m} \times \text{molality of solution}$$

$$\text{molality} = \frac{1.69 \text{ °C}}{37.7 \text{ °C/}m} = 4.48 \times 10^{-2} \text{ mol vapona/kg camphor}$$

But the solution did not involve a whole kilogram of camphor, only 35.0 g or 35.0×10^{-3} kg of camphor. So the actual number of moles of vapona is found by

$$4.48 \times 10^{-2} \frac{\text{mol vapona}}{\text{kg camphor}} \times 35.0 \times 10^{-3} \text{ kg camphor} = 1.57 \times 10^{-3} \text{ mol}$$

Now we can take the ratio of grams of vapona to moles of vapona:

$$\text{molecular mass} = \frac{0.347 \text{ g vapona}}{1.57 \times 10^{-3} \text{ mol vapona}} = 221 \text{ g/mol}$$

Thus the molecular mass of vapona is 221

Answers to Self-Test Questions

1. (a) Both the tendency toward a more probable particle distribution and the intermolecular attractions between the molecules of these two polar liquids assist the formation of their solution.
    (b) The hydration of the K^+ and Br^- ions is the major factor, but the tendency toward a more probable particle distribution is also important.
    (c) When one gas intermixes in another, only the tendency toward a more probable particle distribution is operating.
2. Water molecules are attracted to each other far more strongly than they can be attracted to wax molecules, so the molecules of water remained separated in their own phase.
3. The very strong attractions between Na^+ and Cl^- ions within crystalline NaCl are sharply reduced when each kind of ion becomes surrounded and hydrated by water molecules.
4. Yes, *X* will dissolve in *Y*. There are no forces of attraction to inhibit the operation of the tendency toward a more probable particle distribution.
5. The strong forces of attraction between the Na^+ and NO_3^- ions in the crystalline salt can find no substitutes in new forces of attraction between these ions and the nonpolar molecules of gasoline.
6. (a), because it more closely resembles a hydrocarbon and will have intermolecular attractions that are similar to those in benzene.
7. Enthalpy of solution.
8. The lattice energy is the energy released when separated ions or molecules attract each other and come together to form the crystalline lattice. When a solid dissolves, this energy must be overcome to separate the particles from each other, and is an endothermic contribution to the heat of solution. Therefore, the larger the lattice energy, the more likely the heat of solution will be positive.
9. Increases. Because it must receive outside energy to separate the ions. Hydration. Decreases. Because now energy is released as forces of attraction operate.
10. $\Delta H_{\text{solution}} = -100$ kJ/mol, exothermic. (It costs 400 kJ/mol to break up the crystal and 500 kJ/mol is returned, so the net is 100 kJ/mol released by the system.)
11. The endothermic steps are the separation of the molecules of the individual liquids from each other. The exothermic step is the intermingling of these separated molecules to form the solution.
12. The intermolecular forces of attraction in each of the separated liquids are identical to those in the solution and there is no heat of solution.
13. The forces of attraction between gas particles are zero or extremely small and little energy is needed to make room for the gas molecules in liquid water. Therefore, when a gas dissolves in water only the energy of solvation (which is always exothermic) is a factor. It essentially becomes the heat of solution.
14. (a) Endothermic, (b) Come out of solution
15. $Gas_{\text{undissolved}} \rightleftharpoons Gas_{\text{dissolved}} + heat$

16. $Gas_{undissolved} + heat \rightleftharpoons Gas_{dissolved}$

17. 30.6 mg/L

18. (a) 289 L HCl gas. (b) 0.0169 L N_2 gas. (c) HCl reacts almost completely with water to give H_3O^+ and Cl^-.

19. 80% ethyl alcohol and 20% water

20. 10.4 g

21. $\dfrac{0.250 \text{ mol NaCl}}{1000 \text{ g H}_2\text{O}}$ or $\dfrac{1000 \text{ g H}_2\text{O}}{0.250 \text{ mol NaCl}}$ (One could replace "1000 g H_2O" by 1 kg H_2O.)

22. 400 g of H_2O

23. 4.50 g of glucose

24. Dissolve 2.50 g of sugar in water to make the final mass of the solution equal 500 g by adding water.

25. Weigh out 84.2 g of 10% H_2SO_4 solution.

26. 29.6 % (w/w) C_6H_6

27. For $CaCl_2$, mole fraction = 0.0761; mole percent = 7.61%
    For H_2O, mole fraction = 0.924; mole percent = 92.4%

28. 4.23 *m* LiCl

29. $X_{LiCl} = 0.0710$ or 7.10 mol %, $X_{water} = 0.929$ or 92.9 mol %

30. 22.4% sugar

31. $X_{NaCl} = 0.0203$ or 2.03 mol %

32. 9.98% Na_2CO_3

33. $P_{soln} = 23.4$ torr

34. $P_{total} = 45.1$ torr

35. 137 g mol^{-1}

36. High concentration

37. Dialysis (Both sugar and water pass across the membrane, not just water, so it's dialysis, not osmosis.)

38. The salt forms a very concentrated solution on the surface of the meat and this draws water by dialysis from the less concentrated solution inside the meat and the meat cells.

39. 1.51×10^6 g mol^{-1}

40. −0.93 °C. −0.47 °C

41. Less. At the higher concentration, the ions would interact with each other more and so the solution would behave as if the solute were even less dissociated.

42. law of definite proportions

43. solutions, colloids, suspensions

44. larger than 1000 nm

45. no

46. Tyndall effect

47. They are buffeted too much by the solvent molecules.

48. Colloids show the Tyndall effect and ultimately settle.

49. force of gravity

50. (a) foam, (b) emulsion, (c) gel, (d) sol, (e) emulsion, (f) aerosol, (g) emulsion, (h) emulsion, (i) smoke.

51. their size

52.

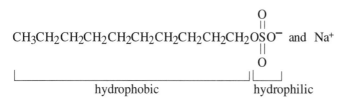

hydrophobic   hydrophilic

The hydrophobic tails of the anions embed themselves in the oily material, leaving the hydrophilic (ionic) heads exposed to the water. As the pincushioned oily matter breaks up into droplets, the droplets all bear the same kind of charge. Since they repel each other, the droplets are emulsified.

Tools you have learned

Consider removing this chart from the Study Guide so you can have it handy when tackling homework problems.

Tool	*How it works*
"Like dissolves like" rule	This rule enables us to use chemical composition and structure to predict whether two substances can form a solution.
Henry's law	We use this law when we wish to calculate the solubility of a gas at a given pressure from its solubility at another pressure.
mass fraction; mass percent	These are tools we use to calculate the mass of a solution that delivers a given mass of solute.
Molal concentration	Molality provides a temperature-independent concentration expression for use with colligative properties.
Raoult's law	This law enables us to calculate the effect of a solute on the vapor pressure of a solution. In its two forms, it applies to a single nonvolatile solute and a volatile solvent, and to mixtures of volatile liquids.
Equations for freezing point depression and boiling point elevation	These allow us to estimate freezing and boiling points or molecular masses. We can also use them to estimate percent dissociation of a weak electrolyte.
Equation for osmotic pressure	This equation permits us to estimate the osmotic pressure of a solution or to calculate molecular masses from osmotic pressure and concentration data.

Summary of Important Equations

Henry's Law

$$\frac{C_1}{P_1} = \frac{C_2}{P_2}$$

mass fraction or mass percent

$$\text{mass fraction} = \frac{\text{mass of component}}{\text{mass of solution}}$$

$$\text{mass percent} = \text{weight fraction} \times 100\%$$

Molality

$$\text{molality} = m = \frac{\text{mol of solute}}{\text{kg of solvent}}$$

Raoult's law

$$P_{\text{solution}} = X_{\text{solvent}} P^{\circ}_{\text{solvent}}$$

Freezing point depression

$$\Delta t = k_f m$$

Boiling point elevation

$$\Delta t = k_b m$$

Osmotic pressure

$$\Pi V = nRT, \text{ or}$$

$$\Pi = MRT$$

van't Hoff factor

$$i = \frac{(\Delta t)_{\text{measured}}}{(\Delta t)_{\text{calcd as nonelectrolyte}}}$$

Chapter 15

Kinetics: The Study of Rates of Reactions

In this chapter we examine the factors that affect reaction rates and some ways of understanding how the factors work.

Learning Objectives

As you study this chapter, keep in mind the following objectives.

1   To learn how the speeds of reactions vary over wide ranges and to learn about the information we gain by the study of speeds of reactions.

2   To learn about the kinds of things that influence how fast a reaction occurs.

3   To learn how rates of reaction are expressed and how they are measured.

4   To learn how the rate of reaction is related to the concentrations of the reactants.

5   To learn how to calculate the concentration of a reactant at any time after the start of a reaction for first- and second-order reactions.

6   To learn how the speed of a reaction is related to the time it takes for half of a reactant to disappear.

7   To learn why the rate of a reaction increases with increasing temperature.

8   To learn how to deal quantitatively with the effect of temperature on the rate of reaction.

9   To learn how the rate law for a reaction can give clues to the sequence of chemical steps that ultimately give the products in a reaction.

10   To learn how substances called catalysts affect the rate of a reaction, and to learn how homogeneous and heterogeneous catalysts work.

15.1   The rate of a reaction is the change in reactant or product concentrations with time

Review

The term *kinetics* implies action or motion. For a chemical reaction, this motion is the speed at which the reaction takes place, the *rate of reaction,* and by that we mean the speed at which the reactants are consumed and the products formed.

The speeds of reactions range from very rapid to extremely slow. One of the benefits of studying reaction rates and the factors that control them is the insights we gain into the *mechanisms* of reactions—the individual chemical steps that produce the net overall change described by a balanced equation.

Self-Test

1. Why is it unlikely that the combustion of octane, C_8H_{18}, a component of gasoline, occurs by a single step as given by the equation

$$2C_8H_{18}(g) + 25O_2(g) \rightarrow 16CO_2(g) + 18H_2O(g)$$

_____

2. Why would chemical manufacturers be interested in studying the factors that affect the rate of a reaction?

_____

New Terms

Write the definitions of the following terms, which were introduced in this section. If necessary, refer to the Glossary at the end of the text.

kinetics   rate of reaction

mechanism of a reaction

15.2   Five factors affect reaction rates

Review

There are five factors that control the rate of a reaction.

1   *The chemical nature of the reactants*: Some substances, because of their chemical bonds, just naturally react faster than others under the same conditions.

2   *Ability of reactants to meet*: Many reactions are classified as *homogeneous reactions* because they are carried out in solution where the reactants can mingle on a molecular level. For *heterogeneous reactions,* at least two phases are present, and particle size is the controlling factor. For a given mass of reactant, the smaller the particle size, the larger the area of contact with the other reactants in other phases, and the faster the reaction is able to proceed.

3   *Concentrations of the reactants*: The rates of most reactions, both homogeneous and heterogeneous, increase with increasing reactant concentrations.

4   *Temperature*: With very few exceptions, reactions proceed faster as their temperature is raised.

5   *Catalysts*: The rates of many reactions are increased by the presence of substances called *catalysts*. A catalyst is a substance that is intimately involved in the reaction, but which is not used up during the reaction.

Thinking It Through

For the following, describe the information needed to answer the question.

1   If you wish to greatly increase the *rate* at which heat can be obtained from the combustion of coal, what might you do? Answer the question in terms of the factors that affect the rate of a reaction.

Self-Test

3. Elemental potassium reacts more rapidly with moisture than does sodium under the same conditions. Which of the factors discussed in this section is responsible for this?

   nature of the reactants

4. Insects move more slowly in autumn than in the summer. Why?

5. What is one reason why coal miners are concerned about open flames during the mining operation?

6. Fire fighters are taught about the "fire triangle."

   Eliminate any one of the three corners, and the fire is extinguished. Which factors discussed in this section affect the fire triangle?

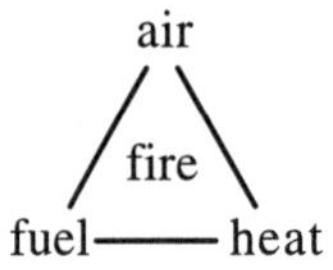

7. Why is pure oxygen more dangerous to work with than air?

New Terms

Write the definitions of the following terms, which were introduced in this section. If necessary, refer to the Glossary at the end of the text.

heterogeneous reaction   homogeneous reaction

catalyst

15.3   Rates of reaction are measured by monitoring change in concentration over time

Review

A *rate* is always a ratio in which a unit of time appears in the denominator. A reaction rate has units of molar concentration in the numerator, so the units of reaction rate are usually mol L^{-1} s^{-1} (mole per liter per second). The rate of a reaction generally changes (decreases) with time as the reactants are used up. The *instantaneous rate* can be measured at any particular instant by determining the slope of the concentration versus time curve as illustrated in Example 15.2.

The rate is usually measured by monitoring the substance whose concentration is most easily determined. Once we know the rate at which one substance is changing, we can calculate the rate for any other substance (study Example 15.1). This is because the relative rates of formation of the products and rates of disappearance of the reactants are related by the coefficients of the balanced equation. (Review Example 15.1 to see how such calculations are made.) Notice that reaction rates are always given as positive quantities

Self-Test

8.   In Figure 15.5 in the text, what is the rate at which HI is reacting at t = 200 seconds?

9.   What would be the rate at which H_2 is forming at t = 200 s? (This refers to Figure 15.5.)

10.   In the combustion of octane

$$2C_8H_{18}(g) + 25O_2(g) \rightarrow 16CO_2(g) + 18H_2O(g)$$

what would be the rate of formation of CO_2 if the concentration of octane was changing at a rate of 0.25 mol L^{-1} s^{-1}?

11.   Referring to the preceding question, if the rate of formation of H_2O is 0.900 mol L^{-1} s^{-1}, what is the rate of reaction of C_8H_{18}?

At what rate will O_2 be reacting under these conditions?

12.   If octane is burning at a constant rate of 0.25 mol s^{-1}, how many moles of it will burn in 15 minutes?

New Terms

Write the definition of the following term, which was introduced in this section. If necessary, refer to the Glossary at the end of the text.

   rate

15.4   Rate laws give reaction rate as a function of reactant concentrations

Review

Concentration and rate are related by the *rate law* for a reaction. For example, for the reaction

$$xA + yB \rightarrow \text{products}$$

in which x and y are coefficients of reactants A and B, respectively, the rate law will be of the form

$$\text{rate} = k[A]^n[B]^m$$

Remember that square brackets, [ ], around a chemical formula stands for the molar concentration (units, mol L^{-1}) of the substance. The proportionality constant, k, is the *rate constant*. The exponents give the *order of the reaction* with respect to each reactant, and their sum is the *overall order of the reaction*. Once you have the rate law for a reaction, it can be used to calculate the rate for any given set of reactant concentrations for the temperature at which k was determined. Keep in mind that k varies with temperature, and therefore, so does the reaction rate.

It is very important to remember that the exponents (n and m in this example) are not necessarily equal to the coefficients x and y, and can only be known for sure if they are determined from experimentally measured data. These data have to show how the rate changes when the concentrations change. In analyzing data like those in Table 15.2 of the text, observe how the rate changes when the concentration of one of the reactants changes while the concentrations of the other reactants are held constant.

Study Table 15.3 in the text and note how it applies to Examples 15.4 to 15.6. After working Practice Exercises 6 and 7, try the following Self-Test.

Self-Test

13. What is the order of the reaction for the dimerization of isoprene in Example 15.5 in the text?

     _____

14. In Practice Exercise 7, what is the order of the reaction with respect to A and B, and what is the overall order of the reaction?

     _____

15. What is the value of the rate constant for the dimerization of isoprene in Example 15.5?

     _____

16. When the concentration of a particular reactant was increased by a factor of 10, the rate of the reaction was increased by a factor of 1000. What is the order of the reaction with respect to that reactant?

     3
     _____

17. For a reaction, $2A + D \rightarrow$ products, the following data were obtained:

Initial Concentration (M)		Initial Rate
A	B	(mol L^{-1} s^{-1})
0.20	0.10	1.0×10^{-2}
0.40	0.10	2.0×10^{-2}
0.20	0.40	1.6×10^{-1}

(a)   What is the rate law for the reaction? _____

(b)   What is the value of the rate constant? (With correct units, too) _____

New Terms

Write the definitions of the following terms, which were introduced in this section. If necessary, refer to the Glossary at the end of the text.

order of a reaction   rate law

overall order of a reaction   rate constant

15.5   Integrated rate laws give concentration as a function of time

Review

For a first-order reaction, the rate constant can be obtained from a graph of the *natural logarithm* of the reactant concentration, ln $[A]$, versus time. The slope of the line equals the rate constant. Alternatively, Equations 15.3 and 15.3a relate concentration and time to the rate constant for a first order reaction.

Working with natural logarithms

To use Equations 15.3 and 15.3a, you have to work with natural logarithms and their antilogarithms. These are easily handled on your calculator. To obtain the natural logarithm of a number, enter the number and then press the key labeled "ln x". (On some calculators, the key may have a different label; check your instruction manual.) The value that appears is the natural logarithm of the number. For example, the natural logarithm of 25.6 is 3.243

$$\ln 25.6 = 3.243$$

Use this to check to see that you're using your calculator correctly. Notice that *the number of digits following the decimal point equals the number of significant figures in the number whose logarithm you're determining.* That's the rule for logarithms and significant figures.

To take the antilogarithm, you use the key labeled e^x on your calculator. (On some calculators, you press an "inverse" key followed by the "ln x" key.) For example, the anti-natural logarithm of 41.69 is 1.3×10^{18}.

$$\text{anti-natural logarithm (41.69)} = e^{41.69} = 1.3 \times 10^{18}$$

Notice that when we take the anti-natural logarithm, the result has only two significant figures — the number of digits after the decimal in 41.69.

Applying the integrated rate laws

Finding the time required for the concentration to drop to some particular value is simple, as shown in Example 15.7 in the text. Just substitute the initial and final concentrations into the concentration ratio, take the logarithm of the ratio, substitute the value of the rate constant, and then solve for t. Notice that the units of t are of the same kind as the units of the rate constant; if k has units of s^{-1}, then t will be in seconds; if k has units of hr^{-1}, then t will be in hours.

Finding the concentration after a specified time is illustrated in Example 15.8. Notice that we calculate the logarithm of the concentration ratio from k and t. Taking the antilogarithm gives the numerical value for the ratio. Then we substitute the known concentration and solve for the concentration we want to find.

For second-order reactions, a graph of the reciprocal of the concentration versus time gives a straight line. The slope of this line equals k. Calculations for second order reactions are done with Equation 15.4. This equation doesn't involve logarithms, so it is easier to manipulate algebraically. Study Example 15.9 to learn how to use it.

Half-lives

The *half-life* of a reaction, $t_{1/2}$, is the time required for half of a given reactant to disappear. For a first-order reaction, $t_{1/2}$ is independent of the initial reactant concentration. The value of $t_{1/2}$ is inversely proportional to the initial reactant concentration for a second-order reaction.

You should be able to use Equations 15.5 and 15.6, and you should be able to use the way $t_{1/2}$ varies with initial concentration to determine whether a reaction is first or second order.

Thinking It Through

For the following, identify the information needed to solve the problem and show (or explain) what must be done with it.

2 The decomposition of SO_2Cl_2 is a first order reaction. In an experiment, it was found that the SO_2Cl_2 concentration was 0.022 M after 300 min. The value of k at the temperature at which the experiment was performed is 2.2×10^{-5} s^{-1}. How many *moles* of the reactant were originally present in the apparatus, which has a volume of 350 mL?

Self-Test

18. A certain reactant disappears by a first-order reaction that has a rate constant $k = 3.5 \times 10^{-3}$ s^{-1}. If the initial concentration of the reactant is 0.500 mol/L, how long will it take for the concentration to drop to 0.200 mol/L?

2.6×10^{2}
_____

19. For the reactant described in the preceding question, what will the concentration of the reactant be after 20.0 min if its initial concentration is 2.50 mol L^{-1}?

_____

20. A certain reaction follows the stoichiometry

$$2D \rightarrow \text{products}$$

It has the rate law Rate = $k[D]^2$. The rate constant for the reaction equals 5.0×10^{-3} L mol^{-1} s^{-1}.

(a) How many seconds will it take for the concentration of D to drop from 1.0 mol/L to 0.60 mol/L?

_____

(b) If the initial concentration of D is 1.0 mol/L, what will its concentration be after 30 min?

_____

21. In a certain reaction, the half-life of a particular reactant is 30 minutes. If the initial concentration of the reactant is 2.0 mol/L, and, if the reaction is first order,

    (a) What will be the concentration of the reactant after 2 hours?

    _____

    (b) How long will it take for the concentration to be reduced to 0.0303 mol/L?

    _____

22. A certain first-order reaction has $t_{1/2}$ = 30 minutes. What is the rate constant for the reaction?

    _____

23. The decomposition of SO_2Cl_2 has $k = 2.2 \times 10^{-5}$ s^{-1} (Example 15.4). What is $t_{1/2}$ for this reaction in seconds and in minutes?

    5.2×10^2 min

    _____

24. At a certain temperature the half-life for the decomposition of N_2O_5 was 350 seconds when the N_2O_5 concentration was 0.200 mol/L. When the concentration was 0.400 mol/L, the half-life was 5.83 minutes. Is this decomposition reaction first or second order?

    _____

New Terms

Write the definition of the following term, which was introduced in this section. If necessary, refer to the Glossary at the end of the text.

half-life ($t_{1/2}$)

15.6 Reaction rate theories explain experimental rate laws in terms of molecular collisions

Review

Collision theory postulates that the rate of a reaction is proportional to the number of effective collisions per second between the reactant molecules or ions. The number of *effective* collisions per second is less than the total number of collisions per second for two principal reasons:

1 For some reactions, it is important that the reactant molecules be in the correct *orientation* when they collide.

2 A minimum kinetic energy, called the *activation energy* (E_a), must be possessed by the reactant molecules in a collision to overcome the repulsions between their electron clouds and thereby permit the electronic rearrangements necessary for the formation of new product molecules.

The activation energy requirement explains why the rate of a reaction increases with increasing temperature. Be sure to study Figure 15.10 and the discussion about it in the text.

In *transition state theory*, we postulate the formation of a high-energy complex formed from the reactant particles by means of a collision, a complex called the *activated complex,* which can collapse into particles of the products as bonds break and form. We follow the energy of the reactants as they are transformed to the products by means of a graph on which the horizontal axis, the *reaction coordinate*, indicates the progress of the reaction and the vertical axis measures the changes in potential energy as the collision occurs. The potential energy at the *transition state* corresponds to a maximum in the curve, and the potential energy change needed to reach the maximum is the *activation energy, E_a*. Study Figures 15.11 to 15.15. On any progress-of-reaction diagram, you should be able to identify the activation energy for both the forward and reverse reactions, the potential energies of the reactants and products, and the *heat of reaction*. You should also be able to locate the transition state on the diagram.

Self-Test

25. Without peeking at the text, sketch and label the potential energy diagram for

     (a)   an exothermic reaction.

     (b)   an endothermic reaction.

26. Where is the transition state located on the energy diagram for a reaction?

     _____

27. One step in the reaction $NO_2(g) + CO(g) \rightarrow NO(g) + CO_2(g)$ is believed to involve the collision of two NO_2 molecules to give NO and NO_3 molecules

$$NO_2 + NO_2 \rightarrow NO + NO_3$$

     What might be a reasonable structure for the activated complex in this collision?

28. On the basis of the transition state theory, explain why the temperature of a collection of molecules rises as an exothermic reaction occurs within it.

     _____

     _____

29. Why do reactions having a low activation energy usually occur faster than ones having a high activation energy (assuming no special requirements for molecular orientations)?

     _____

New Terms

Write the definitions of the following terms, which were introduced in this section. If necessary, refer to the Glossary at the end of the text.

activated complex	reaction coordinate
activation energy	transition state
collision theory	transition state theory

15.7 Activation energies are measured by fitting experimental data to the Arrhenius equation

Review

The activation energy is related to the rate constant by the *Arrhenius equation*,

$$k = A\, e^{-E_a/RT}$$

where A, a proportionality constant, is called the *frequency factor* or *pre-exponential factor*. As described in Example 15.12, plotting the natural logarithm of the rate constant versus the reciprocal of the absolute temperature yields a straight line whose slope is equal to $-E_a/R$. After studying Example 15.12, work Question 34 of the Self-Test at the end of this section. To do that, set up a table with headings of "ln k" and "$1/T$" and compute these values from the data given. Then choose a piece of graph paper, plot the data, measure the slope, and calculate E_a.

To calculate E_a from rate constants at two different temperatures (which is actually less accurate than the graphical procedure) you will need to know Equation 15.9. Examples 15.13 and 15.14 show how this equation is used, and you should study them carefully before beginning the Practice Exercises.

In working with Equation 15.9, it is helpful to note that if a negative value is obtained for E_a when you calculate the activation energy, you probably interchanged the 1 and 2 subscripts on either the rate constants or the temperatures. The only effect that such an error will have on the computed E_a is to change its sign. Since the activation energy must be positive, just change its sign to positive.

When you use the activation energy to calculate a rate constant at some temperature, given the rate constant at some other temperature, keep in mind that the value of k is always larger at the higher temperature. After finishing the calculation, make sure your values of k fit this rule. If they don't, then you switched the subscripts 1 and 2 on the k's in the ratio of rate constants.

A final point to be especially careful about in these calculations is to use Kelvin temperatures in Equation 15.9, not Celsius temperatures.

Self-Test

30. What is the natural logarithm of these quantities? (Express the answers to the correct number of significant figures.)

    (a) 24.6 _____ (b) 0.024 _____

31. What is the natural-antilogarithm of these quantities? (Express the answers to the correct number of significant figures.)

    (a) 1.445 _____ (b) –12.3 _____

32. A certain first-order reaction has a rate constant $k = 1.0 \times 10^{-2}$ s^{-1} at 30 °C. At 40 °C its rate constant is $k = 2.5 \times 10^{-2}$ s^{-1}. Calculate the activation energy for this reaction in kJ/mol.

    _____72.25 kJ/mol_____

33. A certain second-order reaction has an activation energy of 105 kJ/mol. At 25 °C the rate constant for the reaction has a value of 2.3×10^{-3} L mol^{-1} s^{-1}. What would its rate constant be at a temperature of 45 °C?

    _____

34. In the table below are tabulated values of the rate constant for a reaction at 5 °C intervals from 25 °C to 100 °C. Use these data to determine the activation by the graphical method.

Temp (°C)	$k(s^{-1})$	Temp (°C)	$k(s^{-1})$
25	0.0240	65	0.206
30	0.0324	70	0.259
35	0.0433	75	0.325
40	0.0573	80	0.406
45	0.0751	85	0.502
50	0.0978	90	0.619
55	0.126	95	0.757
60	0.162	100	0.922

New Terms

Write the definitions of the following terms, which were introduced in this section. If necessary, refer to the Glossary at the end of the text.

Arrhenius equation   frequency factor

15.8   Experimental rate laws can be used to support or reject proposed mechanisms for reactions

Review

Usually, a net overall reaction occurs by a *mechanism* involving a sequence of simple *elementary processes,* the slowest of which determines how fast the products are able to form. This is the *rate-determining step,* sometimes called the *rate-limiting step,* and the rate law of the overall reaction is the same as the rate law for the rate-determining step.

If we know the stoichiometry for an elementary process, we can predict its rate law; the coefficients of the reactants are equal to their exponents in the rate law. Remember, however, that this works *only* if the elementary process is known. When we first begin to study a reaction we don't know what its mechanism is, so we can't predict with any hope of confidence what the exponents in the rate law will be.

Determining a mechanism involves guessing what the elementary processes are and then comparing the predicted rate law, based on the mechanism, with the rate law determined from experimental data. If the two rate laws match, the mechanism may be correct. If they don't match, the search for a mechanism must continue. Generally, successful collisions involving more than two "bodies," so-called *bimolecular collisions*, are so improbable that chemists almost never postulate them in devising possible mechanisms.

Self-Test

35. Suppose an elementary process is: $2A + M \rightarrow P + Q$. What is the rate law for this step?

_____

36. The following mechanism has been proposed for the reaction

$$(CH_3)_3CBr + OH^- \rightarrow (CH_3)_3COH + Br^-$$

Step 1 $(CH_3)_3CBr \rightarrow (CH_3)_3C^+ + Br^-$ (slow)

Step 2 $(CH_3)_3C^+ + OH^- \rightarrow (CH_3)_3COH$ (fast)

If this mechanism is correct, what is the expected rate law for the overall reaction?

$$rate = k\left[(CH_3)_3CBr\right]$$

37. What would be the rate law for the overall reaction in the preceding question if it occurred in a single step (i.e., if the overall reaction were actually an elementary process)?

$$rate = k\left[(CH_3)_3CBr\right]\left[OH^-\right]$$

New Terms

Write the definitions of the following terms, which were introduced in this section. If necessary, refer to the Glossary at the end of the text.

bimolecular collision

elementary process

mechanism

rate-determining step

rate-limiting step

15.9   Catalysts change reaction rates by providing alternative paths between reactants and products

Review

Catalysts open alternative pathways (mechanisms) for reactions. These paths have lower activation energies than the uncatalyzed mechanisms, so catalyzed reactions occur faster. A *homogeneous catalyst* is in the same phase as the reactants; a *heterogeneous catalyst* is in a different phase than the reactants. A homogeneous catalyst, like the NO_2 used in the lead chamber process, is consumed in one step of the mechanism and then regenerated in a later step. A heterogeneous catalyst functions by *adsorption* of the reactants on its surface where the reactants are able to react with a relatively low activation energy. Be sure you know the difference between adsorption and absorption: adsorption occurs when a substance becomes bound to a surface. Absorption occurs when a substance is taken into the absorbing medium (like a sponge absorbs water).

Self-Test

38. How do catalyst "poisons" work? _____

_____

New Terms

Write the definitions of the following terms. If necessary, refer to the Glossary at the end of the text.

adsorption   catalyst

heterogeneous catalyst   homogeneous catalyst

Solutions to Thinking It Through

1   We could make use of Factors 2 and 3 above. Combustion of coal is a heterogeneous reaction, so we can increase the rate by providing a greater area of contact between the reactants. This could be done by pulverizing the coal before burning it. To further increase the ability of oxygen to reach the coal, we could blow air into the reaction mixture. We could also increase the concentration of the oxygen in the air.

2   Since the reaction is first order, we use Equation 15.3a. The concentration we wish to find corresponds to $[A]_O$ in the equation (i.e., $[SO_2Cl_2]_O$). First we multiply k by t. Then we take the antilogarithm to obtain the ratio of concentrations, Solving for $[SO_2Cl_2]_O$

$$\frac{[SO_2Cl_2]_O}{[SO_2Cl_2]_t} = e^{kt}$$

Solving for $[SO_2Cl_2]_O$ and substituting 0.022 M for $[SO_2Cl_2]_t$ gives us the initial SO_2Cl_2 concentration. Multiplying this by the volume, 0.350 L, gives the number of moles. [The answer is 0.011 mol.]

Answers to Self-Test Questions

1.   It would require the simultaneous collision of 27 molecules, which is an unlikely event.
2.   By adjusting conditions, they can make their reactions go faster and more efficiently.
3.   Nature of the reactants
4.   Their biochemical reactions are slower in the cooler weather.
5.   The possibility of a coal dust explosion.
6.   Ability of reactants to meet; the effect of temperature.
7.   The O_2 is less concentrated in air than in pure O_2.
8.   Approximately 1.2×10^{-4} mol L^{-1} s^{-1}
9. 0.6×10^{-4} mol L^{-1} s^{-1}
10.   2.0 mol L^{-1} s^{-1}
11.   0.100 mol C_8H_{18} L^{-1} s^{-1}, 1.25 mol O_2 L^{-1} s^{-1}
12.   225 mol C_8H_{18}
13.   second order
14.   second order with respect to both A and B, fourth order overall.
15.   7.92 L mol^{-1} s^{-1}
16.   third order
17.   (a) rate $=k[A]^1[B]^2$ (b) $k = 5.0$ L^2 mol^{-2} s^{-1}
18. 2.6×10^2 seconds
19.   0.037 mol/L
20.   (a) 1.3×10^2 seconds, (b) 0.10 mol/L
21.   (a) 0.125 M, (b) 6 half-lives = 3.0 hours
22. 3.8×10^{-4} s^{-1}

23. 3.2×10^4 s = 525 min

24. first order

25. (a) see Figure 13.9 (b) see Figure 13.11

26. At the high-point on the energy curve.

27.
```
    O                 O
     \               /
      N ---O--- N
     /
    O
```

28. As the potential energy decreases, the average kinetic energy rises, which means the temperature rises.

29. When E_a is small, a large fraction of molecules have at least this minimum energy that they need to react.

30. (a) 3.203, (b) –3.73

31. (a) 4.24, (b) 5×10^{-6}

32. $E_a = 72.2$ kJ mol^{-1}

33. 3.3×10^{-2} L mol^{-1} s^{-1}

34. Data to be graphed:

ln k	1/T	ln k	1/T
–3.73	0.00335	–1.58	0.00296
–3.43	0.00329	–1.35	0.00291
–3.14	0.00324	–1.12	0.00287
–2.86	0.00319	–0.902	0.00283
–2.59	0.00314	–0.689	0.00279
–2.33	0.00309	–0.480	0.00275
–2.07	0.00305	–0.278	0.00272
–1.82	0.00300	–0.0809	0.00268

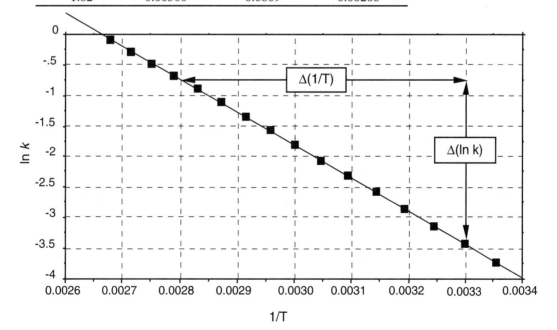

$$\Delta(1/T) = 0.00330 - 0.00280 = 0.00050$$

$$\Delta(\ln k) = -3.40 - (-0.75) = -2.65$$

$$\text{slope} = \frac{\Delta(\ln k)}{\Delta(1/T)} = \frac{-2.65}{5.0 \times 10^{-4} \text{ K}^{-1}} = -5.3 \times 10^3 \text{ K} = \frac{-E_a}{R} \text{ (Note units for the slope.)}$$

Using $R = 8.314$ J mol^{-1} K^{-1}, $E_a = -8.314(-5.3 \times 10^3) = 44 \times 10^3$ J/mol $= 44$ kJ/mol

The activation energy is approximately 44 kJ/mol.

35. rate $= k[A]^2[M]$
36. rate $= k[(CH_3)_3CBr]$
37. rate $= k[(CH_3)_3CBr]\ OH^-]$
38. The poison becomes attached to the catalyst's surface and prevents adsorption of the reactants.

Tools you have learned

Consider removing this chart from the Study Guide so you can have it handy when tackling homework problems.

Tool	How it Works
Rate law of a reaction	The rate law permits us to calculate the rate of reaction for any given set of concentrations, provided we know the rate law.
Half-lives	For a first order reaction, we can calculate a half-life from the rate constant. For a second order reaction, we need the rate constant and the concentration of the reactant.
Arrhenius equation	We use this equation to determine an activation energy graphically and to interrelate rate constants, activation energy, and temperature.

Summary of Important Equations

Concentration versus time, first order reaction

$$\ln \frac{[A]_0}{[A]_t} = kt$$

Concentration versus time, second order reaction

$$\frac{1}{[B]_t} - \frac{1}{[B]_0} = kt$$

Half-life, first order reaction

$$t_{1/2} = \frac{\ln 2}{k}$$

Half-life, second order reaction

$$t_{1/2} = \frac{1}{k[B]_0}$$

Arrhenius equation

$$k = A\, e^{-E_a/RT}$$

$$\ln\left(\frac{k_2}{k_1}\right) = \frac{-E_a}{R}\left(\frac{1}{T_2} - \frac{1}{T_1}\right)$$

Remember, if you mix up the 1's and 2's, the only effect will be to change the sign of E_a. However, E_a must be positive. Also, k is always larger at the higher temperature.

Chapter 16

Chemical Equilibrium— General Concepts

In Chapter 15 you learned the factors that determine how fast reactions are able to proceed. In this chapter we explore the fate of most chemical reactions, namely, dynamic equilibrium. For most reactions, the concentrations eventually level off at constant values, which are the equilibrium concentrations in the reaction mixture.

The principles we develop here will be used in the next several chapters as we study equilibria in solutions of acids and bases and solubility equilibria. For this reason, it is important that you learn well the topics discussed here, especially how we use the balanced chemical equation to construct the *equilibrium law* for the reaction. Also, study carefully the approach to solving equilibrium problems.

Learning Objectives

As you study this chapter, keep in mind the following objectives.

1   To become totally familiar with the concept of dynamic equilibrium.

2   To learn how the composition of an equilibrium mixture is independent of where we begin along the path from reactants to products, provided that the overall composition of the system is the same each time.

3   To learn how to construct an equation called the *equilibrium law* that relates the equilibrium concentrations of the reactants and products in a chemical system to a constant called the equilibrium constant, K_c.

4   To learn how the equilibrium law for gaseous reactions can be written using partial pressures and related to an equilibrium constant K_p.

5   To be able to use the magnitude of the equilibrium constant to make a qualitative estimate of the extent of reaction.

6   To learn how to convert between K_p and K_c for gaseous reactions.

7   To be able to write the equilibrium law for a heterogeneous reaction.

8   To learn how to apply Le Châtelier's principle to chemical equilibria. You should be able to predict the effects of adding or removing a reactant or product, changing the volume of a gaseous reaction, changing the temperature, adding a catalyst, and adding an inert gas at constant volume.

9   To learn how to calculate K_c from data related to equilibrium concentrations and to learn how to use the value of K_c and initial concentrations to calculate equilibrium concentrations.

16.1   Dynamic equilibrium is achieved when the rates of two opposing processes are equal

Review

Equilibrium is established in a chemical system when the rate at which the reactants combine to form the "products" is equal to the rate at which the products react to form the "reactants." We call it a dynamic equilibrium because the reaction hasn't ceased; instead there are two opposing reactions occurring at equal rates.

When equilibrium is reached in a chemical system, the concentrations of the reactants and products attain steady, constant values that do not change with time. Because the reaction is proceeding in both directions simultaneously, the terms *reactants* and *products* no longer have their usual meanings. Instead, we use the term *reactants* to mean the substances on the left of the equilibrium arrows and the term *products* to mean the substances on the right of the arrows.

Self-Test

1.   If we were able to follow a particular carbon atom in a solution of $HC_2H_3O_2$, part of the time it would exist in an acetate ion, $C_2H_3O_2^-$ and part of the time it would exist in a molecule of acetic acid, $HC_2H_3O_2$. Why is this so?

Weak acid

New Terms

None

16.2   Closed systems reach the same equilibrium concentrations whether we start with reactants or products

Review

In the example used in this section, 0.0350 mol N_2O_4 and 0.0700 mol NO_2 each contain the same total amount of nitrogen and oxygen. The two "initial" systems differ in how the nitrogen and oxygen atoms are distributed among the molecules. When these gases are allowed to come to equilibrium from either direction, the same equilibrium composition of NO_2 and N_2O_4 is reached. This demonstrates that for a given *overall* composition, the same equilibrium composition will always be reached regardless of whether we begin with reactants, products, or a mixture of them. The reaction can go in either direction, and the direction in which it proceeds is determined by how the concentrations must change in order to become equal to the appropriate equilibrium concentrations.

New Terms

None

16.3 A law relating equilibrium concentrations can be derived from the balanced chemical equation for a reaction

Review

As you learned in Chapter 15, the molar concentration of a substance is represented symbolically by placing the formula for the substance between brackets. Thus, $[CO_2]$ stands for the molar concentration (in units of mol L^{-1}) of CO_2.

The *mass action expression* for a reaction is a fraction. Its numerator is constructed by multiplying together the molar concentrations of the *products*, each raised to a power that is equal to its coefficient in the balanced chemical equation for the equilibrium. The denominator is obtained by multiplying together the molar concentrations of the *reactants*, each raised to a power equal to its coefficient. For example, for the reaction

$$2CO(g) + O_2(g) \rightleftharpoons 2CO_2(g)$$

the mass action expression is

$$\frac{[CO_2]^2}{[CO]^2[O_2]}$$

The numerical value of the mass action expression is called the *reaction quotient*, Q, and at equilibrium, the reaction quotient has a value that we call the *equilibrium constant* K_c. Remember that the "c" in K_c means that the mass action expression is written with molar concentrations. Equating the mass action expression to the equilibrium constant gives the *equilibrium law* for the reaction. Thus, for the reaction above, the equilibrium law is

$$\frac{[CO_2]^2}{[CO]^2\,[O_2]} = K_c$$

For any given reaction, the numerical value of K_c varies with temperature. But at a given temperature, the reaction quotient will always be the same when the system is at equilibrium, regardless of the individual concentrations. An important point is that there are no restrictions on individual equilibrium concentrations. They can have *any* values as long as they satisfy the equilibrium law when the system is at equilibrium.

Manipulating equations for chemical equilibria

The following rules apply to calculating the equilibrium constant when we manipulate chemical equilibria. Notice that the rules are different than the ones we used when we applied Hess's law. (Study the examples that begin on page 703.)

- Change the direction of a reaction:
    Take the reciprocal of K

- Multiply by a factor
    Raise K to a power equal to the factor

- Add two equations
    Multiply their K's

Self-Test

2. Write the equilibrium law for the following reactions:

(a) $C_2H_4(g) + H_2(g) \rightleftharpoons C_2H_6(g)$

$$K_c = \frac{[C_2H_6]}{[C_2H_4][H_2]}$$

(b) $2N_2O(g) + 3O_2(g) \rightleftharpoons 4NO_2(g)$

_____

(c) $2HCrO_4^-(aq) \rightleftharpoons Cr_2O_7^{2-}(aq) + H_2O(l)$

_____

(d) $CH_3OH(l) + CH_3CO_2H(l) \rightleftharpoons CH_3CO_2CH_3(l) + H_2O(l)$

_____

3. Write the equilibrium law for the following reaction:

$$4NH_3(g) + 7O_2(g) \rightleftharpoons 4NO_2(g) + 6H_2O(g)$$

$$\frac{[H_2O]^6[NO_2]^4}{[NH_3]^4[O_2]^7} = K_c$$

4. We saw that at 440 °C, $K_c = 49.5$ for the reaction,

$$H_2(g) + I_2(g) \rightleftharpoons 2HI(g).$$

Which of the following mixtures are at equilibrium at 440 °C?

(a) $[H_2] = 0.0122\ M$, $[I_2] = 0.0432\ M$, $[HI] = 0.154\ M$

(b) $[H_2] = 0.708\ M$, $[I_2] = 0.0115\ M$, $[HI] = 0.635\ M$

(c) $[H_2] = 0.0243\ M$, $[I_2] = 0.0226\ M$, $[HI] = 0.165\ M$

_____

5. We saw that at 440 °C, $K_c = 49.5$ for the reaction,

$$H_2(g) + I_2(g) \rightleftharpoons 2HI(g).$$

(a) What is K_c for the reaction $2HI(g) \rightleftharpoons H_2(g) + I_2(g)$?

_____

(b) What is K_c for the reaction $\frac{1}{2} H_2(g) + \frac{1}{2} I_2(g) \rightleftharpoons HI(g)$?

$\sqrt{K_c}$ or $K_c^{1/2}$

_____

(c) Write the equation for the reaction we would obtain if we added the equations in parts a and b. What is the value of the equilibrium constant for this final equation?

_____

New Terms

Write the definitions of the following terms, which were introduced in this section. If necessary, refer to the Glossary at the end of the text.

mass action expression   equilibrium law

reaction quotient K_c

equilibrium constant

16.4   Equilibrium laws for gaseous reactions can be written in term of concentrations or pressures

Review

The partial pressure of a gas in a mixture is proportional to its concentration. Therefore, the mass action expression for reactions involving gases can be written using partial pressures in place of concentrations. When partial pressures are used in the mass action expression, the equilibrium constant is designated as K_p. In general K_p does not have the same value as K_c.

Thinking It Through

For the following, identify the information needed to solve the problem and show (or explain) what must be done with it.

1   What is the molar concentration of nitrogen in air that has a total pressure of 760 torr at 25 °C? Air is composed of 79% N_2 by volume.

$$K_p = K_c (RT)^{\Delta n_{gas}}$$

Self-Test

$PV = nRT$

6.   Write the K_p expression for the reaction

$$2N_2O(g) + 3O_2(g) \rightleftharpoons 4NO_2(g)$$

$$\frac{P_{NO_2}^4}{P_{O_2}^3 \cdot P_{N_2O}^2}$$

7.   A chemical reaction has the following equilibrium law.

$$K_p = \frac{P_{BrF_3}^2}{P_{F_2}^3 \, P_{Br_2}}$$

(a)   What is the chemical equation for the reaction?

$$3F_2 + Br_2 \rightleftharpoons 2BrF_3$$

(b)   What is the expression for K_c?

$$[BrF_3]^2 / [F_2]^3 [Br_2]$$

8. The partial pressure of O_2 in air is approximately 160 torr. What is the concentration of O_2 in air at 25 °C expressed in mol L^{-1}?

$$\underline{\quad .21 \;=\; K_c \left(.0821 \cdot 298 \right)^1 \quad}$$

New Terms

Write the definition of the following term, which was introduced in this section. If necessary, refer to the Glossary at the end of the text.

K_p

16.5   A large K means a product-rich equilibrium mixture; a small K means a reactant rich mixture at equilibrium

Review

When the equilibrium constant (either K_p or K_c) is large, the reaction goes far toward completion by the time equilibrium is reached. We express this by saying that the "position of equilibrium lies far to the right." When K is small, only small amounts of products are formed when equilibrium is reached, so we say the position of equilibrium lies far to the left. See the summary on page 706 of the text.

By comparing K's for reactions of similar stoichiometry, we are able to compare the extent to which the reactions proceed toward completion. Even if the stoichiometries differ, comparisons are still valid if the K's are vastly different.

Self-Test

9. For the reaction

$$CH_4(g) + H_2O(g) \rightleftharpoons CO(g) + 3H_2(g)$$

$K_c = 1.78 \times 10^{-3}$ at 800 °C, $K_c = 4.68 \times 10^{-2}$ at 1000 °C, and $K_c = 5.67$ at 1500 °C. From these data, should more or less CO and H_2 form as the temperature of this equilibrium is increased? Explain.

$\underline{\qquad more\ is\ formed \qquad}$

$\underline{\qquad\qquad\qquad\qquad\qquad}$

10. Hypochlorite ion, OCl^-, is the active ingredient in liquid laundry bleach. It has a tendency to react with itself as follows (although the reaction is slow at room temperature).

$$3OCl^- \rightleftharpoons 2Cl^- + ClO_3^- \qquad K_c = 10^{27}$$

If a bottle of bleach is allowed to come to equilibrium, how should the relative concentrations of the reactants and products compare?

$\underline{\qquad\qquad\qquad\qquad\qquad}$

New Terms

None

16.6 A simple expression relates K_p and K_c

Review

For reactions in which there is a change in the number of moles of gas going from the reactants to the products, the numerical values of K_p and K_c are not the same, and in this section you learn how to convert between them. The equation that you need to remember in order to do this is Equation 16.4.

$$K_p = K_c \, (RT)^{\Delta n_{gas}} \qquad\qquad (16.4)$$

In applying this equation, be careful to compute Δn_{gas} correctly; it is the difference in the total number of moles of *gas* between the product side and the reactant side of the equation. It is also important to remember to use $R = 0.0821$ L atm mol^{-1} K^{-1}. Any other value of R will give incorrect answers.

Thinking It Through

Explain how you would solve the following problem.

2   The reaction $2H_2(g) + O_2(g) \rightleftharpoons 2H_2O(g)$ has $K_c = 9.1 \times 10^{80}$ at 25 °C. What is the value of K_p for the following reaction?

$$H_2(g) + \tfrac{1}{2} O_2(g) \rightleftharpoons H_2O(g)$$

Self-Test

11. For which of the following reactions will K_p be numerically equal to K_c?

(a) $PCl_5(g) \rightleftharpoons PCl_3(g) + Cl_2(g)$

(b) $H_2(g) + Cl_2(g) \rightleftharpoons 2HCl(g)$

(c) $2CO(g) + O_2(g) \rightleftharpoons 2CO_2(g)$ _____

12. For the reactions in the preceding question, which will have

(a) $K_p > K_c$ at 25 °C?   _____

(b) $K_p < K_c$ at 25 °C?   _____

13. The reaction $2H_2(g) + O_2(g) \rightleftharpoons 2H_2O(g)$ has $K_c = 9.1 \times 10^{80}$ at 25 °C. What is the value of K_p for this reaction?

_____

14. At 300 °C, the reaction $2SO_2(g) + O_2(g) \rightleftharpoons 2SO_3(g)$ has $K_p = 1.0 \times 10^8$. What is the value of K_c for this reaction at this temperature?

_____

New Terms

None

16.7 Heterogeneous equilibria involve reaction mixtures with more than one phase

Review

A heterogeneous equilibrium is one in which not all of the reactants and products are in the same phase. The mass action expression for a heterogeneous reaction is normally written without terms for the concentrations of pure liquids or solids. This is because the number of moles per liter for such substances does not depend on the amount of the substance in the reaction mixture. Here are two examples. For the reaction

$$Cl_2(g) + 2NaBr(s) \rightleftharpoons Br_2(l) + 2NaCl(s)$$

$$K_c = \frac{1}{[Cl_2]} \quad \text{and} \quad K_p = \frac{1}{P_{Cl_2}}$$

For the reaction

$$NH_4Cl(s) \rightleftharpoons NH_3(g) + HCl(g)$$

$$K_c = [NH_3]\,[HCl] \qquad \text{and} \qquad K_p = P_{NH_3}\,P_{HCl}$$

Self-Test

15. Write the equilibrium law for the following reactions in terms of K_c.

(a) $PCl_3(l) + Cl_2(g) \rightleftharpoons PCl_5(s)$

$\frac{1}{[Cl_2(g)]} = K_c$

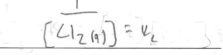

(b) $CaCO_3(s) + SO_2 \rightleftharpoons CaSO_3(s) + CO_2(g)$

$\frac{[CO_2(g)]}{[SO_2]} = K_c$

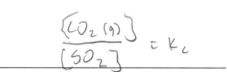

16. Write the equilibrium law for K_c the reaction: $PbCl_2(s) \rightleftharpoons Pb^{2+}(aq) + 2Cl^-(aq)$

$[Cl^-]^2\,[Pb^{2+}]$

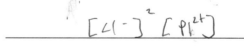

17. Write the equilibrium law for K_c the reaction: $3Zn(s) + 2Fe^{3+}(aq) \rightleftharpoons 2Fe(s) + 3Zn^{2+}(aq)$

$[Zn^{2+}]^3 / [Fe^{3+}]^2$

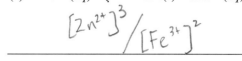

New Terms

Write the definitions of the following terms, which were introduced in this section. If necessary, refer to the Glossary at the end of the text.

homogeneous reaction   heterogeneous reaction

homogeneous equilibrium   heterogeneous equilibrium

16.8   When a system at equilibrium is stressed, it reacts to relieve the stress

Review

Le Châtelier's principle states that if an equilibrium system is *stressed*, meaning it is subjected to an outside disturbance that upsets the equilibrium, the position of equilibrium shifts in a direction that counteracts the disturbance and, if possible, returns the system to equilibrium.

1   A reaction shifts in a direction away from the side of the reaction to which a substance is added. It shifts toward the side from which a substance is removed.

2   A decrease in the volume of a gaseous reaction mixture increases the pressure and shifts the equilibrium toward the side with the fewer number of molecules of *gas*. If both sides have the same number of molecules of gas, a volume change will not affect the equilibrium. Solids and liquids are virtually incompressible, so pressure changes have no effect on them.

3   An increase in temperature shifts an equilibrium in a direction that absorbs heat. The value of K increases with increasing temperature for a reaction that is endothermic in the forward direction. If you remember this, it's easy to figure out what happens for an exothermic reaction as well. 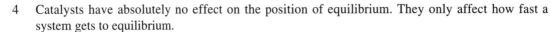 Remember that temperature is the *only* thing that affects the value of K for a reaction.

4   Catalysts have absolutely no effect on the position of equilibrium. They only affect how fast a system gets to equilibrium.

5   Adding an inert gas, without simultaneously changing the volume, will have no effect on the position of equilibrium.

Self-Test

18.   Self-Test Question 9 gives values of K_c at three different temperatures for the reaction

$$CH_4(g) + H_2O(g) \rightleftharpoons CO(g) + 3H_2(g)$$

Is this reaction, as read from left to right, exothermic or endothermic? Explain.

endo

_____

_____

_____

19. For the reaction in the preceding question, state how the amount of CO at equilibrium will be affected by

    (a) adding CH_4 _shift right_ ↑

    (b) adding H_2 _shift left_ ↓

    (c) removing H_2O _shift left_ ↓

    (d) decreasing the volume   _shift left_ ↓

    (e) adding helium at constant volume

         nothing

20. For the reaction, $F_2 + Cl_2 \rightleftharpoons 2ClF$ + heat, describe how the amount of Cl_2 will be affected by

    (a) adding F_2 ↓

    (b) adding ClF   ↑

    (c) decreasing the volume   _nada_

    (d) raising the temperature   ↑

    (e) adding a catalyst   _nada_

21. How will the value of K_p change for the reaction in the preceding question if the temperature is lowered?   ↑

New Terms

None

16.9 Equilibrium concentrations can be used to predict equilibrium constants, and vice versa

Review

As described in the text, the calculations that you learn how to perform in this section can be divided into two categories—calculating K_c from equilibrium concentrations, and calculating equilibrium concentrations from K_c and initial concentrations. These are not really very difficult *if you approach them systematically*. Until you become experienced at solving these problems, it's important that you not attempt to take shortcuts; that's where many students make mistakes or become lost and can't finish the problem.

In working equilibrium problems there are some important rules to follow, and the concentration table that we construct under the chemical equation helps you follow them.

1   The concentrations that you substitute into the equilibrium law *must* correspond to equilibrium concentrations. These are the *only* quantities that satisfy the equilibrium law.

2   Since we are working with K_c, quantities in the concentration table should have the units of molar concentration (mol L^{-1}). This means that if you are not given the molar concentration, but instead a certain number of moles in a certain volume, you must immediately change the data to mol L^{-1}. For example, if you are told that 5.0 mol of a certain reactant is in a volume of 2.0 L, change the data to 5.0 mol/2.0 L = 2.5 mol L^{-1} before entering it into the appropriate place in the concentration table.

3   An important difference between the two kinds of calculations that you are learning to perform is that when you are asked to calculate K_c, the equilibrium constant is the unknown and you must therefore find *numerical values* for *all* the entries in the "Equilibrium concentration" row of the table. On the other hand, when K_c is known and you are being asked to find an equilibrium concentration, then an x (or other symbol) appears in some way in the "Equilibrium concentration" row.

4   The initial concentrations in a reaction mixture are determined by the person doing the experiment. When you are given the composition of a mixture in a problem, these are initial concentrations, unless stated otherwise. In filling in this line of the table, we imagine that the reaction mixture can be prepared before any chemical changes take place. Then we let the reaction proceed to equilibrium and see how the concentrations change.

5   The changes in concentration are controlled by the stoichiometry of the reaction. When we want to compute K_c and have to calculate what the equilibrium concentrations are, the entries in this column are *numbers* whose ratios are the same as the ratios of the coefficients. When you are given K_c and initial concentrations, this is where the x's first appear. The following points are useful to remember:

   • The coefficients of x can be the same as the coefficients in the balanced equation; this ensures that they are in the right ratio.

   • The "changes" for the reactants all must have the same algebraic sign, and these signs must be the opposite of the algebraic signs of the "changes" for the products. (In other words, if the changes for the reactants are positive, the changes for the products are negative.)

   • If the initial concentration of some reactant or product is zero, its change *must* be positive.

6   Equilibrium concentrations are obtained by algebraically adding the "Change in concentration" to the "Initial concentration." In cases where K_c is *very* small, the extent of reaction from left to right will also be *very* small. This allows equilibrium concentration expressions such as $(0.200 + x)$ or $(0.100 - x)$ to be simplified. The initial concentrations of the reactants hardly change at all as the reaction proceeds, so x can be expected to have a very small value. In this case, a quantity such as $(0.200 + x)$ will be very nearly equal to 0.200 after the very small value of x is added to 0.200.

$$(0.200 + x) \cong 0.200$$

This kind of simplifying assumption often makes a difficult problem very easy. If you are setting up a problem and find the algebra very complicated, it is likely you will be able to make simplifying assumptions.

7   Look for ways of simplifying the algebra in solving for x. Sometimes this isn't possible, as we see in Example 16.11, where the quadratic formula is used.

Thinking It Through

Explain how you would solve the following problem.

3   The reaction $2HI(g) \rightleftharpoons H_2(g) + I_2(g)$ has $K_c = 1.6 \times 10^{-2}$. A mixture is prepared containing $0.100\ M$ HI, $0.010\ M$ H_2, and $0.020\ M$ I_2. When this mixture reaches equilibrium, what will be the molar concentration of HI?

Self-Test

22.   At 25 °C a mixture of Br_2 and Cl_2 in carbon tetrachloride reacted according to the equation,

$$Br_2 + Cl_2 \rightleftharpoons 2BrCl.$$

The following equilibrium concentrations were found: $[Br_2] = 0.124\ M$, $[Cl_2] = 0.237\ M$, $[BrCl] = 0.450\ M$. What is the value of K_c for this reaction?

_____

23.   Referring to Question 22, what is the value of K_c for the reaction

$$2BrCl \rightleftharpoons Br_2 + Cl_2 \text{ at 25 °C?}$$

_____

24.   A mixture of $N_2O(g)$, $O_2(g)$, and $NO_2(g)$ was prepared in a 4.00-liter container and allowed to come to equilibrium according to the equation

$$2N_2O(g) + 3O_2(g) \rightleftharpoons 4NO_2(g)$$

The mixture originally contained 0.512 mol N_2O.

(a)   As the reaction came to equilibrium, the concentration of N_2O decreased by 0.0450 mol/L. What was the equilibrium concentration of N_2O?

_____

(b)   By how much did the O_2 concentration change?

_____

(c)   By how much did the NO_2 concentration change?

_____

25.   At a certain temperature, a mixture of $N_2O(g)$ and $O_2(g)$ was prepared having the following initial concentrations: $[N_2O] = 0.0972\ M$, $[O_2] = 0.156\ M$. The mixture came to equilibrium following the equation,

$$2N_2O(g) + 3O_2(g) \rightleftharpoons 4NO_2$$

At equilibrium, the NO_2 concentration was found to be $0.0283\ M$. What is K_c for this reaction?

_____

26. The reaction, $2BrCl \rightleftharpoons Br_2 + Cl_2$, has $K_c = 0.145$ in CCl_4 at 25 °C. If 0.220 mol of BrCl is dissolved in 250 mL of CCl_4, what will be the concentrations of BrCl, Cl_2 and Br_2 when the reaction reaches equilibrium?

27. At 25 °C, $K_c = 1.6 \times 10^{-17}$ for the reaction,

$$N_2(g) + 2O_2(g) \rightleftharpoons 2NO_2(g)$$

In air, the concentrations of N_2 and O_2 are: $[N_2] = 0.0320\ M$, $[O_2] = 0.00860\ M$. Taking these as initial concentrations, what should be the equilibrium concentration of NO_2 in air?

New Terms

None

Solutions to Thinking It Through

1   The molar concentration of a gas is given by the equation

$$M = \frac{n}{V} = \frac{P}{RT}$$

where P is the pressure, in this case, the partial pressure of N_2. Air is 79% N_2 by volume. This means that if we had 1 liter of air, 79% of it would be N_2; when expanded to the full 1 liter volume, it would exert 79% of the total pressure. Therefore, the partial pressure of the N_2 is 79% of 760 torr. We can change the calculated partial pressure of N_2 to atmospheres by multiplying by the factor (1 atm/760 torr). We use $R = 0.0821$ L atm mol^{-1} K^{-1} to make the units cancel correctly, and we must express the temperature in kelvins. Substituting values into the equation above gives the molarity of the N_2. [The answer is 0.0323 M]

2   We can calculate the value of K_p for the reaction in which 2 moles of H_2O are formed. For this we use the equation

$$K_p = K_c (RT)^{\Delta n_{gas}}$$

We use $R = 0.0821$ L atm mol^{-1}K^{-1}, $T = 298$ K, and $\Delta n_{gas} = (2 - 3) = -1$. Substituting

$$K_p = (9.1 \times 10^{80})\ [(0.0821)(298)]^{-1}$$

$$K_p = 3.7 \times 10^{79}$$

Notice, however, that the equation for which we want K_p is for the formation of 1 mole of H_2O. We obtain this from the original equation by dividing its coefficients by 2, which means we take the square root of the equilibrium constant (which is the same as raising K_p to the $\frac{1}{2}$ power) to obtain the new K_p. [The answer is $K_p = 6.1 \times 10^{39}$]

3   First we write the equilibrium law for the reaction

$$K_c = \frac{[H_2][I_2]}{[HI]^2} = 1.6 \times 10^{-2}$$

We also must set up the concentration table, which is shown below. Since we don't know which way the reaction will proceed to equilibrium, let's assume the reaction will go to the left, to form more HI.

	$2HI(g)$ $\rightleftharpoons$	$H_2(g)$ +	$I_2(g)$
Initial concentration (*M*)	0.100	0.010	0.020
Change in concentration (*M*)	$+2x$	$-x$	$-x$
Equilibrium concentrations (*M*)	$0.100 + 2x$	$0.010 - x$	$0.020 - x$

Next, we substitute equilibrium concentrations into the mass action expression.

$$K_c = \frac{(0.10 + 2x)^2}{(0.010 - x)\,(0.020 - x)}$$

The equation is not a perfect square, but the highest order term in x is x^2. Therefore, we can expand the equation and substitute values into the quadratic formula. Two values will be obtained for x, but only one of them will make sense physically. This is the one we use to substitute into the expression for [HI], namely, [HI] = $(0.100 + 2x)$.

Answers to Self-Test Questions

1.   The $C_2H_3O_2^-$ ion that the carbon atom is in can pick up an H^+ and become an $HC_2H_3O_2$ molecule. Later this molecule can lose H^+ to become a $C_2H_3O_2^-$ ion again. This can be repeated over and over.

2.   (a) $\dfrac{[C_2H_6]}{[C_2H_4]\,[H_2]} = K_c$ (b) $\dfrac{[NO_2]^4}{[N_2O]^2\,[O_2]^3} = K_c$

     (c) $\dfrac{[Cr_2O_7{}^{2-}]\,[H_2O]}{[HCrO_4{}^-]^2} = K_c$ (d) $\dfrac{[CH_3CO_2CH_3]\,[H_2O]}{[CH_3OH]\,[CH_3CO_2H]} = K_c$

3. $\dfrac{[NO_2]^4\,[H_2O]^6}{[NH_3]^4\,[O_2]^7} = K_c$

4.   Only (b) and (c) are at equilibrium.

5.   (a) $K_c = 0.0202$,
     (b) $K_c = 7.04$,
     (c) $HI(g) \rightleftharpoons \frac{1}{2} H_2(g) + \frac{1}{2} I_2(g)$, $K_c = 0.142$

6. $K_p = \dfrac{P_{NO_2}^4}{P_{N_2O}^2 P_{O_2}^3}$

7. (a) $3F_2(g) + Br_2(g) \rightleftharpoons 2BrF_3(g)$

    (b) $K_c = \dfrac{[BrF_3]^2}{[F_2]^3 [Br_2]}$

8. 8.60×10^{-3} mol L^{-1}

9. More CO_2 and H_2 should form, because K_c increases with increasing temperature.

10. At equilibrium the concentrations of Cl^- and ClO_3^- should be large compared to the concentration of OCl^-.

11. Reaction b

12. (a) Reaction a (b) Reaction c

13. $K_p = 3.7 \times 10^{79}$

14. $K_c = 4.7 \times 10^9$

15. (a) $K_c = \dfrac{1}{[Cl_2]}$ (b) $K_c = \dfrac{[CO_2]}{[SO_2]}$

16. $K_c = [Pb^{2+}][Cl^-]^2$

17. $K_c = \dfrac{[Zn^{2+}]^3}{[Fe^{3+}]^2}$

18. Endothermic, because K_c increases with increasing temperature.

19. (a) increase (b) decrease (c) decrease (d) decrease (e) no change

20. (a) decrease (b) increase (c) no change (d) increase (e) no change

21. K_p increases as the temperature is lowered.

22. $K_c = 6.89$

23. $K_c = 0.145$

24. (a) 0.083 M (The initial N_2O concentration was 0.512 mol/4.00 L = 0.128 M.)   (b) decreased by 0.0675 mol/L (c) increased by 0.0900 mol/L

25. $K_c = 0.0378$

26. $[BrCl] = 0.500$, $[Cl_2] = 0.190$ M, $[Br_2] = 0.190$ M

27. $[NO_2] = 3.1 \times 10^{-12}$ M

Tools you have learned

Consider removing this chart from the Study Guide so you can have it handy when tackling homework problems.

Tool	*How it Works*
Mass action expression and the equilibrium law	It defines the relationship that exists between the equilibrium concentrations or partial pressures of the reactants and products in a chemical reaction. Be sure you can write the equilibrium law from the balanced chemical equation for the equilibrium.
Manipulating equilibrium equations	We apply the rules on pages 703 and 704 to determine K for a reaction by manipulating a chemical equilibrium or by combining other chemical equilibria.
Magnitude of the equilibrium constant	It lets us determine qualitatively the extent of reaction at equilibrium. The larger the value of K, the farther the reaction proceeds toward completion when equilibrium is reached.
$K_p = K_c (RT)^{\Delta n_{gas}}$	The equation is used when we have to convert between K_p and K_c.
Le Châtelier's Principle	It enables us to analyze the effects of disturbances (stresses) on the position of equilibrium.
Concentration table	These are constructed when we are solving problems in chemical equilibria. It incorporates initial concentrations, changes in concentration, and equilibrium concentration expressions.
Simplifying assumptions	These are used to simplify the algebra when working equilibrium problems for reactions for which K is very small.

Summary of Important Equations

The equilibrium law from the coefficients of a balanced equation

For the general chemical equation

$$dD + eE \rightleftharpoons fF + gG$$

$$\frac{[F]^f [G]^g}{[D]^d [E]^e} = K_c$$

Relationship between K_p and K_c

$$K_p = K_c (RT)^{\Delta n_{gas}}$$

Chapter 17
Acids and Bases: A Second Look

This chapter continues our study, begun in Chapter 5, of *acid–base reactions*, which are among most common and important kinds of chemical reactions. Chemists most often use the Brønsted view of acids and bases, because in the overwhelming majority of all acid–base reactions, *proton–transfer* occurs. However, many reactions with the look and feel of an acid–base reaction do not involve proton transfers but involve nothing more, in the Lewis view, than the formation of a coordinate covalent bond.

The Brønsted view is nearly always used when working with aqueous solutions, but Brønsted acids (proton donors) vary widely in strength. We learn here how the periodic table and the concept of electronegativity help us to understand the trends in acid strengths and to organize them for study and learning.

The chapter includes a shorthand way for describing small molar concentrations (less than $1 M$) of hydrogen ions in water, the pH concept. It's used throughout all of chemistry, biochemistry, and any fields of medicine that focus on the acid–base balance of a fluid of a living system. That's why it must be learned well.

Learning Objectives

In this chapter, you should keep in mind the following objectives.

1   To learn the Brønsted concept of acids and bases.

2   To be able to write the formula for the conjugate acid or conjugate base of a substance.

3   To be able to identify the Brønsted acids and bases in an acid-base reaction.

4   To use the periodic table to organize information about the relative strengths of acids.

5   To use the concept of electronegativity to predict trends in the strengths of acids.

6   To learn the Lewis concept of acids and bases.

7   To learn how the periodic table "stores" information about the acidities and basicities of the oxides of the elements.

8   To be able to write the equation for the autoionization of water and its ion product constant, K_w.

9   To learn how to calculate the molarity of H_3O^+, given that of OH^-, or the molarity of OH^-, given that of H_3O^+.

10   To learn the equations that define pH and pOH, and to calculate either given the other.

11   To be able to convert $[H_3O^+]$ or $[OH^-]$ into pH or into pOH.

12   To learn what the terms acidic solution, basic solution, and neutral solution mean in terms of the relative concentrations of $[H_3O^+]$ and $[OH^-]$ and in terms of pH.

13   To learn how the pH of a solution can be measured.

14   To learn how to calculate the concentration of $[H_3O^+]$ as well as the pH (or pOH) of a dilute solution of a strong acid or a strong base.

17.1   Brønsted-Lowry acids and bases exchange protons.

Review

Water is not a required solvent for acid–base reactions if we define an acid as any proton (H^+) donor and a base as any proton acceptor. Acids and bases thus defined are sometimes referred to as *Brønsted-Lowry acids and bases,* or simply *Brønsted acids and bases,* after the chemists who had the fundamental insight. A Brønsted acid–base reaction simply involves the transfer of a hydrogen ion (which is the same as a proton) from the acid to the base.

Brønsted acid–base reactions can always be viewed as reversible, so that in *both* the forward reaction and the reverse there is an acid and a base. An acid or proton donor occurring to the left of the equilibrium arrows becomes a base on the right; the proton acceptor or base on the left thereby becomes a Brønsted acid on the right.

Two chemical species that differ by only one proton are called an acid–base *conjugate pair.* The conjugate acid member has one more hydrogen and one additional positive charge (or one less negative charge) than the conjugate base. For example, consider the following pair,

$$H_2PO_4^- \longleftrightarrow HPO_4^{2-}$$

<table>
<tr><td>This is the conjugate acid because it has one more hydrogen and one less negative charge.</td><td>This is the conjugate base because it has one less hydrogen and one more negative charge.</td></tr>
</table>

If you need to write the formula for the conjugate acid of something, add a hydrogen and a positive charge (i.e., add one H^+). For example, to find the conjugate acid of PH_3, we add one H^+.

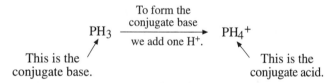

Similarly, to write the formula for the conjugate base of something, we take away one H^+. Suppose we want the formula for the conjugate base of PH_3:

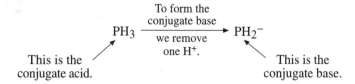

Be sure to study Examples 17.1 and 17.2.

In any Brønsted acid–base reaction there are two conjugate pairs. Be sure to study Example 17.3 so that you can quickly identify the conjugate pairs in an equation.

Two particularly important conjugate pairs are H_3O^+ and H_2O, and H_2O and OH^-. The hydronium ion is the conjugate acid of water; water is the conjugate base of the hydronium ion. Similarly, water is the conjugate acid of the hydroxide ion, and the hydroxide ion is the conjugate base of water. Notice that water is the base in one pair and the acid in the other. Any species that can be either an acid or a base, depending on the circumstances, is said to be *amphoteric* or *amphiprotic*. In the illustrations above, for instance, we would describe PH_3 as amphoteric if it is able to form both PH_2^- and PH_4^+.

Thinking It Through

Explain how you would answer the following questions.

1 Nitrogen monoxide, NO, reacts with oxygen to give nitrogen dioxide, NO_2, but this reaction is not a Brønsted acid–base reaction. Why not?

2 Consider the *possibility* that H_2O and S^{2-} might give a Brønsted acid–base reaction. If they do, what would be the likeliest products: Identify each of them as Brønsted acids or bases.

Self–Test

1. Write the formula of the conjugate acid of each of the following.

   (a) CH_3NH_2 $CH_3NH_3^+$ (c) ClO_3^- $HClO_3$

   (b) NH_2OH $NH_2OH_2^+$ (d) $HC_2O_4^-$ $H_2C_2O_4$

2. Write the formula of the conjugate base of each of the following.

   (a) $HC_2O_4^-$ $C_2O_4^{2-}$ (c) $HCHO_2$ CHO_2^-

   (b) $H_2AsO_4^-$ $HAsO_4^{2-}$ (d) H_2S HS^-

3. Identify the conjugate acid–base pairs in the following reaction. In each pair, state which is the acid and which is the base.

$$H_2SO_4 + Cl^- \rightarrow HSO_4^- + HCl$$
acid   base   Base   acid

4. Identify the conjugate acid–base pairs in the following reaction. In each pair, state which is the acid and which is the base.

$$NH_3 + NH_3 \rightarrow NH_4^+ + NH_2^-$$
acid   base   acid   base

5. Acetic acid, $HC_2H_3O_2$, is a Brønsted base in concentrated sulfuric acid and it is a Brønsted acid in water. Write chemical equations that illustrate these reactions.

$$HC_2H_3O_2 + H_2O \rightarrow H_3O^+ + C_2H_3O_2^-$$

$$HC_2H_3O_2 + H_2SO_4 \rightarrow H_2C_2H_3O_2^+ + HSO_4^-$$

Explain why acetic acid is said to be *amphoteric*.

acts as both acid and base

New Terms

amphiprotic	Brønsted base	conjugate acid–base pair
amphoteric	conjugate acid	
Brønsted acid	conjugate base	

17.2 Strengths of Brønsted acids and bases follow periodic trends

Review

Measuring acid-base strength

To compare the strengths of a series of acids, we use a reference base and compare the extents to which the base is protonated by the acids. If we represent an acid by the formula HA and the base by B, we examine the equilibrium

$$\text{H}A + B \rightleftharpoons A^- + B\text{H}^+$$

The farther the reaction proceeds toward completion, the strong is the acid.

In water, the strongest acid that can exist is H_3O^+; any stronger acid reacts with water to form H_3O^+ and the corresponding anion. This means we can't use water as a reference to compare the strengths of very strong acids such as HNO_3 and $HClO_4$. Both react completely with water, so differences in their strengths cannot be observed.

The strongest base that can exist in water is OH^-. Stronger bases, such as NH_2^-, react completely with water to give OH^- and the corresponding conjugate acid (the reaction is given on page 743).

The position of equilibrium favors the weaker acid and base

This is the theme of the discussion that begins on page 743. If we look at the position of equilibrium in a Brønsted acid-base reaction, we can tell the relative acid and base strengths of the substances involved. For example, consider the equilibrium

$$\text{HCO}_3^- + \text{H}_2\text{O} \rightleftharpoons \text{H}_3\text{O}^+ + \text{CO}_3^{2-}$$

Position of equilibrium lies to the left. ⟵

The position of equilibrium tells us that CO_3^{2-} is a stronger base than H_2O, and that H_3O^+ is a stronger acid than HCO_3^-.

Reciprocal relationships

Be sure to learn the inverse relationship between the strengths of acids and their conjugate bases. This is one of the most important concepts to come from this section.

The stronger the acid is, the weaker is its conjugate base. Thus, very strong acids, like HCl, $HClO_4$, and H_3O^+, have *very* weak conjugate bases, Cl^-, ClO_4^- and H_2O, respectively.

The weaker the acid is, the stronger is its conjugate base. Thus, H_2O, a very weak acid, has a very strong conjugate base, namely, OH^-. Acetic acid, $HC_2H_3O_2$, is a stronger acid than H_2O, but still a weak acid among all the acids. Thus, its conjugate base, the acetate ion, $C_2H_3O_2^-$, is not as strong a base as OH^- but is much stronger as a base than Cl^-.

Binary acids: relative strengths and the periodic table

Binary acids have molecules made of just two elements, one being hydrogen joined to another atom. Examples are HCl and H_2S. Let's represent their formulas as H_nX. *The acids become stronger as the location of atom X in the periodic table moves from left to right in a period or from top to bottom in a group.*

The left-to-right variation is caused by increases in the electronegativity of *X*, which also increases from left to right in a period.

The top to bottom trend, however, defies what we would expect on the basis of changes in electronegativity. Electronegativity *decreases* as we move *down* a vertical column (group), but the acidities of the binary acids *increase* as we move down. There is a weakening of the H—X from top to bottom in a group. The atom *X* holding H increases in size as we move down a group, and the greater the radius of *X*, the weaker the H—X bond. The decrease in bond strength outweighs the decrease in electronegativity, so the acids H_nX become better proton donors as we descend a group.

Oxoacids: Relative strengths and the periodic table

Oxoacids are those in which a central atom, other than O or H, holds one or more OH groups and, sometimes, extra O atoms, called *lone oxygens.* Because O is very electronegative, the OH group is polar. It's made even more polar when it's attached to another electronegative atom, like a halogen, or S, or N. This combination makes the $\delta+$ on H positive enough so that H can be transferred as H^+ from the oxoacid's OH group to an acceptor, like a molecule of H_2O. *All of the correlations in this section between the acidities of oxoacids and their structures or their locations in the periodic table ultimately come down to the magnitude of $\delta+$ on H. Whatever makes this $\delta+$ greater makes the oxoacid stronger.*

Oxoacids with central atoms in the same *group* holding identical numbers of O atoms. Examples are H_2SO_4 and H_2SeO_4. Such acids increase in strength as the central atom changes from bottom to top in the group. S is more electronegative than Se (which is below it in the table), and so the $\delta+$ on H in an OH group of H_2SO_4 is greater than it is when in H_2SeO_4. As a result, H_2SO_4 is stronger than H_2SeO_4.

Oxoacids with central atoms in the same *period* holding identical numbers of O atoms. Examples are H_2SO_4 and H_3PO_4. Such acids increase in strength as the central atom moves from left to right in a period. This trend parallels the left-to-right trending increase in electronegativity. Thus, S is to the right of P in Period 3 of the table and S is more electronegative than P. As a result, H_2SO_4 is a stronger acid than H_3PO_4.

Oxoacids with the same central atom but different numbers of O atoms. Examples are H_2SO_3 and H_2SO_4. O atoms over and above those in OH groups add electron withdrawing ability and increase the sizes of $\delta+$ on the H atoms of the OH groups. So the more O atoms that are joined to the same central atom, the greater is the acidity of the oxoacid. Thus, H_2SO_4 is stronger than H_2SO_3.

When extra O atoms are *lone oxygens*, those not holding H atoms, they not only increase the electron withdrawal from OH groups but also help to disperse and stabilize the negative charge in the conjugate base.

Thinking It Through

Explain how you would answer the following question.

3   An oxoacid of the general formula H_xZO_y has been discovered and found to be a stronger acid than H_2SeO_4. (a) If $y = 4$, which location in the periodic table is more likely for *Z*, to the *left* of Se or to the *right* of Se in the same row? (b) If $y = 4$ and $x = 2$, where does Z most likely lie in the periodic table, *above* or *below* Se in the same group?

Self–Test

6. Consider the following equilibria and their equilibrium constants;

$$HCHO_2 + H_2O \rightleftharpoons H_3O^+ + CHO_2^- \quad K = 1.8 \times 10^{-4}$$

$$HOCl + H_2O \rightleftharpoons H_3O^+ + OCl^- \qquad K = 3.0 \times 10^{-8}$$

Which is the stronger acid? _____HCHO$_2$_____

7. The position of equilibrium in the following reaction lies to the left:

$$HOCl + NO_2^- \rightleftharpoons HNO_2 + OCl^-$$

   (a) Which is the stronger acid? _____HNO$_2$_____

   (b) Which is the stronger base? _____OCl$^-$_____

8. NH_2^- is a stronger Brønsted base than OH^-. Which is the stronger acid, NH_3 or H_2O? __H$_2$O__

9. Which is the stronger acid, H_2Te or H_2Se? _____H$_2$Te_____. Explain.
   _____whichever is below or to the right_____

10. Which is the stronger acid, H_2Te or HI? _____HI_____. Explain.
    _____to the right of Te_____

11. Which is the stronger acid, H_3PO_3 or H_3PO_4? _____H$_3$PO$_4$_____. Explain.
    _____one more O_____

12. Which is the stronger acid, H_3AsO_4 or H_3PO_4? _____H$_3$P$_3$O$_4$_____. Explain.
    _____above and to ... more electro_____

17.3 Lewis acids and bases involve coordinate covalent bonds

Review

Ordinarily, a covalent bond forms between two atoms when each atom supplies one electron for the shared pair. When both electrons for the pair come from one of the atoms, the bond is called a *coordinate covalent bond*. (Once formed, of course, it's just like any other covalent bond.)

A species furnishing an electron pair for a covalent bond is a *Lewis base*. The species accepting the pair is a *Lewis acid*. In the Lewis acid–base system, a neutralization reaction is the formation of a coordinate covalent bond. Hydrogen ions are not necessarily involved, although H$^+$ is a Lewis acid. (It's also what transfers in the Brønsted system.)

Lewis acids are species with less than an octet of electrons on a central atom (for example, BF_3), or they are species that can develop a vacancy open to more electrons by shifting electrons to other atoms around the central atom (like the C in CO_2).

Lewis bases are species that have unshared electron pairs on a central atom that can be donated into a coordinate covalent bond (like the O in H_2O or in OH^-).

A Brønsted acid-base reaction can be viewed as the transfer of a Lewis acid (H^+) from one Lewis base to another. The position of equilibrium favors the H^+ being attached to the stronger Lewis base.

Self–Test

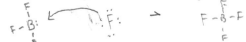

13. Boron trifluoride, BF_3, can react with a fluoride ion to form the tetrafluoroborate ion, BF_4^-. Diagram this reaction using Lewis symbols to show the formation of a coordinate covalent bond.

14. BF_3 reacts with organic chemicals called ethers to form addition compounds. Use Lewis formulas to diagram the reaction of BF_3 with dimethyl ether. The structure of dimethyl ether is

$$CH_3 - \overset{\cdot\cdot}{\underset{\cdot\cdot}{O}} - CH_3$$

15. In the preceding question, which substance is the Lewis acid? _____BF_3_____

   Which is the Lewis base? _____dimethyl ether_____

16. Which are the two Lewis bases in Question 7? Which is the stronger of them?

   _____OCl^-_____

New Terms

Lewis acid   Lewis acid–base neutralization

Lewis base

17.4   Elements and their oxides demonstrate acid–base properties

Review

As you learned in Section 5.5, some metal oxides, particularly those of metals in Groups IA and IIA of the periodic table, are called *basic anhydrides*, because they react with water to give hydroxide ions. The oxides of most nonmetals react with water to give acidic solutions and so are *acidic anhydrides.*

In water, metal ions with sufficiently high *charge densities*, which is the ratio of ionic charge to ionic volume, form proton–donating hydrates, and their solutions are acidic. A charge density large enough to yield an acidic species usually requires both a charge of at least 2+ and a small radius. Generally, the cations of Groups IA and IIA, except for Be^{2+}, do not have high enough charge densities to make their cations acidic. But the water soluble cations of other groups form slightly acidic solutions.

Thinking It Through

Explain how you would answer the following questions.

4   A white solid readily dissolves in water to form a solution that turns red litmus paper blue. Which of the following compounds could this solid be?

$$CO_2, \quad Na_2O, \quad HNO_3, \quad KOH, \quad SO_3, \quad HC_2H_3O_2$$

(handwritten annotations in top margin: $Na_2O + H_2O \rightleftharpoons 2NaOH$; $Z^{2+} + O^{2-} \rightarrow ZO$; blue; hydroxide; $Z^{2+} + O \rightarrow \boxed{ZO}$; basic)

5 Consider the oxide of an element in Group IIA of the periodic table. (a) Write the formula of this oxide, using *Z* as the element's symbol. (b) Will a solution of this oxide in water give a blue or a red color to litmus paper? (c) Is this oxide very soluble in water or is it likely to be sparingly soluble?

(handwritten: Sparingly soluble)

Self–Test

17. Why is Na_2O called a *basic anhydride*? Write the equation for its reaction with water.

(handwritten: $Na_2O + H_2O \rightarrow 2NaOH$ base OH⁻)

18. Why is SO_3 called an *acidic anhydride*? Write the equation for its reaction with water.

(handwritten: $SO_3 + H_2O \rightarrow H_2SO_4$ acid)

19. Which salt, if either, will form the more acidic solution in water, $MgCl_2$ or $AlCl_3$? Explain.

(handwritten: $AlCl_3$ because higher density (Al^{3+} Mg^{2+}))

20. Is the oxide OsO_4 more likely to be acidic or basic? Explain.

(handwritten: acidic (high density))

New Terms

None

17.5   pH is a measure of the acidity of a solution

Review

Water undergoes *autoionization* as follows, and you should learn this equilibrium equation.

$$H_2O \rightleftharpoons H^+ + OH^-$$

Actually, because H^+ never exists in water unattached to H_2O, a better way to express water's autoionization is as follows, but chemists nearly always "shorten" the equation to the above.

$$2H_2O \rightleftharpoons H_3O^+ + OH^-$$

It is very common to use H^+ and $[H^+]$ as "stand-ins" for H_3O^+ and $[H_3O^+]$.

The *ion product* of the autoionization equilibrium is $[H^+][OH^-]$, and at 25 °C, the value of this ion product, called the *ion product constant*, K_w, is 1.0×10^{-14}. It is very important that you learn the following.

$$[H^+][OH^-] = K_w = 1.0 \times 10^{-14} \text{ (at 25 °C)} \tag{1}$$

(Virtually all calculations involving the value of K_W in this chapter as well as in quizzes and examinations assume a temperature of 25 °C and a value of K_W of 1.0×10^{-14}.)

It's important to remember that Equation 1 relates $[H^+]$ and $[OH^-]$ *in any aqueous solution, regardless of the solutes.* You should be able to use the Equation to calculate $[H^+]$ from $[OH^-]$ and vice versa. Study Example 17.5 in the text and work Practice Exercise 11.

Regardless of the solute and regardless of the temperature of the solution, $[H^+]$ and $[OH^-]$ *equal each other in a neutral solution.* In fact, *only* in a neutral solution do $[H^+]$ and $[OH^-]$ equal each other. At 25 °C, their values happen to equal 1.0×10^{-7} M, because at this temperature $K_W = 1.0 \times 10^{-14}$.

When $[H^+]$ is greater than $[OH^-]$, the solution is acidic. When $[H^+]$ is less than $[OH^-]$, the solution is basic. Learn the colors that a few indicators have, like phenolphthalein, bromothymol blue, and thymol blue, in more acidic and more basic solutions (see Table 17.5).

When the value of $[H^+]$ is small (less than 1 M), it is convenient to represent acidity on a logarithmic scale called pH. Equations 17.5 and 17.6 in the text are alternative definitions of pH, Equation 17.5 being used to calculate pH from $[H^+]$ and Equation 17.6 being used to find $[H^+]$ from pH. Similar equations apply to pOH and values of $[OH^-]$.

$$pH = -\log [H^+]$$

$$pOH = -\log [OH^-]$$

$$pH + pOII = 14.00 \text{ (at 25 °C)}$$

*Be careful to note that we use common logarithms in calculations involving pH and pOH, **not natural** logarithms.*

Remember, the lower the pH, the more acidic the solution is. Similarly, the higher the pH, the more basic the solution is. Be sure to study Examples 17.6, 17.7, and 17.8 and to work Practice Exercises 12-14 before tackling the following problems. Once you've done a few and have gotten used to your pocket calculator for doing them, these problems are not hard. Get as good at doing them as you can.

Thinking It Through

Explain how you would solve the following problem.

6   An aqueous solution with a pH of 3.0 has how many times the molar concentration of H^+ as an aqueous solution with a pH of 6.0?

1000 times more

Self–Test

21.   When $[H_3O^+]$ equals 3.80×10^{-6} M at 25 °C,

(a)   What is the value of $[OH^-]$?

2.63×10^{-9}

(b)   Is the solution acidic, basic, or neutral?

Acidic

22.   At 5 °C, $K_W = 1.85 \times 10^{-15}$. If an aqueous solution has $[H_3O^+]$ equal to 4.30×10^{-8} M, is the solution acidic, basic, or neutral?

neutral

23.   Calculate the concentration of H^+ in a solution in which the hydroxide ion concentration is

(a) 2.0×10^{-5} M

5×10^{-10}

(b) 4.0×10^{-8} M

2.5×10^{-7}

24. An aqueous solution at 25 °C with a pOH of 10.00 has a hydrogen ion concentration equal to

    (a) $1.00 \times 10^{-10}\ M$ (c) $10.00\ M$

    (b) $1.00 \times 10^{-4}\ M$ (d) $4.00\ M$ *b*

25. When [H⁺] equals 4.8×10^{-6}, the pH is

    (a) 5.20   (b) 6.48   (c) 5.32   (d) 4.80   *c*

26. When the pH equals 9.65 at 25 °C, the value of [H⁺] is

    (a) $2.24 \times 10^{-10}\ M$ (c) $4.35 \times 10^{-10}\ M$

    (b) $9.65 \times 10^{-14}\ M$ (d) $3.50 \times 10^{-9}\ M$ *a*

27. When pOH equals 7.35 at 25 °C, what is [H⁺]? 2.2×10^{-7}

28. What color does each of the following indicators have in a strongly acidic and in a strongly basic solution? (See Table 17.5 to check your answers.)

	Color in	
Indicator	**Acid**	**Base**
phenolphthalein	X	r~l
bromothymol blue	yellow	blue
thymol blue	yellow	blue

New Terms

Write definitions of the following terms, which were introduced in this section. If necessary, refer to the Glossary at the end of the text.

    acidic solution (both in terms of relative molarities of H⁺ and OH⁻ and in pH at 25 °C)

    basic solution (both in terms of relative molarities of H⁺ and OH⁻ and in pH at 25 °C)

    ion product constant of water, $K_w = 10^{-M} = [H^+][OH^-]$

    neutral solution (both in terms of relative molarities of H⁺ and OH⁻ and in pH at 25 °C)

    pH

    pOH

17.6   Strong acids and bases are fully dissociated in solution

Review

When we have dilute solutions of strong acids and bases, the calculation of [H⁺] or pH is particularly simple because these solutes are 100% ionized in solution. Remember, *we do not write the ionization reaction of a*

strong acid or a strong base as an equilibrium. For strong, monoprotic acids, for example, the value of $[H^+]$ is identical with the molar concentration of the acid. Thus, 0.10 M HCl has $[H^+] = 0.10\ M$.

The autoionization of water contributes essentially nothing to the value of $[H^+]$ except in extremely dilute solutions ($10^{-6}\ M$ or less).

If you have not done so yet, be sure to memorize the names and formulas of the following strong, monoprotic acids:

$HClO_4$, perchloric acid	HCl, hydrochloric acid
$HClO_3$, chloric acid	HBr, hydrobromic acid
HNO_3, nitric acid	HI, hydriodic acid

Sulfuric acid, H_2SO_4, is also a strong acid, at least in its first ionization, but calculating the pH of its solutions is not as simple as for strong monoprotic acid. Also be sure to remember that all of the Group IA metal hydroxides are water-soluble strong bases and fully dissociated in water; the Group IIA metal hydroxides are strong bases too, but they are not as soluble as the hydroxides of the Group IA metals.

For a metal hydroxide from Group IIA, remember that you need to take into account the stoichiometry of the dissociation reaction when calculating the OH^- concentration. For example,

$$Ba(OH)_2 \rightarrow Ba^{2+} + 2OH^-$$

Two moles of hydroxide ion are
released for each mole of $Ba(OH)_2$.

Self–Test

29. Calculate the pH of each of the following solutions.

    (a)   0.25 M HCl   *0.6*

    (b)   0.25 M NaOH   *13.39*

    (c)   0.0010 M Ca(OH)$_2$ ✗ 2   *11.3*

30. A solution of KOH has a pH of 12.45. What is the molarity of the KOH solution?

    0.0282

31. Suppose 25.0 mL of 0.45 M HCl is added to 30.0 mL of 0.35 M NaOH.

    (a)   Write the equation for the chemical reaction that takes place in the solution.

    HCl + NaOH → H₂O + NaCl

    (b)   After these substances have reacted, what is the pH of the solution? (Hint: be sure to find the H^+ concentration in the total final volume of the solution.)

    _____

.055 L

$$\frac{.45\ mol\ HCl}{L} \times .025\ L = \frac{.01125\ mol\ HCl}{.055}$$

New Terms

None

Solutions to Thinking It Through

1   In the Brønsted theory, an acid–base reaction is the transfer of a proton, and neither reactant is able to donate H^+.

2   The likeliest products are OH^- and HS^-, the results of a proton transfer from H_2O to S^{2-}.

$$H_2O + S^{2-}(aq) \rightarrow OH^-(aq) + HS^-(aq)$$

In this particular equation, the two Brønsted acids are H_2O and HS^-. The two Brønsted bases are S^{2-} and OH^-.

3   (a) To the right. For oxoacids with the same number of oxygens, acid strength increases from left to right in the same row.

(b) Above. For oxoacids of elements in the same group and so having the same general formula, acid strength increases from bottom to top within the group.

4   Go through the list and identify the *kinds* of substances and recall their general acid–base properties. Carbon dioxide is a gas and a *nonmetal oxide* that dissolves in water to make a weakly acidic solution. Sodium oxide is the oxide of a group IA element; *all the oxides of the Group IA elements react with water to give solutions of their hydroxides.* HNO_3 is nitric acid, a strong acid. KOH is the hydroxide of one of the group IA metals, *all of which are strongly basic.* SO_3 is a *nonmetal oxide (and a gas) which gives an acid in water* (H_2SO_4). $HC_2H_3O_2$ is acetic acid, a weak acid. Thus only the metal oxide and the metal hydroxide are both solids and give a basic solution in water.

5   (a)   ZO, because Group IIA metals have 2+ charges, which would balance the 2– charge on O.

(b)   Blue; oxides of Group IA and IIA metals all do this.

(c)   Sparingly soluble, like all of the Group IIA oxides.

6   The definition of pH needs to be recalled and translated into $[H^+]$. When pH = 3, $[H^+] = 1 \times 10^{-3}$ M; when pH = 6, $[H^+] = 1 \times 10^{-6} M$. The ratio of these two molar concentrations of H^+ is 1000:1, because,

$$\frac{1 \times 10^{-3}\ M}{1 \times 10^{-6}\ M} = 1000$$

So the solution with a pH of 3 has 1000 times the H^+ concentration as the solution with a pH of 6.

Answers to Self–Test Questions

1.   (a) $CH_3NH_3^+$, (b) NH_3OH^+, (c) $HClO_3$, (d) $H_2C_2O_4$.
2.   (a) $C_2H_4^{2-}$, (b) $HAsO_4^{2-}$, (c) CHO_2^-, (d) HS^-
3.   **H_2SO_4**, HSO_4^-; Cl^-, **HCl** (acid in bold type)
4. NH_3 (base), NH_4^+ (acid); NH_3 (acid), NH_2^- (base)
5. $HC_2H_3O_2 + H_2SO_4 \rightarrow H_2C_2H_3O_2^+ + HSO_4^-$
     $HC_2H_3O_2 + H_2O \rightarrow C_2H_3O_2^- + H_3O^+$
     In these reactions, acetic acid can be either an acid or a base.

6. $HCHO_2$

7. (a) HNO_2, (b) OCl^-

8. H_2O

9. H_2Te. The H–Te bond is weaker, because Te is below Se in the same group in the table.

10. HI. The element I is more electronegative than Te, standing to the right of Te in the same period in the table.

11. H_3PO_4 is stronger. It has one more O.

12. H_3PO_4 is stronger. P is more electronegative than As, standing above As in the table.

13.

14.

15. BF_3 is the Lewis acid; $(CH_3)_2O$ is the Lewis base.

16. NO_2^- and OCl^-; the stronger base is OCl^-.

17. Na_2O reacts with water to give sodium hydroxide, NaOH
    $$Na_2O(s) + H_2O \rightarrow 2NaOH(aq)$$

18. SO_3 reacts with water to give sulfuric acid, H_2SO_4
    $$SO_3(g) + H_2O \rightarrow H_2SO_4(aq)$$

18. $AlCl_3$. Al^{3+} has a greater charge density than Mg^{2+}, so the hydrated aluminum ion will be more acidic.

20. Acidic. The large charge on an Os^{8+} ion would give the ion a very large charge density.

21. (a) $2.63 \times 10^{-9}\ M$ (b) acidic because $[H_3O^+]$ is greater than $[OH^-]$.

22. The solution is neutral because $[H_3O^+] = [OH^-] = 4.30 \times 10^{-8}\ M$.

23. (a) $5 \times 10^{-10}\ M$ (b) $2.5 \times 10^{-7}\ M$

24. b. When pOH = 10, pH = 4, and $[H^+] = 1.0 \times 10^{-4} M$.

25. c. When $[H^+] = 4.68 \times 10^{-6}\ M$, pH $= -\log 4.68 \times 10^{-6}$.

26. a. When pH = 9.65, $[H^+] = 1 \times 10^{-9.65}\ M. = 2.24 \times 10^{-10}\ M$.

27. $2.2 \times 10^{-7}\ M$

28.

| | **Color in** | |
Indicator	**Acid**	**Base**
phenolphthalein	none	red
bromothymol blue	yellow	blue
thymol blue	yellow	blue

29. (a) 0.60, (b) 13.40, (c) 11.30

30. 0.028 M

31. (a) $HCl(aq) + NaOH(aq) \rightarrow NaCl(aq) + H_2O$
    (b) pH = 1.90

Tools you have learned

Consider removing this chart from the Study Guide so you can have it handy when tackling homework problems.

Tool	How it Works
List of strong monoprotic acids	When you are working with solutions of solutes and one is a strong acid, you know it is 100% ionized. This enables you to determine the $[H^+]$ in the solution.
Periodic trends in strengths of binary acids.	You can use the trends to predict the relative acidities of X–H bonds, both for the binary hydrides themselves and for molecules that contain X–H bonds.
Periodic trends in strengths of oxoacids.	You use the trends to compare the relative acidities of oxoacids according to the nature of the central nonmetal as well as the number of oxygens attached to a given nonmetal. The principles involved also let you compare acidities of compounds containing different electronegative elements.
Ion-product constant for water. $[H^+][OH^-] = K_W$	Use this equation to calculate $[H^+]$ if you know $[OH^-]$, and vice versa. Be sure you have learned the value of K_W.
Defining equations for pX $pX = -\log X$ and $X = 10^{-pX}$	Use the equations of this type to calculate pH and pOH from $[H^+]$ and $[OH^-]$, respectively. You also use them to calculate $[H^+]$ and $[OH^-]$ from pH and pOH, respectively.
Relationship between pH and pOH: **pOH = 14.00**	This relationship is used to calculate pH if you know pOH, and vice versa.

Summary of Important Equations

Ion product constant of water

$$[H^+][OH^-] = K_W$$

pH and pOH and the relationship between pH and pOH

$$pH = -\log[H^+] \qquad pOH = -\log[OH^-] \qquad pH + pOH = 14.00 \text{ (at 25 °C)}$$

General equation for pX

$$pX = -\log X$$

Chapter 18

Equilibria in Solutions of Weak Acids and Bases

The list of the strong acids and bases available in chemistry, as you now know, is very short. Most acids and bases are weak, and there are thousands of them. Those that are somewhat soluble in water do not come even close to being 100% ionized in solution. But they vary widely in how weak they are, so special equilibrium constants for these substances—acid or base ionization constants—have been devised whose values tell us at a glance how *relatively* strong or weak such acids and bases are.

Weak acids or bases and their conjugates have life-and-death roles in nature because they help to maintain the pH values of the fluids of living systems within very narrow limits. Our study of buffers explains how this all works. If you are planning to enter any one of the health sciences, buffers might be the most important single topic in this chapter, given their importance in human health and disease.

Learning how to do the various calculations in this chapter is of paramount importance. If you can do the calculations, and doing them is not just mechanical, you almost certainly understand the concepts. The large number of worked examples have been very carefully prepared. If you master one kind before going on to the next, the way to success will be smooth. The trickiest parts involve making simplifying assumptions about what we can safely ignore in certain calculations. It's particularly important that you understand these assumptions and can judge when to make them, because they really do simplify the calculations.

Good luck in your study of this vital chapter. There are very few places in either the chemical or the biological sciences where the equilibria of weak acids and bases in water are not important.

Learning Objectives

Throughout your study of this chapter, keep in mind the following objectives.

1. To learn to write the ionization equilibrium for a weak acid and use the equation to write the appropriate expression for the acid ionization constant, K_a.

2. To learn to write the ionization equilibrium for a weak base and use the equation to write the appropriate expression for the acid ionization constant, K_b.

3. To be able to calculate the value of pK_a or pK_b given K_a or K_b, respectively, for a weak acid or base.

4. To learn how the values of K_a and K_b for a conjugate acid–base pair are related and how one can be calculated given the other.

5. To learn how pK_a and pK_b are related for members of an acid-base conjugate pair.

6. For a solution containing a single solute, to learn how to calculate the value of K_a or K_b given the initial molarity and the measured pH.

7. To learn how to use the percentage ionization of an acid or base to calculate K_a or K_b, respectively.

8   For a solution containing a single solute, to learn how to calculate equilibrium concentrations, the solution's pH, and the solute's percentage ionization from the K_a or K_b value and the initial concentration of the solute.

9   To learn the conditions under which simplifying assumptions enable the simplification of the algebra in acid-base equilibrium calculations.

10   When simplifying assumptions fail, to be able to use either a quadratic solution or the method of successive approximations.

11   To be able to use the formula of a salt to predict if its cation or its anion can affect the pH of a solution, and if it can, to carry out pH calculations involving such salts.

12   To learn how to select solutes to prepare a buffer and to write the equations that describe how the buffer works.

13   To be able to calculate the pH of a buffer.

14   To be able to calculate the concentrations of the buffer components needed to achieve a specified pH.

15   To be able to calculate the pH of a buffered solution after the addition of a strong acid or base.

16   To learn the concept of buffer capacity.

17   To be able to calculate [H$^+$] (or the pH), [HA^-], and [A^{2-}] at equilibrium in a solution of any weak, diprotic acid, H_2A, given its values of K_{a_1} and K_{a_2} , as well as [HA]$_{initial}$.

18   To be able to estimate the pH of a solution of a salt of a polyprotic acid, given its molarity and the relevant ionization constants.

19   To calculate how the pH values of solutions of various kinds of acids or bases change during a titration as a titrant is added and neutralization occurs.

20   To learn the principles involved in the choice of a good acid–base indicator.

18.1   Ionization constants can be defined for weak acids and bases

Review

Weak acids

The first thing you should learn here is the general chemical equation for the ionization of a weak acid.

$$HA \rightleftharpoons H^+ + A^- \tag{1}$$

The same equation applies regardless of the formula of the acid. Thus, HA could be a neutral species, like $HC_2H_3O_2$, an anion, like HSO_4^-, or a cation, like NH_4^+. From this equilibrium equation comes the corresponding *acid ionization constant,* K_a.

$$K_a = \frac{[H^+] [A^-]}{[HA]} \tag{2}$$

Sometimes, it's convenient to express K_a in logarithmic form as pK_a, which is defined as follows.

$$pK_a = -\log K_a \tag{3}$$

You must be sure to master the skills for preparing equations (1) – (3) for any specific acid before you go on to Section 18.2. You can test your progress by working the Self-Test below.

Weak bases

The general equation for the ionization of a weak base is

$$B + H_2O \rightleftharpoons BH^+ + OH^- \tag{4}$$

The corresponding *base ionization constant, K_b* is

$$K_b = \frac{[BH^+]\,[OH^-]}{[B]} \tag{5}$$

The logarithmic expression for pK_b is

$$pK_b = -\log K_b \tag{6}$$

Be sure that you can construct such equations for any specific weak base before you go to Section 18.2. You can test your progress by working the Self-Test below.

Conjugate acid–base pairs

For any acid–base conjugate pair, the following relationships apply.

$$K_a \times K_b = K_w \tag{7}$$

$$pK_a + pK_b = 14.00 \text{ (at } 25 \text{ °C)} \tag{8}$$

These useful equations mean that when we know K_a (or pK_a) for a weak acid we can calculate the values of K_b (or pK_b) for its conjugate base. Most references supply tables only for molecular weak acids or bases, so when we want the K_a or K_b for an ionic species that's a weak acid or base, we usually have to calculate it from the corresponding values for its conjugate base or acid. For example, if we need the K_a for $N_2H_5^+$, we calculate it from the K_b for its conjugate base, N_2H_4. The K_b is tabulated for N_2H_4, but the K_a for $N_2H_5^+$ is not.

Be sure to study Examples 18.1 – 18.4 and work Practice Exercises 1 – 4; then try the following Self–Test.

Self–Test

1.   For each of the following acids, write both the equilibrium equation for the ionization and the expression for K_a.

(a)   HCN $HCN \rightleftharpoons CN^- + H^+$ $K_a = \dfrac{[H^+][CN^-]}{[HCN]}$

(b) $HClO_2$ $HClO_2 \rightleftharpoons H^+ + ClO_2^-$ $K_a = [ClO_2^-][H^+] / [HClO_2]$

(c) $N_2H_5^+$ $N_2H_5^+ \rightleftharpoons N_2H_4 + H^+$ $K_a = [H^+][N_2H_4] / [N_2H_5^+]$

2. For each of the following bases, write both the equilibrium equation for the ionization and the expression for K_b.

   (a) $(CH_3)_2NH$ $(CH_3)_2NH + H_2O \rightleftharpoons (CH_3)_2NH_2^+ + OH^-$ $K_b = \dfrac{[OH^-][(CH_3)_2NH_2^+]}{[(CH_3)_2NH]}$

   (b) N_2H_4 $N_2H_4 + H_2O \rightleftharpoons N_2H_5^+ + OH^-$ $[OH^-][N_2H_5^+] / [N_2H_4] = K_b$

   (c) CN^- $CN^- + H_2O \rightleftharpoons HCN + OH^-$ $K_b = \dfrac{[OH^-][HCN]}{[CN^-]}$

3. What is the value of pK_a for the bicarbonate ion if its K_a is 4.7×10^{-11}?

   10.33

4. Which is the stronger acid, nitrous acid ($K_a = 7.1 \times 10^{-4}$) or formic acid ($K_a = 1.8 \times 10^{-4}$)?

   nitrous acid

5. Which is the stronger acid, hydrocyanic acid, HCN ($pK_a = 9.20$) or the ammonium ion, NH_4^+ ($pK_a = 9.24$)?

   HCN

6. The K_b for the fluoride ion is 1.5×10^{-11}. What is its pK_b?

   10.82

7. Nitrous acid, HNO_2, has $K_a = 7.1 \times 10^{-4}$. What is the value of K_b for NO_2^-?

   1.41×10^{-11}

8. The azide ion, N_3^-, has $pK_b = 9.26$. What is K_a for hydrazoic acid, HN_3?

   1.82×10^{-5}

New Terms

Write definitions of the following terms, which were introduced in this section. If necessary, refer to the Glossary at the end of the text.

acid ionization constant, K_a pK_a

base ionization constant, K_b pK_b

18.2   Calculations can involve finding or using K_a and K_b

Review

We deal with two kinds of calculations in this section.

- Calculating K_a or K_b from an initial concentration of a weak acid or base and either the pH of the solution or some other information about the extent of ionization, like the percent ionization.

$$\text{percent ionization} = \frac{\text{moles ionized per liter}}{\text{moles available per liter}} \times 100\%$$

- Calculating the extent of ionization of a weak acid or base from K_a or K_b and the initial concentration of the solute. The results of such a calculation are often expressed as the calculated pH of the solution.

The first type of calculation is illustrated by Examples 18.5 and 18.6 in the text. Example 18.5 gives a pH value, and Example 18.6 gives a percent ionization. Be sure to study the reasoning involved in obtaining the entries in the concentration tables. We use such tables often.

Calculating equilibrium concentrations from K_a or K_b

The one thing all problems of this type have in common is the necessity *at the outset* of writing the correct equation for the equilibrium. Then you can write the appropriate equilibrium law. And then you'll have something into which to substitute quantities and to calculate equilibrium concentrations of products of the equilibrium. Writing the correct equilibrium equation is easy because there are only three options.

1. ***The only solute is a weak acid.*** Write the equilibrium equation for the ionization of the acid and from it write the equation for the acid ionization constant, K_a. Next, use this equation and the value of K_a (either given or from a table) to solve the problem. If the ionization constant you're given is not K_a but is K_b for the acid's conjugate base, then you must use Equation 7, above, to find K_a.

2. ***The only solute is a weak base.*** In this case, you must write the equation for the ionization of the base, modeling the equation after (4), above. Then you write the expression for K_b and use this equation and the value of K_b to solve the problem. If you've been given, instead, the value of K_a for the base's conjugate acid, you have to use equation (7), above, to find K_b.

3. ***The solution contains both a weak acid and a salt of its conjugate base.*** Such solutions are called *buffers* and are discussed in Section 18.5. With a solution of this type, you can use either K_a or K_b to solve the problem. It depends on the information supplied. If you have a value of K_a, then work with the acid and write the equation for its ionization and for the equilibrium law, K_a. On the other hand, if you have a value of K_b for the acid's conjugate base, develop the ionization equation for the base, prepare the equilibrium law, K_b, and proceed to solve the problem.

Simplifications that apply to most weak acid-base calculations

Example 18.7 takes you through a calculation where we're given the concentration of a weak acid and its K_a, and it illustrates a very important simplification: when K_a is very small, the change in the concentration of the weak acid or base that occurs when the ionization takes place doesn't appreciably change the initial concentration. As a result, the equilibrium concentration of the acid is effectively the same as the initial concentration. Let's review the reasoning.

Suppose we have a 0.10 M HA solution and K_a for the acid is very small. When the acid ionizes, a tiny amount of HA will be lost per liter, which we can call x. In other words, the concentration of HA decreases by x mol per liter, and at equilibrium the concentration of HA equals $(0.10 - x)$ M. But we can anticipate that x will be very small compared to the initial concentration, so we make the approximation

$$(0.10 - x)\,M \approx 0.10\,M$$

In all the calculations you encounter in this section, this approximation is valid, so you can use the *initial* concentration of the acid or base as its *equilibrium* concentration. This makes these equilibrium computations much simpler.

The problem solving strategy, then, boils down to the following.

- First, determine which of the three categories, above, fits the problem. (In other words, is the solute a weak acid, a weak base, or a mixture of an acid and its conjugate base?)

- Next, write the appropriate chemical equilibrium equation.

- Construct the equilibrium law (an equation either for K_a or K_b) from the equilibrium equation.

- If the only solute is a weak acid or a weak base, the algebra works out as follows.

$$\frac{x^2}{[\text{initial conc.}]} = K_a \quad \text{or} \quad \frac{x^2}{[\text{initial conc.}]} = K_b$$

depending on whether the solute is an acid or a base.

- If the solute is an acid, the value of x obtained is both the H^+ concentration and the concentration of the conjugate base formed by the ionization of the acid.

- If the solute is a base, the value of x obtained is both the OH^- concentration and the concentration of the conjugate acid formed by the ionization of the base.

Study Examples 18.7 and 18.8 carefully to see how this problem strategy works. Then work the Practice Exercises following the Examples. When you've done this, you're ready for the problems given next.

Thinking It Through

Describe how you would solve the following problem.

1   Consider a solution of acetic acid with a concentration of 0.620 *M*. Describe how you would follow the strategy above to calculate $[H^+]$, $[C_2H_3O_2^-]$, and the pH of the solution. (Use $K_a = 1.8 \times 10^{-5}$.)

Self–Test

9.   A new organic acid was discovered and at 25 °C a 0.115 *M* solution of the acid in water had a pH of 3.42. Calculate the value of K_a for the acid.

1.26×10^{-6}

10.   A new drug obtained from the seeds of a Uruguayan bush was found to be a weak organic base. A 0.100 *M* solution in water of the drug had a pH of 10.80. What is the K_b of the drug?

4.02×10^{-6}

11.   Calculate the pH of a 0.425 *M* solution of NH_3 in water. (Use $K_b = 1.8 \times 10^{-5}$.)

11.44

12.   Calculate the percentage ionization of HCN in a solution that is 0.10 *M* HCN. Obtain the K_a value from Table 18.1 in the text.

0.00787%

New Term

Write the definition of the following term, which was introduced in this section. If necessary, refer to the Glossary at the end of the text.

percentage ionization

$\dfrac{(1.58 \times 10^{-11})^2}{.1 - (1.58 \times 10^{-11})}$

$1.8 \times 10^{-5} = \dfrac{x^2}{.425 - x}$

18.3   Salt solutions are not neutral if the ions are weak acid or bases

Review

The questions in this section are "How can we tell if a salt, dissolved in water, affects the pH of the solution?" And, "How can we calculate the pH of such a solution?" If neither the anion nor the cation of a salt influences the pH, there is no calculation to carry out.

Does a salt affect the pH?

To answer this question, you need to know whether either (or both) of the ions of the salt are able to change the pH of the solution. In other words, are the ions weak acids or bases?

You can tell if a salt's anion affects a solution's pH very simply. First, determine the formula for the conjugate acid of the anion of the salt. Then, ask yourself whether this acid is on the list of strong acids. If it is not on the list, the conjugate acid is a weak acid, meaning that the anion in question is at least a weak base and it *will* affect the pH of the solution. On the other hand, if the conjugate acid of the salt's anion is strong, the anion will be such a weak base that it will not affect the pH. For example, suppose the anion is $C_3H_5O_2^-$.

$$C_3H_5O_2^- \xrightarrow{\text{Add } H^+} HC_3H_5O_2 \xrightarrow{\substack{\text{Is the acid on} \\ \text{the list of} \\ \text{strong acids?}}} No$$

Anion of salt   Conjugate acid

The acid isn't a strong acid, so the anion, $C_3H_5O_2^-$, must be a weak base that is capable of affecting the pH. This anion should tend to make the solution basic.

A somewhat similar procedure is applied to the cation of the salt. First, recall from Chapter 17 that the cations of the metals of Groups IA and IIA (except for Be^{2+}) do not affect the pH. The cations of other metals, however, form hydrates that can release protons to water molecules to form extra hydronium ions and thus make a solution acidic, *but we'll not deal with calculations involving them*. The only kinds of cations that could potentially affect pH and that we will work with are proton-donating cations, such as the ammonium ion and a few others like it. Therefore, if the cation of the salt is from Group IA or IIA, it won't affect the pH. However, *if the cation of the salt is not a metal, you can anticipate that it be acidic and tend to lower the pH.*

With some salts, like ammonium cyanide, both the cation and the anion of a salt react to some extent with water, with the cation tending to lower the pH and the anion tending to raise it. (See Example 18.12 and Practice Exercise 15.) But *we are not dealing with calculations of pH for such systems.*

Summary

1   If neither the cation nor the anion of a salt can affect the pH, the solution should be neutral. (Example: $NaNO_3$)

2   If only the anion is basic, the solution will be basic. (Example: $NaC_2H_3O_2$)

3   If only the cation is acidic, the solution will be acidic. (Example: NH_4Cl)

4   If the cation is acidic and the anion is basic, the pH outcome depends on their strengths relative to each other. We'll not encounter calculations dealing with such systems. (Example: NH_4CN)

Study Example 18.9 and work Practice Exercises 10 – 12.

How can we calculate the pH of a salt solution when the salt does affect the pH?

If you determine that a salt's anion (but not its cation) affects pH, the pH calculation is simply the same kind that you do for any Brønsted base, just as you learned in Section 18.2.

If the salt involves a proton-donating cation, like the ammonium salt of a strong acid, the pH calculation is nothing more than what you do for any weak acid, just as you learned in Section 18.2.

The calculations, in other words, involve essentially nothing new, once it's determined that a calculation is warranted. The problem solving strategy mirrors the one described in the previous section, above:

1   Determine the nature of the salt. Which ion, if either, affects the pH?

2   Write the appropriate chemical equilibrium equation for the reaction of that ion with water.

3   Construct the equilibrium law (an equation either for K_a or K_b) from the equilibrium equation.

4   Get the value of the ionization constant, K_a or K_b. Seldom for any *ion* will this value be found directly in a table. If you need the K_b of an *anion*, you'll have to use the K_a of its conjugate acid and than calculate K_b from $K_a \times K_b = K_w$. If you need the K_a of a cation, like NH_4^+, you may have to find the K_b of its conjugate base, NH_3, and use the equation $K_a \times K_b = K_w$ to calculate the K_a that you need.

From here on, the calculation is just like that for weak molecular acids or molecular bases. Once again, the algebra works out to be

$$\frac{x^2}{[\text{initial conc.}]} = K_a \quad \text{or} \quad \frac{x^2}{[\text{initial conc.}]} = K_b$$

depending on whether the reactant is the cation (an acid) or the anion (a base).

With the above before you, study Examples 18.10 and 18.11 and work the Practice Exercises that follow them. Then tackle the following problems.

Thinking It Through

Describe how you would solve the following problem.

2   Is a 0.100 *M* solution of KNO_2 acidic, basic, or neutral? If it is not neutral, what is its calculated pH? Outline the strategy you would use to answer these questions.

Self–Test

13.   Decide if each ion is a weak base that's capable of affecting the pH of a solution.

(a) $C_6H_5O^-$ _____ yes _____

(b) OCN^- _____ yes _____

(c) NO_3^- _____ no _____

14.   Predict the behavior of each salt in water by stating if it will cause an aqueous solution to be acidic, basic, or neutral.

(a) KNO_3 ____neutral____   (d) $NaClO_4$ ____neutral____

(b) $NaClO_2$ ____basic____   (e) C_5H_5NHCl →Cl⁻ ____acidic____
                                                      acidic

(c)   KF ____basic____   (f) $NaNO_2$ ____basic____

15. Will either ion from potassium butyrate, $KC_4H_7O_2$, react with water? _____

    If so, calculate the pH of a 0.120 M solution. (For $HC_4H_7O_2$, $K_a = 1.5 \times 10^{-5}$.) _____

16. Will either ion from ammonium bromide, NH_4Br, react with water? _____

    If so, calculate the pH of a 0.240 M solution. _____

17. Will either ion from potassium iodide, KI, react with water? _____

    If so, calculate the pH of a 0.100 M solution. _____

New Terms

None

18.4   Simplifications fail for some equilibrium calculations

Review

If your instructor is covering this section in your course, the first step that you will have to take in solving *any* acid–base equilibrium problem is to determine if the simplifications we've used so far can be employed.

You *can* use the simplifications if the following conditions are met.

- For a solution of a weak acid: $[HA]_{initial} \geq 400 \times K_a$

- For a solution of a weak base: $[B]_{initial} \geq 400 \times K_b$

If these conditions are not met, the extent of ionization will not be negligible and equilibrium concentration of the solute will differ by a significant amount from initial concentration. As a result, when you substitute quantities from the "Equilibrium concentration" row of the concentration table into the mass action expression, the denominator cannot be simplified. For a weak acid, for example, the expression that you will have to solve for x is

$$\frac{x^2}{[\text{initial conc.}] - x} = K_a$$

To obtain x, you will have to use either the quadratic formula (see Example 18.13) or the method of successive approximations (see Figure 18.3). The method of successive approximations is quicker and easier, so we recommend that you learn it if you have to choose.

Self–Test

18. What is the minimum concentration of chloroacetic acid, $HC_2H_2O_2Cl$, in an aqueous solution that would allow us to reasonably use the simplifying assumption about the equilibrium concentration of the acid? The K_a of this acid is 1.4×10^{-3}. _____

$x = .00379$

$$\frac{(x)^2}{(.01-x)} = 1.4 \times 10^{-3}$$

19. Calculate the pH of a 0.010 M solution of chloroacetic acid. (Refer to the previous problem for its formula and K_a value.)

$[H^+] = 3.1 \times 10^{-3} \quad pH = 2.5$

New Terms

None

HA^*

18.5   Buffers enable the control of pH

conj. acid + OH^- → H_2O + conj. base

conj. base + H^+ → conjugate acid

Review

A *buffer* is a solution that maintains a nearly constant pH even when small amounts of strong acid or strong base are added to it. Be sure to note, however, that a buffer doesn't necessarily hold a solution to a *neutral* pH of 7.00 (at 25 °C). Rather, the pH is held reasonably steady at whatever pH value the buffer is prepared. Thus, the buffered solution could be acidic, basic, or neutral.

Note further that a buffer doesn't necessarily maintain an absolutely *constant* pH. It only prevents large changes of pH when strong acids or bases challenge the system.

A buffer is a solution that contains both a weak acid and a weak base. Often, the *buffer pair* consists of a weak acid, HA, and one of its salts, e.g., NaA. The anion A^- is the Brønsted base in the buffer; its conjugate acid, HA, is the acid component of the buffer. A typical pair is sodium acetate and acetic acid. The base, acetate ion, has the ability to react with extra protons that enter. The acid component, acetic acid, can neutralize extra hydroxide ions that get into the solution. This prevents large changes in the concentrations of H^+ or OH^- ions, and therefore, large changes in pH.

Another kind of buffer pair consists of a weak base and its salt, for example, ammonia plus the ammonium salt of a strong acid, such as ammonium chloride. The cation, NH_4^+, can neutralize any OH^- ion that enters the solution, and NH_3 can react with any protons that get in.

To perform buffer calculations, you must understand how the components of a buffer function. In other words, you should be able to write equations that show how they neutralize H^+ or OH^-. Here's a summary:

When a strong acid is added to a buffered solution, it supplies H^+, which is neutralized by the reaction:

$$(\text{conjugate base}) + H^+ \rightarrow (\text{conjugate acid})$$

$$C_2H_3O_2^- + H^+ \rightarrow HC_2H_3O_2$$

$$NH_3 + H^+ \rightarrow NH_4^+$$

When a strong base is added to a buffered solution, it supplies OH^-, which is neutralized by the reaction:

$$(\text{conjugate acid}) + OH^- \rightarrow (\text{conjugate base}) + H_2O$$

$$HC_2H_3O_2 + OH^- \rightarrow C_2H_3O_2^- + H_2O$$

$$NH_4^+ + OH^- \rightarrow NH_3 + H_2O$$

Study these equations as they apply to the acetic acid/acetate buffer and the ammonium ion/ammonia buffer.

Factors that determine the pH that a buffer pair will hold

There are two.

1   The pK_a of the Brønsted acid of the pair. (If the buffer is the ammonium ion/ammonia buffer or one like it, then this factor is the pK_b of the Brønsted base.)

2   The logarithm of the *ratio* of the *initial* values of the molarities of the buffer's components.

It's safe to use *initial* molarities for *equilibrium* molarities in all of our buffer calculations.

Buffer calculations start with the chemical equations and equilibrium expressions that define K_a or K_b. When the buffer consists of a weak acid, HA, and a salt, like NaA or KA, we use the equation for K_a.

$$K_a = \frac{[H^+][A^-]}{[HA]}$$

The text explains how, in buffer calculations, we can use *initial* molar concentrations as equilibrium concentrations for both $[A^-]$ and $[HA]$. Solving this equation for $[H^+]$ and switching to logarithms, gives us the pH of the buffer in terms of the value of pK_a and the log of the ratio of acid to anion. Assuming the acid is monoprotic and can be represented by HA, we have

$$pH = pK_a - \log \frac{[HA]_{init}}{[A^-]_{init}} \tag{9}$$

This equation is known as the Henderson-Hasselbalch equation. It is frequently used in biology and other life sciences.

When the buffer is made up of a weak base, B, and its conjugate acid, BH^+, (example, the ammonia/ammonium ion buffer), then we use the defining equation for K_b for the weak base.

$$K_b = \frac{[BH^+][OH^-]}{[B]}$$

Again, *initial* molar concentrations of cation and base can be used as equilibrium values. Solving for $[OH^-]$ and taking the negative logarithm gives pOH from which pH is easily obtained.

Factors that determine the choice of a specific buffer pair

There are two, and they are essentially identical with the two given above that are involved in the pH that a buffer pair will hold. As Equation 9 above shows, for a buffer that is to hold an acidic pH, these factors are the pK_a of the weak acid and the ratio of the molar concentrations of the weak acid and its conjugate base. In experimental work, these factors govern the selection of a buffer pair.

For a buffer meant to hold an acidic pH, we pick a weak acid with a value of pK_a as close to our preselected pH as we can get. We do this because when the ratio of molarities, anion to acid, equals 1, then pH = pK_a. {Notice that when $[A^-] = [HA]$ in Equation 9, these two molarities cancel, leaving $K_a = [H^+]$ and so pK_a = pH].

When we want a buffer for a pH not identical to, but still not too far from the acid's pK_a, we calculate the ratio of the molarities of anion to acid that will adjust the system to the desired pH. But the choice of the acid for its pK_a value is of first importance. For maximum effectiveness, the ratio of the initial molar concentrations of the anion to the acid should be somewhere between 10:1 and 1:10.

The text does not deal with choosing buffer pairs for basic pH values in general; it only develops the ammonium ion/ammonia buffer. However, the same principles would apply.

Calculating the effect of strong acid or base on the pH of a buffer

In calculations that deal with the results of adding strong acid or base to a given buffer, we need to apply the equations given above that describe how a buffer works. Remember that when H^+ is added to the buffer, it reacts with the conjugate base. This reduces the amount of conjugate base by the number of moles of H^+ added, and increases the amount of conjugate acid by the identical amount. Similarly, when OH^- is added to a buffer, by reaction it reduces the amount of conjugate acid by the number of moles of OH^- added, and increases the amount of conjugate base by the same amount. Study Example 18.17 to see how these calculations proceed.

Factors that determine actual quantities of the members of the buffer pair

This consideration has to do with buffer *capacity*, that is, with just how much acid or base we expect the prepared buffer will have to absorb during some experiment while holding the pH within a predetermined range. For a large capacity buffer, we would want relatively high concentrations of each component, remembering that it is the *ratio* of the two molarities that, together with the pK_a, determines the initial pH. In biological applications of buffers, however, the potential toxicities of the buffer components at high concentrations must be considered.

Thinking It Through

$\dfrac{.25mol}{L} \times .2 L$

Identify the information needed to solve the following problem and explain what must be done with it.

3   You are a member of a team of scientists that decides that the lab should have on hand 1.0 L of an aqueous buffer for a pH of 4.5. What questions or problems have to be answered before the chemicals can be assembled, measured, and mixed?

Self–Test

20. Calculate the pH of a buffered solution made up of $0.125\ M\ KC_2H_3O_2$ and $0.100\ M\ HC_2H_3O_2$. The pK_a of acetic acid is 4.74.

    _____4.84_____

21. The pK_a of formic acid, $HCHO_2$, is 3.74. What mole ratio of sodium formate, $NaCHO_2$, to formic acid is needed to prepare a solution buffered at a pH of 3.00?

    _____.18_____

22. How many grams of NH_4Cl have to be added to 200 mL of $0.25\ M\ NH_3$ to make a solution that is buffered at pH 9.10? For NH_3, $K_b = 1.8 \times 10^{-5}$.

    _____

23. Suppose 0.050 mol of HCl were added to 1.00 L of the buffer in Question 20. By how much and in which direction will the pH change?

    _____0.9 decrease_____

$pH = 4.74 + log \dfrac{.075}{.15}$

New Term

Write the definition of the following term, which was introduced in this section. If necessary, refer to the Glossary at the end of the text.

buffers

18.6   Polyprotic acids ionize in two or more steps

Review

Be sure you learn to write the equations for the step-wise ionizations of polyprotic acids. Follow the example in this section for carbonic acid, H_2CO_3. Also learn to write the acid ionization constant for each step. These are designated as K_{a_1}, K_{a_2}, and so forth.

The heart of the section is in the simplifications that greatly reduce the work of calculations. They are possible because the numerical values of the K_a's generally differ very much from one ionization step to the next. The simplifications are as follows.

1   We can use just the first ionization step and its K_{a_1} to calculate $[H^+]$ of a solution of the acid.

2   The value of the anion formed in the first step of the ionization is equal to $[H^+]$.

3   The equilibrium concentration of the polyprotic acid equals its initial concentration.

4   The molarity of the ion formed in the second ionization step numerically equals K_{a_2} for the acid, provided that there is no other solute.

In other words, by simplifications 1 and 2 we can treat the acid as if it were monoprotic to find $[H^+]$. The third simplification is made possible by the very small sizes of the first ionization constants of weak polyprotic acids; very little ionization occurs to lower the concentration of the un-ionized species.

Self–Test

24.   On a separate sheet of paper, write the equilibrium equations and the expressions for K_{a_1} and K_{a_2} for the step-wise ionization of H_2SeO_3.

25.   What are the values of $[H^+]$, pH, and $[Asc^{2-}]$ in a solution of ascorbic acid, which we represent as H_2Asc, with a concentration of 0.100 M? $K_{a_1} = 6.8 \times 10^{-5}$ and $K_{a_2} = 2.7 \times 10^{-12}$.

$[H^+] =$ ___.00261___   pH = ___2.54___ $[Asc^{2-}] =$ ___2.7×10^{-12}___

26.   What is the concentration of $[HAsc^-]$ in the solution of the previous question?

___.00261___

New Terms

None

18.7   Salts of polyprotic acids give basic solutions

Review

The anions of polyprotic acids are weak Brønsted bases. Those of diprotic acids are able to react with water in two steps, each step producing some OH^-. Section 18.7 in the text illustrates this using the carbonate ion.

Just as with diprotic acids, there are two equilibrium constants, one for each of the two steps, designated as K_{b_1} and K_{b_2}. In general, K_{b_1} is much larger than K_{b_2}. Therefore, almost all of the OH^- produced by the two successive reactions with water comes from the first step. We are able to ignore the contribution of OH^- by the second step. Therefore, we can use K_{b_1} and its associated equilibrium expression to calculate $[OH^-]$ and ultimately the pH of the solution.

You may have already noticed the parallel between the simplifications that apply to solutions of bases such as CO_3^{2-} and those that apply to solutions of diprotic acids, such as H_2CO_3. In fact, the simplifications are identical, except that in one case we are calculating $[OH^-]$ and in the other $[H^+]$.

The one complicating factor in dealing with problems involving salts of polyprotic acids is that we must calculate the K_b values from the corresponding K_a values for the conjugate acids. But we now know the relationship among K_a, K_b, and K_w for any acid–base conjugate pair. With polyprotic acids, this relationship works itself out with slightly different labels.

$$K_{b_1} = \frac{K_w}{K_{a_2}} \quad \text{and} \quad K_{b_2} = \frac{K_w}{K_{a_1}}$$

Note carefully the subscripts: K_{b_1} comes from K_{a_2} and K_{b_2} comes from K_{a_1}.

After studying Example 18.19 and working Practice Exercises 22 and 23, take the Self-Test below.

Self–Test

27. Calculate the pH of a 0.20 M solution of sodium oxalate, $Na_2C_2O_4$. Use Table 18.3 in the text for the successive ionization constants of oxalic acid.

     8.75

28. Calculate the pH of a 0.10 M solution of Na_2X at 25 °C. The first ionization constant of H_2X is 9.5×10^{-8} and the second is 1.3×10^{-14}. (In this case, $[X^{2-}]_{initial} < 400 \times K_{b_1}$, so you will have to use either the quadratic equation or the method of successive approximations.)

29. In the solution described in Question 27, what is the concentration of $H_2C_2O_4$ at equilibrium?

     $1.53 \times 10^{-13} M$

30. In the solution described in Question 27, what is the concentration of $C_2O_4^{2-}$?

     .2 M

New Terms

None

18.8   Acid–base titrations have sharp changes in pH at the equivalence point

Review

The nature and appearance of an acid–base titration curve is a function of the relative strengths of the acid and the base, because these determine how the ions being produced by the neutralization reaction react (if at all) with water as Brønsted acids or bases.

- A strong acid is titrated with a strong base. Example: HCl and NaOH

    The titration curve is symmetrical about pH 7, the equivalence point.

- A weak acid is titrated with a strong base. Example: $HC_2H_3O_2$ and NaOH

    The equivalence point is greater than 7 (because the newly forming anion, a Brønsted base, reacts with water to give some OH^- and a slightly alkaline pH).

- A weak base is titrated with a strong acid. Example: NH_3 and HCl.

    The equivalence point is less than 7 (because the newly forming cation, a Brønsted acid, reacts with water to give some H^+ and a slightly acidic pH).

    The calculations required to obtain points for a titration curve are identical in kind to those studied earlier in the chapter.

- Before any titrant is added.

    Use molarity and an acid (or base) ionization constant to find $[H^+]$ (or $[OH^-]$, depending on what exactly is the substance about to be titrated).

- As soon as some titrant is added, you have to take into account three factors.

    (1)   The *total* volume changes, so concentrations change for this reason alone.

    (2)   The concentration of the substance being titrated, which is needed in the calculation of pH, decreases because it is being consumed in the neutralization.

    (3)   *If the salt being formed is one with an ion that is basic or acidic,* there is a buffer effect requiring a buffer-type calculation.

    Choice of Indicator   Ideally, the midpoint of the pH range of the indicator's color change should be the same as the pH of the equivalence point for the titration. What this means is that if the acid ionization constant of the indicator is K_{In}, then pK_{In} should equal the pH of the titration's equivalence point.

Self–Test

31.   Calculate the pH of the solution that exists in a titration after 22.00 mL of 0.10 M NaOH has been added to 25.00 mL of 0.10 M $HC_2H_3O_2$.

_____ 5.61 _____

32.   Calculate the pH of the solution that forms in a titration in which 24.00 mL of 0.20 M KOH is added to 25.00 mL of 0.20 M HCl.

_____ 2.14 _____

33. Calculate the pH of the solution that forms during a titration when 26.00 mL of 0.10 M NaOH has been added to 25.00 mL of 0.10 M $HC_2H_3O_2$.

11.3

34. Calculate the pH of the solution that has formed during a titration when 28.00 mL of 0.10 M HCl has been added to 25.00 mL of 0.10 M NH_3.

2.25

35. If you know that there are two indicators, X and Y, that undergo their color changes in the same pH range, and that X changes from colorless to blue and Y from orange to red, which indicator would be the better choice and why?

whichever is more dramatic

36. If you were asked to carry out a titration of dilute hydrobromic acid with potassium hydroxide, which indicator would be the better choice (and why)—thymol blue or bromothymol blue?

whichever has a change a pH near the equivalency point

New Terms

Write the definitions of the following terms, which were introduced in this section. If necessary, refer to the Glossary at the end of the text.

end point   equivalence point

Solutions to Thinking It Through

1   First we note that the only solute in the solution is acetic acid, a weak acid. This tells us we need to use K_a and that we must write the chemical equation for the ionization of a weak acid. The equilibrium equation is

$$HC_2H_3O_2 \rightleftharpoons H^+ + C_2H_3O_2^-$$

and the equilibrium law is

$$K_a = \frac{[H^+][C_2H_3O_2^-]}{[HC_2H_3O_2]} = 1.8 \times 10^{-5}$$

Because H^+ and $C_2H_3O_2^-$ form in equal numbers, we let their concentrations be equal to x. So,

$$[H^+] = [C_2H_3O_2^-] = x$$

The concentration of $HC_2H_3O_2$ equals the initial concentration reduced by the amount that ionized.

$$[HC_2H_3O_2] = (0.60 - x)\,M$$

But, we expect x to be small compared to 0.60, so we simplify by dropping the x to give

$$[HC_2H_3O_2] = 0.60\ M$$

Next, we substitute these values into the equation for K_a and get

$$K_a = \frac{(x)(x)}{(0.620)} = 1.8 \times 10^{-5}$$

$$x = 3.3 \times 10^{-3} \, M$$

Therefore, $[H^+] = [C_2H_3O_2^-] = 3.3 \times 10^{-3} \, M$. The pH equals $-\log(3.3 \times 10^{-3}) = 2.48$.

2   We have to study the ions that are present in KNO_2, which are K^+ and NO_2^-. The cation is of a Group IA metal, so it cannot affect the pH of water. The anion has HNO_2 as its conjugate acid, an acid not on the list of strong acids. Therefore, we expect that HNO_2 must be a weak acid. Reasoning in this way, we infer that NO_2^- must be a weak Brønsted base and we expect it to affect the pH. So in this way we discover that the solution is not neutral and that we have to make a calculation.

We next write the equilibrium equation for the reaction of the weak base NO_2^- with water and then write the corresponding expression for K_b.

$$NO_2^- + H_2O \rightleftharpoons HNO_2 + OH^- \qquad K_b = \frac{[HNO_2][OH^-]}{[NO_2^-]}$$

To get the value of K_b, we look up the value of K_a of the conjugate acid of NO_2^-, namely, HNO_2, in an appropriate table (for example, Table 18.1), and convert it to the value of K_b for NO_2^-. The calculation is

$$K_b = \frac{K_w}{K_a} = \frac{1.0 \times 10^{-14}}{7.1 \times 10^{-4}} = 1.4 \times 10^{-11}$$

The ionization equilibrium tells us that $[HNO_2] = [OH^-]$, both of which we can set equal to x. The equilibrium value of $[NO_2^-]$ is essentially the same as the initial value of the salt, $0.100 \, M$, because $0.100 - x \approx 0.100$. So, after constructing a concentration table, it will be apparent that

$$\frac{(x)(x)}{0.100} = K_b$$

from which we solve for x. The value we obtain is $x = 1.2 \times 10^{-6} \, M = [OH^-]$. From this we calculate the pOH and then subtract from 14.00 to obtain the pH. The pH equals 8.08. As a check, we note that the pH is above 7.00, which is what we expect for a solution of a weak base.

3   The desired pH is on the acidic side of 7, so a combination of a weak *acid* and its (perhaps sodium) salt should be used. But which acid? This is decided by finding an acid with a pK_a value that lies in the range of 4.5 ± 1.0, because the best pK_a is one that is related to the desired pH by the equation $pH = pK_a \pm 1$.

If a biological system were to be buffered, the research team would have to ask at this point in the evaluation, "how toxic is the proposed buffer pair to the biological system?"

The next question is "what ratio of acid molarity to anion (or salt) molarity would bring the pH of the buffer to 4.5?" The "tool" we need for this question is the equation that relates pH to pK_a and the ratio of acid to anion (salt).

$$pH = pK_a - \log \frac{[HA]_{init}}{[A^-]_{init}}$$

So we would next insert the values of pH and pK_a into this equation and solve for the molar concentration ratio.

The ratio does not tell us "what actual number of moles of the acid and the salt should be used to satisfy the ratio?" This depends on how much neutralizing *capacity* is needed. And this depends on an estimate by the research group of how much incursion of strong acid or base into the buffer solution can be tolerated. Remember, as discussed in the text, you need about 10 times as many moles of a given buffer component as are the estimated number of moles of acid (or base) that are expected to invade the system.

Answers to Self–Test Questions

1. (a) $HCN \rightleftharpoons H^+ + CN^-$, $K_a = \dfrac{[H^+][CN^-]}{[HCN]}$

   (b) $HClO_2 \rightleftharpoons H^+ + ClO_2^-$, $K_a = \dfrac{[H^+][ClO_2^-]}{[HCClO_2]}$

   (c) $N_2H_5^+ \rightleftharpoons H^+ + N_2H_4$, $K_a = \dfrac{[H^+][N_2H_4]}{[N_2H_5^+]}$

2. (a) $(CH_3)_2NH + H_2O \rightleftharpoons (CH_3)_2NH_2^+ + OH^-$, $K_b = \dfrac{[(CH_3)_2NH_2^+][OH^-]}{[(CH_3)_2NH]}$

   (b) $N_2H_4 + H_2O \rightleftharpoons N_2H_5^+ + OH^-$, $K_b = \dfrac{[N_2H_5^+][OH^-]}{[N_2H_4]}$

   (c) $CN^- + H_2O \rightleftharpoons HCN + OH^-$, $K_b = \dfrac{[HCN][OH^-]}{[CN^-]}$

3. 10.33
4. Nitrous acid
5. HCN
6. 10.82
7. 1.4×10^{-11}
8. 1.8×10^{-5}
9. $K_a = 1.26 \times 10^{-6}$
10. $K_b = 4.01 \times 10^{-6}$
11. pH = 11.44
12. 0.0079%
13. (a) $C_6H_5O^-$ is a Brønsted base, (b) OCN^- is a Brønsted base, (c) NO_3^- is not a Brønsted base.
14. (a) neutral, (b) basic, (c) basic, (d) neutral, (e) acidic (contains $C_5H_5NH^+$ and Cl^- ions), (f) basic
15. Yes. $C_4H_7O_2^-$. pH = 8.95
16. Yes. NH_4^+. pH = 4.93
17. No
18. 0.56 M
19. By successive approximations: $[H^+] = 3.1 \times 10^{-3}$ M; pH = 2.51
20. pH = 4.84
21. $[CHO_2^-]/[HCHO_2] = 0.18$, so the desired mole ratio is (0.18 mol $NaCHO_2$)/(1 mol $HCHO_2$).
22. 3.7 g NH_4Cl

23. The pH will decrease by 0.40 units.

24. $H_2SeO_3 \rightleftharpoons H^+ + HSeO_3^-$ $\quad K_{a_1} = \dfrac{[H^+][HSeO_3^-]}{[H_2SeO_3]}$

    $HSeO_3^- \rightleftharpoons H^+ + SeO_3^{2-}$ $\quad K_{a_2} = \dfrac{[H^+][SeO_3^{2-}]}{[HSeO_3^-]}$

25. $[H^+] = 2.6 \times 10^{-3}$ M; pH $= 2.58$; $[Asc^{2-}] = 2.7 \times 10^{-12}$ M

26. $[HAsc^-] = 2.6 \times 10^{-3}$ M
27. pH $= 8.76$
28. pH $= 12.95$
29. $[H_2C_2O_4] = K_{b_2} = 1.5 \times 10^{-13}$ M
30. 0.20 M
31. pH $= 5.61$
32. pH $= 2.39$
33. pH $= 11.30$ (caused solely by the OH^- ion provided by the excess NaOH)
34. pH $= 2.25$ (caused solely by the H^+ ion provided by the excess HCl)
35. Indicator X, because a very dramatic color change is easier to notice.
36. Bromothymol blue, because the equivalence point for the titration of a strong acid and a strong base is at pH 7, so we need an indicator whose color change occurs at or near pH 7.

Tools you have learned

Consider removing this chart from the Study Guide so you can have it handy when tackling homework problems.

Tool	How it Works
Acid ionization equilibria and K_a	Following the general form, you can write the correct chemical equation for the ionization of any weak acid in water and construct the correct equilibrium law to be used in calculations.
Base ionization equilibria and K_b	Following the general form, you can write the correct chemical equation for the ionization of any weak base in water and construct the correct equilibrium law to be used in calculations.
percent ionization $= \dfrac{\text{amount ionized}}{\text{amount available}} \times 100\%$	We use the equation to calculate the percentage ionization from the initial concentration of acid (or base) and the change in the concentrations of the ions. If the percentage ionization and initial concentration of solute are known, the K_a (or K_b) can be calculated.

Relationship among K_a, K_b, and K_w for a conjugate acid–base pair	Enables you to calculate K_a given K_b, or vice versa.
Identification of the nature of the solute as **(1) only a weak acid (2) only a weak base** **(3) a mixture of a weak acid and its conjugate base**	This identification points the way in solving problems by enabling you to write the correct equilibrium chemical equation, and therefore the correct equilibrium law.
Recognition of acidic cations and basic anions	Enables you to determine whether a salt solution is acidic, basic, or neutral. Also, the first step in calculating the pH of a salt solution. Use it in conjunction with the preceding tool to establish the correct direction in analyzing the problem.
Simplifications in calculations:	When the concentration of the solute is larger than $400 \times K$, initial concentrations can be used as though they are equilibrium values in the mass action expression.
Method of Successive Approximations (Figure 18.3)	Provides a quick procedure for calculating equilibrium concentrations when the usual simplifications are inappropriate.
Reactions when H+ or OH− are added to a buffer	Use these equations to determine the effect of strong acid or strong base on the pH of a buffer. Adding H^+ lowers $[A^-]$ and raises $[HA]$, adding OH^- lowers $[HA]$ and raises $[A^-]$.

Summary of Important Equations

Percent ionization of a weakly ionized solute

$$\text{percent ionization} = \frac{\text{moles ionized per liter}}{\text{moles available per liter}} \times 100\%$$

Acid ionization constant: K_a and pK_a for a weak acid: $HA \; \rightleftharpoons \; H^+ + A^-$

$$K_a = \frac{[H^+][A^-]}{[HA]} \qquad pK_a = -\log K_a$$

Base ionization constant: K_b and pK_b for a weak base: $B + H_2O \; \rightleftharpoons \; BH^+ + OH^-$

$$K_b = \frac{[BH^+][OH^-]}{[B]} \qquad pK_b = -\log K_b$$

Summary of Important Equations (*continued*)

Relationship among K_a, K_b, and K_w for conjugate acid–base pairs

$$K_a \times K_b = K_w$$

$$pK_a + pK_b = 14.00 \text{ (at 25 °C)}$$

Buffers

$$pH = pK_a - \log \frac{[HA]_{init}}{[A^-]_{init}}$$

For the best buffer action:

$$pH = pK_a \pm 1$$

Successive acid ionization constants, polyprotic acids, illustrated by H_2A

$$H_2A \;\rightleftharpoons\; H^+ + HA^- \quad K_{a_1} = \frac{[H^+][HA^-]}{[H_2A]}$$

$$HA^- \;\rightleftharpoons\; H^+ + A^{2-} \qquad K_{a_2} = \frac{[H^+][A^{2-}]}{[HA^-]}$$

Concentration of A^{2-} in a solution of H_2A (when H_2A is the *only* solute)

$$[A^{2-}] = K_{a_2}$$

Chapter 19

Solubility and Simultaneous Equilibria

In this chapter we extend our discussion of ionic equilibria to deal with salts that until now we have considered insoluble. As you learn here, such compounds do dissolve slightly and their saturated solutions represent situations in which the solid salt is in equilibrium with its ions in solution. Among the topics we examine are the effects of one salt on the solubility of another and how relative degrees of insolubility permit separation of metal ions by selective precipitation. We also study equilibria in the formation of substances called complex ions and how such substances can influence the solubilities of salts. This is a topic that has many practical applications.

Learning Objectives

Throughout your study of this chapter, keep in mind the following objectives.

1   To learn how to write the special equilibrium law for a *sparingly soluble* (i.e., very slightly soluble) salt, namely, an equation for a *solubility product constant* or K_{sp}.

2   To learn how to calculate the value of K_{sp} from solubility data.

3   To learn how to calculate the solubility of a sparingly soluble salt, given its K_{sp} value.

4   To learn the *common ion effect* and how it affects the solubility of some solute.

5   To learn how to calculate the solubility of a sparingly soluble salt in the presence of a common ion.

6   To learn how to use K_{sp} to find out if the precipitation of a sparingly soluble salt will take place when a solution develops a certain composition.

7   To learn what happens chemically when metal sulfides or oxides dissolve in aqueous solutions.

8   To learn how a special solubility product constant, the *acid solubility product constant* or K_{spa}, works best with sparingly soluble sulfides of metal ions.

9   To learn how to adjust the pH of a solution so that *selective precipitation* can be used to separate one metal ion from another.

10   To learn the vocabulary of complex ions and how to represent the equilibria involving metal ion complexes in solution.

11   To learn how to write appropriate chemical equations and equilibrium laws that correspond to formation constants and instability constants for complexes

12   To be able to apply Le Châtelier's principle to explain how complex ion formation affects the solubility of a sparingly soluble salt.

13   To learn how to use a formation constant (or its reciprocal, an instability constant) to calculate the effect of complex ion formation on the solubility of a sparingly soluble salt.

19.1 An insoluble salt is in equilibrium with the solution around it

Review

The equilibrium law for the solubility equilibrium of a salt involves the salt's *ion product,* which is the product of the ion molarities raised to powers obtained from the subscripts in the salt's formula. The sparingly soluble salt calcium phosphate, $Ca_3(PO_4)_2$, for example, gives the following equilibrium in its saturated aqueous solution.

$$Ca_3(PO_4)_2(s) \;\rightleftharpoons\; 3Ca^{2+}(aq) + 2PO_4^{3-}(aq)$$

The ion product for $Ca_3(PO_4)_2$ is $[Ca^{2+}]^3 \, [PO_4^{3-}]^2$. Notice that the exponents come from the coefficients of the ions in the equilibrium. When the solution is *saturated,* and only then, the value of the ion product is a constant (at a given temperature) and is called the *solubility product constant,* K_{sp}. Thus, for calcium phosphate, the equation for its K_{sp} is

$$K_{sp} = [Ca^{2+}]^3 \, [PO_4^{3-}]^2 \quad \text{(saturated solution)}$$

The calculations in this section fall into three major categories.

- Calculating K_{sp} from the known or measured solubility of a salt.

- Calculating solubility from K_{sp}.

- Using the calculated ion product for a solution of a salt to determine whether a precipitate will form.

   For the first two kinds of calculations, the concentration table is helpful in organizing our thinking. To have a consistent approach that we can apply to a variety of situations, we imagine the saturated solution being formed in a stepwise fashion. First, we examine what's in the solvent to which the sparingly soluble salt will be added and determine concentration values for any of the salt's ions that might already be present (say, from some other compound already dissolved). These go in the "initial concentration" row. If the solvent is pure water, then all the values in this top row are zero.

   The next step is to consider how the concentrations change as the salt dissolves. If we're calculating K_{sp} from solubility data, then the entries in the "changes in concentration" row will have known numerical values. When we're calculating solubility from K_{sp}, the quantities in the "change" row will be our unknown quantities expressed as multiples of *x*.

   As usual, the equilibrium concentrations in the last row are obtained by adding the changes to the initial values.

   In the worked examples in this section, you will notice that we place no entries under the formula for the sparingly soluble salt. The concentration of the solid doesn't appear in the mass action expression (the ion product).

Calculating a salt's K_{sp} from its solubility

When we are dealing with a salt dissolved in water, there are no ions of the salt present in the solvent before the salt is added, so the entries in the "initial concentrations" row are zero. This is what we see in Examples 19.1 and 19.2. In Example 19.3, the solvent already contains an ion found in the sparingly soluble salt ($PbCl_2$ in this case). Notice how we enter the value of the Cl^- concentration in the solvent into the first row.

   Next, we consider what happens when the salt dissolves and forms the saturated solution. The entries in the *change* row of the concentration table are obtained from the *molar solubility,* which is the solubility expressed in moles per liter. This is the number of moles of the salt that dissolve in one liter to make a saturated solution. If the solubility isn't given in units of moles per liter, it should be changed to these units. The *change* in the concentrations of the salt's ions depends on relatively how many are produced as the salt dissociates, as illustrated in Examples 19.1 to 19.3. (For example, 1 mol of $AgCl$ can give one mole each of Ag^+ and Cl^-, but 1 mol of Ag_2CrO_4 yields *two* moles of Ag^+ ion and one of CrO_4^{2-} ion.)

The entries in the "equilibrium concentration" row are simply the sums to the columns holding entries of the previous rows. These entries are now substituted into the equation for the salt's K_{sp} and the calculation of K_{sp} is carried out.

Calculating a salt's molar solubility from its K_{sp}

To set up a concentration table for a specific salt, we must first obtain three pieces of information.

1   The equation for the equilibrium in the saturated solution of the solute.
2   The ion product equation for the salt's K_{sp} (which we figure out from the salt's equilibrium equation).
3   The value of K_{sp} for the salt.

Again a concentration table has to be set up. When the solute is the only one present, the row for the *initial concentrations* of the ions carries only zeros. The row for the *changes in concentration* uses x to stand for the molar solubility, but for each ion, x must be multiplied by the number of ions each formula unit releases, as illustrated in Examples 19.4 and 19.5. This assures us that the changes in concentration will be in the same ratio as the coefficients in the balanced equilibrium equation. The individual *equilibrium concentrations* are the sums, column by column, of the initial concentration and the change in concentration already developed.

The equilibrium values expressed in units of x are finally substituted into the K_{sp} equation, and then the equation is solved for x, the molar solubility.

Common ion effect

A salt is less soluble in a solution that contains one of its ions than it is in pure water, a phenomenon called the *common ion effect.* This is really nothing more than an application of Le Châtelier's principle.

To calculate a salt's solubility from its K_{sp} under a common ion situation, the only thing that changes in the strategy described above is that the *initial concentration* row will have a non-zero entry, namely, the molar concentration of one of the ions already provided by some other solute. The unknown is still the molar solubility, x. Entries into the change row are made exactly as before, with the coefficients of x equal to the coefficients in the equation for the solubility equilibrium. The sums that you next enter into the *equilibrium concentration* row might include an expression such as $(0.10 + 2x)$, as in Example 17.6. Because we're dealing with *sparingly* soluble salts, we can safely assume that x or $2x$ is very small compared to 0.10, so we can simplify a $(0.10 + 2x)$ term to 0.10. This makes the arithmetic much simpler without sacrificing accuracy.

Predicting if a precipitate will form

A precipitate can form only if a solution is supersaturated, and this condition is fulfilled only when the ion product of the salt in question exceeds the value of K_{sp} for the salt. This simple fact will help you to remember the summary just preceding Example 19.7 in the text. (See page 842.) Be sure you know this summary.

When trying to predict what might happen when two solutions of different salts are mixed, you first have to consider which combinations of their ions might produce a sparingly soluble salt. If you find one, you have to do a solubility-product type of calculation. When the problem describes the mixing of two solutions, be sure to take into account the dilution of the ion concentrations, as illustrated in Example 19.8. Then, with the correct molarities, compute the value of the ion product of the salt. Finally, compare the ion product with the salt's K_{sp} to see if a precipitate should form.

Thinking It Through

Describe the steps you would follow to obtain the solution to the following problem.

1 Will a precipitate of $CaSO_4$ form if 40.0 L of 2.0×10^{-3} M $CaCl_2$ is mixed with 60.0 mL of 3.0×10^{-2} M Na_2SO_4?

Self–Test

1. The molar solubility of $MnCO_3$ in pure water is 2.24×10^{-5} mol/L. What is the K_{sp} for $MnCO_3$?

   _____

2. The molar solubility of lead iodate, $Pb(IO_3)_2$, is 4.0×10^{-5} M. What is the K_{sp} of this salt?

   _____

3. In 100.0 mL of a saturated solution of $PbBr_2$ in water there are 0.30 g of $PbBr_2$. What is K_{sp} for this salt?

   _____

4. The molar solubility of $Cu(OH)_2$ in 0.10 M NaOH is 4.8×10^{-18} mol/L. What is the K_{sp} of $Ca(OH)_2$?

   _____

5. For $PbCO_3$, $K_{sp} = 7.4 \times 10^{-14}$. What is the molar solubility of $PbCO_3$ in pure water?

   _____

6. What is the molar solubility of $PbCO_3$ in 0.020 M Na_2CO_3? For $PbCO_3$, $K_{sp} = 7.4 \times 10^{-14}$.

   _____

7. What is the molar solubility of $PbCl_2$ in 0.30 M $CaCl_2$? For $PbCl_2$, $K_{sp} = 1.7 \times 10^{-5}$.

   _____

8. Referring to Question 7, what is the molar solubility of $PbCl_2$ in 0.30 M NaCl solution?

   _____

9. How many grams of Ag_2CrO_4 will dissolve in 100 mL of water to give a saturated solution?

   _____

10. Will a precipitate of $PbCl_2$ form in a solution that contains 0.20 M Pb^{2+} and 0.030 M Cl^-? For $PbCl_2$, $K_{sp} = 1.7 \times 10^{-5}$.

   _____

11. Will a precipitate of $CaSO_4$ form in a solution that contains 0.0030 M Ca^{2+} and 0.0010 M SO_4^{2-}? Check a table in the chapter for the needed K_{sp}.

   _____

New Terms

Write the definitions of the following terms, which were introduced in this section. If necessary, refer to the Glossary at the end of the text.

solubility product constant, K_{sp} ion product   molar solubility

common ion   common ion effect

19.2   Solubility equilibria of metal oxides and sulfides involve reactions with water

Review

The sparingly soluble oxides and sulfides of metal ions cannot be treated like other sparingly soluble ionic compounds because the O^{2-} and S^{2-} ions cannot exist in water. Once either is released from a crystalline substance, it instantly reacts essentially 100% with water to give OH^- and HS^-, respectively. The rest of this section deals only with metal sulfides.

For a metal sulfide of the general formula MS, the solubility equilibrium is

$$MS(s) + H_2O \; \rightleftharpoons \; M^{2+}(aq) + HS^-(aq) + OH^-(aq)$$

and the solubility product constant is

$$K_{sp} = [M^{2+}] [HS^-] [OH^-]$$

If the solution is made acidic, we have to rewrite the equilibrium, because both HS^- and OH^- are neutralized by acids. In dilute acid, the equilibrium involving the general sulfide, MS, is

$$MS(s) + 2H^+(aq) \; \rightleftharpoons \; M^{2+}(aq) + H_2S(aq)$$

The solubility product expression based on this equilibrium gives a different constant, the *acid solubility product constant,* or K_{spa}.

$$K_{spa} = \frac{[M^{2+}] [H_2S]}{[H^+]^2}$$

(When the metal sulfide is not of the form MS, these equations must be altered accordingly.)

Acid-insoluble sulfides

One group of metal sulfides, the *acid-insoluble sulfides*, have values of K_{spa} so small that no acid exists that can be made concentrated enough in water to bring them into solution simply by converting the sulfide ion to H_2S. Their K_{spa} values range from about 10^{-32} to 10^{-5}.

Acid-soluble sulfides

When K_{spa} of a metal sulfide is about 10^{-4} or greater, then it will dissolve in acid. These are called the *acid-soluble sulfides.*

Self–Test

12. Consider lead(II) sulfide.

(a) Write its solubility equilibrium and the K_{sp} equation for a saturated aqueous solution.

_____

(b) Write its solubility equilibrium and the K_{spa} equation for a saturated solution in aqueous acid.

_____

New Term

Write the definition of the following term, which was introduced in this section. If necessary, refer to the Glossary at the end of the text.

acid solubility product, K_{spa}

19.3   Metal ions can be separated by selective precipitation

Review

Because of the huge differences in K_{spa} values among the metal sulfides, the adjustment of the pH of an aqueous solution containing two metal ions that can form insoluble sulfides can sometimes permit the precipitation of one metal ion but not the other when H_2S is bubbled into the solution.

Some metal carbonates also differ widely enough in solubility that an adjustment of the pH of a solution of metal ions that are able to form insoluble carbonates can sometimes permit the precipitation of one metal ion but not the other when CO_2 is bubbled into the solution.

This section uses equations developed in the previous section to illustrate how selective precipitation works.

Selective precipitation of a metal sulfide

To achieve a separation of metal ions that can form insoluble sulfides, hydrogen sulfide is bubbled into their solution until it is saturated in hydrogen sulfide (0.1 M. in H_2S). But before this is done, the pH of the solution is adjusted according to the results of a calculation, as explained in Example 19.9.

Selective precipitation of a metal carbonate

The solubility products of the sparingly soluble metal carbonates do not differ nearly as widely as do those of the metal sulfides. Yet it's still possible to use pH to control the carbonate ion concentration so that one of two metal carbonates will precipitate and the other will not as CO_2 is bubbled into the solution of the ions. Using MCO_3 as a general formula, the relevant ionization equilibrium is

$$MCO_3(s) \rightleftharpoons M^{2+}(aq) + CO_3{}^{2-}(aq)$$

How the molarity of $CO_3{}^{2-}$ responds to pH control is seen in the ionization equilibrium of carbonic acid, shown next in a form that combines the two step-wise ionizations. Adding the two ionization equilibria gives the combined equation.

$$H_2CO_3 \rightleftharpoons H^+ + HCO_3^-$$

$$HCO_3^- \rightleftharpoons H^+ + CO_3^{2-}$$

Sum: $H_2CO_3 \rightleftharpoons 2H^+ + CO_3^{2-}$

The combined equilibrium constant, K_a, equals the product of K_{a_1} and K_{a_2}.

$$K_a = \frac{[H^+]^2 [CO_3^{2-}]}{[H_2CO_3]}) = 2.4 \times 10^{-17}$$

Recalling Le Châtelier's principle, you can see that if we decrease the pH by adding acid to this system, the combined equilibrium will shift to the left, reducing the availability of CO_3^{2-}. And if we increase the pH by adding something to neutralize H^+, like OH^-, the equilibrium will shift to the right, increasing the availability of CO_3^{2-}.

The source of carbonic acid for the selective precipitation of carbonates is simply the bubbling into the solution of CO_2 until the solution is saturated, where its concentration becomes approximately $0.030\ M$. We use this value for $[H_2CO_3]$ in calculations. Example 19.10 describes a calculation for separating ions by selective precipitation of their carbonates.

Thinking It Through

Describe in detail how you would go about solving the following problem.

2   A solution contains both $AgNO_3$ and $Co(NO_3)_2$ at concentrations of $0.010\ M$ each. It is to be saturated with hydrogen sulfide. Is it possible to adjust the pH of the solution so as to prevent one of the cations from precipitating when hydrogen sulfide gas is bubbled into the system? If so, which cation remains dissolved and what pH should be used? Consult Table 19.2 as needed.

Self–Test

13. What value of pH permits the selective precipitation of the sulfide of just one of the two metal ions in a solution that is 0.020 molar in Pb^{2+} and 0.020 molar in Fe^{2+}?

14. A solution contains $Mg(NO_3)_2$ and $Sr(NO_3)_2$, each at a concentration of $0.050\ M$. Can one of the metal ions be precipitated as its insoluble carbonate without the other? If so, which one? The pH of the solution must be within what range of values for this to work as CO_2 is bubbled into the solution?

New Terms

None

19.4   Complex ions participate in equilibria in aqueous solutions

Review

This section delves into equilibria involving complex ions and how the ability to form complex ions can be used to affect the solubility of a salt. Only an overview is given in this chapter. Chapter 23 has several sections on complex ions that go into more detail, including the rules of nomenclature and the description of bonding in these substances.

Metal ions other than those in Groups IA and IIA are strong enough Lewis acids to draw electron-rich species, whether neutral or negatively charged, into Lewis acid–base interactions, forming coordinate covalent bonds. The products are called *complex ions* or simply *complexes*. Compounds made up of complex ions and ions of opposing charge are *coordination compounds* (or *coordination complexes*). The metal ion is called the *acceptor*, and a Lewis base to which the metal becomes attached is called a *ligand*. The specific atom within the ligand that provides the electron pair for the new bond is called the *donor atom*. Common ligands include neutral species, like H_2O and NH_3, and anions, like OH^- or any halide ion.

Complex ion equilibria

There is no uniform agreement among chemists concerning the way to represent the equilibrium between a complex ion and the metal and ligands that form it. Two approaches to writing the equilibria are used, one being for the formation of the complex from its constituents and the other being for the decomposition of the complex. Fortunately, one is simply the inverse of the other.

Formation constants

When the equilibrium equation shows the formation of the complex ion (rather than its decomposition), the associated equilibrium constant is called the *formation constant* or the *stability constant* and is given the symbol K_{form}. The equation corresponding to K_{form} has the complex in the numerator and the ligand and donor in the denominator. For example, the *formation* of the complex of the cadmium ion with the cyanide ion is represented as follows:

$$Cd^{2+}(aq) + 4CN^-(aq) \rightleftharpoons Cd(CN)_4{}^{2-}(aq) \qquad K_{form} = \frac{[Cd(CN)_4{}^{2-}]}{[Cd^{2+}][CN^-]^4}$$

When K_{form} is relatively large, it must mean that the value of $[Cd(CN)_4{}^{2-}]$ is high, relative to the terms in the denominator. In other words, the larger that K_{form} is, the more stable is the complex, which is why K_{form} is sometimes called the *stability constant* of a complex ion. Some chemists prefer to view complexes from such a perspective.

Instability constants

Other chemists prefer the reciprocal view. They write the equilibrium equation to show the decomposition of the complex and the equation for its associated equilibrium constant as follows.

$$Cd(CN)_4{}^{2-}(aq) \rightleftharpoons Cd^{2+}(aq) + 4CN^-(aq) \qquad K_{inst} = \frac{[Cd^{2+}][CN^-]^4}{[Cd(CN)_4{}^{2-}]}$$

The equilibrium constant is now called the *instability constant* because this view puts the emphasis on the instability of a complex. Thus, the larger the value of K_{inst}, the more unstable is the complex and the greater its tendency to undergo decomposition into the metal ion and free ligands. Notice that the equilibrium equation for K_{form} is the reverse of that for K_{inst}. This means that K_{inst} is merely the reciprocal of K_{form}.

Complex ions and the solubilities of salts

Example 19.11 shows how AgBr is 4000 times more soluble in 1.0 M NH_3 than it is in pure water. Let's look at another example to be sure you understand the thinking involved.

Suppose we wish to anticipate the effect of adding sodium cyanide on the solubility of cadmium hydroxide. We saw above that Cd^{2+} forms a complex with cyanide ion. When CN^- is added to a saturated solution of $Cd(OH)_2$, the following two equilibria must be considered.

$$Cd(OH)_2(s) \rightleftharpoons Cd^{2+}(aq) + 2OH^-(aq) \qquad K_{sp} = [Cd^{2+}][OH^-]^2$$

$$Cd^{2+}(aq) + 4CN^-(aq) \rightleftharpoons Cd(CN)_4{}^{2-}(aq) \qquad K_{form} = \frac{[Cd(CN)_4{}^{2-}]}{[Cd^{2+}][CN^-]^4}$$

Thus any Cd^{2+} ion that is in the solution from the first equilibrium will begin to form the complex ion when CN^- is added, which upsets the first equilibrium. The lowering of $[Cd^{2+}]$ as its complex forms must cause the first equilibrium to shift to the right; it's the Le Châtelier's principle response. This is just another way of saying that forming a complex with the cadmium ion increases the solubility of $Cd(OH)_2$. It's as if we're dealing with the following overall equilibrium.

$$Cd(OH)_2(s) + 4CN^-(aq) \rightleftharpoons Cd(CN)_4{}^{2-}(aq) + 2OH^-(aq) \qquad K_c = K_{sp} \times K_{form}$$

Note: it is assumed when working problems involving a system like this that *all* of the Cd^{2+} ion is tied up either in $Cd(OH)_2$ or in $Cd(CN)_4{}^{2-}$; that there is a totally negligible amount of free $Cd^{2+}(aq)$.

The text shows how the equilibrium constant, K_c, is the product of two other equilibrium constants.

$$K_c = K_{sp} \times K_{form}$$

Self–Test

15. The zinc ion forms a complex with four OH^- ions. Write the formula of the complex ion.

                             _____

16. Fe^{3+} and SCN^- form a complex ion with a net charge of 3–. Write the formula of this complex ion.

                             _____

17. Write the equilibria that are associated with the equations for K_{form} for each of the following complex ions. Write also the equations for the K_{form} of each.

(a) $Hg(NH_3)_4{}^{2+}$ _____

(b) $SnF_6{}^{2-}$ _____

(c) $Fe(CN)_6{}^{3-}$ _____

18. Cobalt(II) ion, Co^{2+}, forms a complex with ammonia, $Co(NH_3)_6^{2+}$. Write the chemical equation involving this complex for which the equilibrium constant would be referred to as

(a) K_{form}

_____

(b) K_{inst}

_____

19. Silver ion forms a complex ion with cyanide, $Ag(CN)_2^-$. Its formation constant equals 5.3×10^{18}. What is the value of the instability constant?

_____

20. Consider AgI in the presence of aqueous NaCN. For AgI, $K_{sp} = 8.3 \times 10^{-18}$; for $Ag(CN)_2^-$ $K_{form} = 5.3 \times 10^{18}$.

(a) Calculate the molar solubility of AgI in 0.010 M NaCN.

_____

(b) Calculate the molar solubility of AgI in pure water.

_____

(c) By what factor has the solubility of AgI in water been increased by complex formation?

_____

New Terms

acceptor

complex

complex ion

coordination compound

coordination complex

donor atom

formation constant, K_{form}

instability constant, K_{inst}

Solutions to Thinking It Through

1   What we need is the value of the ion product, $[Ca^{2+}] [SO_4^{2-}]$, for $CaSO_4$ in the given mixture of ions in the final solution. If this product *exceeds* K_{sp} for $CaSO_4$, a precipitate must form. To find values for $[Ca^{2+}]$ and $\{SO_4^{2-}\}$, we have to calculate the ratio of the moles of each ion to the *final* volume, assuming that no precipitate forms. For the calcium ion,

$$40 \text{ mL of } CaCl_2 \text{ soln} \times \frac{2.0 \times 10^{-3} \text{ mol } Ca^{2+}}{1000 \text{ mL } CaCl_2 \text{ soln}} = 8.0 \times 10^{-5} \text{ mol } Ca^{2+}$$

The final volume is 40.0 mL + 60.0 mL = 100.0 mL or 0.100 L, so the concentration of Ca^{2+} in the final solution is

$$[Ca^{2+}] = \frac{8.0 \times 10^{-5} \text{ mol } Ca^{2+}}{0.100 \text{ L}} = 8.0 \times 10^{-4} \ M$$

Doing the same kind of calculation for the sulfate ion gives $[SO_4^{2-}] = 1.8 \times 10^{-2} \ M$. The ion product is then found by

$$[Ca^{2+}] [SO_4^{2-}] = (8.0 \times 10^{-4})(1.8 \times 10^{-2}) = 1.4 \times 10^{-5}$$

The ion product, therefore, is *less* than the value of K_{sp} (2.4×10^{-5}), so no precipitate forms.

2   We first have to assemble the equilibrium equations and their acid solubility products (Table 19.2)

For Ag_2S:

$$Ag_2S(s) + 2H^+(aq) \ \rightleftharpoons \ 2Ag^+(aq) + H_2S(aq) \qquad K_{spa} = \frac{[Ag^+]^2 [H_2S]}{[H^+]^2} = 6 \times 10^{-30}$$

For CoS:

$$CoS(s) + 2H^+(aq) \ \rightleftharpoons \ Co^{2+}(aq) + H_2S(aq) \qquad K_{spa} = \frac{[Co^{2+}] [H_2S]}{[H^+]^2} = 5 \times 10^{-1}$$

Now we check to see if there is an upper limit to the value of $[H^+]$ above which the less soluble sulfide, Ag_2S, would dissolve. The value of $[Ag^+]$ is given as 0.010 M, and from the text the value of $[H_2S]$ in a saturated solution is 0.1 M, so

$$K_{spa} = \frac{(0.010)^2 \times (0.1)}{[H^+]^2} = 6 \times 10^{-30}$$

Solving this for $[H^+]$ gives $[H^+] = 1.3 \times 10^{12}$ mol/L. No acid solution can be this concentrated, so there is no upper limit to the acidity. Ag_2S is truly an acid-insoluble sulfide.

Next we check the lower limit; $[Co^{2+}] = 0.010 \ M$, so

$$K_{spa} = \frac{(0.010) \times (0.1)}{[H^+]^2} = 5 \times 10^{-1}$$

Solving for $[H^+]$ gives $[H^+] = 0.04 \ M$, and a pH of 1.4. So if the value of $[H^+]$ is made greater than 0.04 M, (or the pH is made less than 1.4), then cobalt sulfide will not precipitate.

Answers to Self–Test Questions

1. $K_{sp} = 5.0 \times 10^{-10}$
2. $K_{sp} = 2.6 \times 10^{-13}$
3. $K_{sp} = 2.2 \times 10^{-6}$
4. $K_{sp} = 4.8 \times 10^{-20}$
5. $2.7 \times 10^{-7}\ M$
6. $3.7 \times 10^{-12}\ M$
7. $4.7 \times 10^{-5}\ M$
8. $1.9 \times 10^{-4}\ M$
9. $2.2 \times 10^{-3}\ g$
10. Yes. The ion product is equal to 1.8×10^{-5}, which is greater than K_{sp}.
11. No. The ion product is less than K_{sp}.
12. (a) $PbS(s) + H_2O \rightleftharpoons Pb^{2+}(aq) + OH^-(aq) + HS^-(aq)$

     (b) $PbS(s) + 2H^+(aq) \rightleftharpoons Pb^{2+}(aq) + H_2S(aq)$

13. $pH = 0.7$
14. Yes. The Sr^{2+} ion can be selectively precipitated. The pH of the solution must be in the range of 5.2 and 6.0.
15. $Zn(OH)_4{}^{2-}$
16. $Fe(SCN)_6{}^{3-}$
17. (a) $Hg^{2+}(aq) + 4NH_3(aq) \rightleftharpoons Hg(NH_3)_4{}^{2+}(aq)$ $K_{form} = \dfrac{[Hg(NH_3)_4{}^{2+}]}{[Hg^{2+}]\,[NH_3]^4}$

     (b) $Sn^{4+}(aq) + 6F^-(aq) \rightleftharpoons SnF_6{}^{2-}(aq)$ $K_{form} = \dfrac{[SnF_6{}^{2-}]}{[Sn^{4+}]\,[F^-]^6}$

     (c) $Fe^{3+}(aq) + 6CN^-(aq) \rightleftharpoons Fe(CN)_6{}^{2+}(aq)$ $K_{form} = \dfrac{[Fe(CN)_6{}^{2+}]}{[Fe^{3+}]\,[CN^-]^6}$

18. (a) $Co^{2+}(aq) + 6NH_3(aq) \rightleftharpoons Co(NH_3)_6{}^{2+}(aq)$

     (b) $Co(NH_3)_6{}^{2+}(aq) \rightleftharpoons Co^{2+}(aq) + 6NH_3(aq)$

19. $K_{inst} = 1.9 \times 10^{-19}$
20. (a) $1.9 \times 10^{-2}\ M$

     (b) $2.8 \times 10^{-9}\ M$

     (c) 6.6×10^{-6} (AgI is 6.6 million times more soluble in the NaCN solution.)

Tools you have learned

Consider removing this chart from the Study Guide so you can have it handy when tackling homework problems.

Tool	How it Works
Solubility product constant, K_{sp}	We can use K_{sp} to calculate the molar solubility of a salt in pure water or in a solution containing a common ion.
Ion product for a salt	Calculating the ion product and comparing it with K_{sp} enables us to determine whether a precipitate should form in a solution. Remember: If ion product $> K_{sp}$, precipitate will form. If ion product $\leq K_{sp}$, precipitate will not form.
Acid solubility product constant, K_{spa}	We use K_{spa} to calculate the solubility of a metal sulfide at a given pH. We can also use it to determine the conditions under which metal ions can be separated by selective precipitation of their sulfides.
Formation constants (stability constants) for complex ions	In this chapter you learned to use formation constants to determine how complex ion formation affects the solubility of sparingly soluble salts in which the metal ion is able to form stable complex ions.

Summary of Important Equations

Equilibrium equation for a sparingly soluble salt, M_mA_a

$$M_mA_a(s) \rightleftharpoons mM^{a+}(aq) + aA^{m-}(aq)$$

Solubility product constant for a sparingly soluble salt, M_mA_a

$$K_{sp} = [M^{a+}]^m[A^{m-}]^a$$

For a sparingly soluble metal sulfide, MS, in acid

The solubility product equilibrium:

$$MS(s) + 2H^+(aq) \rightleftharpoons M^{2+}(aq) + H_2S(aq)$$

The solubility product constant, K_{spa}:

$$K_{spa} = \frac{[M^{2+}][H_2S]}{[H^+]^2}$$

Chapter **20**

Thermodynamics

In this chapter we turn our attention to what it is that causes events of any kind, whether they be physical or chemical, to occur spontaneously—that is, by themselves without outside assistance. Spontaneous events are crucial for our existence, because without them nothing would ever happen and we wouldn't exist. Understanding the factors that contribute to spontaneity is important because such events provide the driving force for all natural changes. You will also learn that there is a close connection between thermodynamics and chemical equilibrium.

Learning Objectives

As you study this chapter, keep in mind the following objectives:

1 To learn what thermodynamics means and to discover the kinds of questions it seeks to answer.

2 To learn how thermodynamics deals with the exchange of energy between a system and its surroundings.

3 To review the concepts of internal energy and enthalpy and the heats of reaction at constant volume and at constant pressure.

4 To learn how to convert between ΔE and ΔH for a chemical system.

5 To learn what a spontaneous change is and how everything that happens can be traced to some spontaneous change somewhere.

6 To learn what influence energy changes have on the tendency for an event to occur spontaneously.

7 To learn the meaning of the term entropy, S, and to see how an entropy increase favors a spontaneous change.

8 To learn how entropy changes can be calculated from absolute entropies.

9 To learn how the Gibbs free energy, G, interrelates the energy and entropy factors in determining the spontaneity of a chemical or physical change.

10 To learn how to calculate standard free energy changes from $\Delta H°$ and $\Delta S°$ and from standard free energies of formation.

11 To learn what relationship exists between the free energy change and the work that is available from a chemical reaction.

12 To see how free energy is related to equilibrium.

13 To learn how $\Delta G°$ can be used to predict the outcome of a chemical reaction.

14 To learn to compute equilibrium constants from thermodynamic data, and vice versa.

20.1   Internal energy can be transferred as heat or work, but it cannot be created or destroyed

Review

Thermodynamics is the study of energy changes, with a focus on the way energy flows between system and surroundings. Studying thermodynamics allows us to understand why spontaneous events occur and gives us insight into answers to many questions that plague modern society.

In this section, we review some topics introduced in Chapter 7. The *first law of thermodynamics* is given by the equation

$$\Delta E = q + w$$

where q is the heat *absorbed* by the system and w is the work done *on the system*. Study the signs of q and w in relation to the direction of energy flow.

A system can do work by expanding against an opposing pressure, and the amount of work can be calculated as $w = -P\Delta V$, where ΔV is the change in volume. (Here w is expressed as a negative quantity because the system loses energy when it does work.) If a system cannot change volume during a reaction, then the heat of reaction is equal to ΔE for the change.

$$\Delta E = q_v \text{ (where } q_v \text{ is the heat of reaction at constant volume)}$$

The enthalpy, H, is defined to deal with changes that occur at constant pressure. If the only kind of work a system can do is expansion work against the opposing atmospheric pressure, then ΔH is equal to the heat of reaction, and it is called the heat of reaction at constant pressure.

$$\Delta H = q_p \text{ (where } q_p \text{ is the heat of reaction at constant pressure)}$$

For most reactions, the difference between ΔE and ΔH is very small and can be neglected. To convert between ΔE and ΔH, we can use the equation

$$\Delta H = \Delta E + \Delta n_{gas} RT$$

where Δn_{gas} is the change in the number of moles of *gas* on going from the reactants to the products. (If the energy is to be in joules or kilojoules, remember to use $R = 8.314$ J mol^{-1} K^{-1}.) Study Example 20.1 to see how this equation is applied.

Self-Test

1.   Give the definition of enthalpy in terms of the internal energy, the pressure, and the volume.

   _____

2.   The product of pressure in pascals (Pa, which has the units newtons per square meter, N/m^2) times volume in cubic meters (m^3) gives work in units of joules. In Chapter 8, the standard atmosphere was defined in terms of the pascal as 1 atm = 101,325 Pa. With this information, calculate the amount of work, in joules, done by a gas when it expands from a volume of 500 mL to 1500 mL against a constant opposing pressure of 5.00 atm.

   (Remember that 1 mL = 1 cm^3.)   _____

3.   Suppose that during an exothermic reaction at constant volume, a gas is consumed, so that the pressure decreases. How would ΔE for this reaction compare to its ΔH? Why?

_____

_____

4.   How would the values of ΔE and ΔH compare for the reaction:

$$CO_2(g) + H_2(g) \rightarrow CO(g) + H_2O(g)$$

Explain your answer.   _____

_____

5.   At 25 °C and a pressure of 1 atm, the reaction

$$2NO_2(g) \rightarrow N_2O_4(g)$$

has $\Delta H = -57.9$ kJ. What is the value of ΔE for this reaction at the same temperature?

_____

New Terms

Write the definitions of the following terms, which were introduced in this section. If necessary, refer to the Glossary at the end of the text.

thermodynamics	first law of thermodynamics	enthalpy
internal energy	heat of reaction at constant volume	
heat of reaction	heat of reaction at constant pressure	

20.2   A spontaneous change is a change that continues without outside intervention

Review

A spontaneous change is one that takes place all by itself without continual assistance. Once conditions are right, it proceeds on its own. A nonspontaneous change needs continual help to make it happen, and some spontaneous change must occur first to drive the nonspontaneous one. Everything we see happen is ultimately caused by a spontaneous event of some kind.

When a change is accompanied by a decrease in the potential energy of the system (i.e., when the change is exothermic), it tends to occur spontaneously. Some examples are shown in the photographs in Figure 18.2. A potential energy decrease is a factor in *favor* of spontaneity. It is not the sole factor, however, because there are changes that are endothermic and nevertheless spontaneous.

In chemical reactions at constant pressure and temperature, the energy change is given by ΔH. For a pressure of 1 atm and a temperature of 25 °C, it is $\Delta H°$, the standard enthalpy change. When $\Delta H°$ is negative, the process occurs by a decrease in its potential energy and it *tends* to be spontaneous.

Self-Test

6. Which of the following do you recognize as spontaneous events?

    (a) Ice melts on a hot day.   _____

    (b) Water boils in a home in Alaska.   _____

    (c) A scratched fender on an automobile rusts.   _____

    (d) Dirty clothes become clean.   _____

7. Which of the following chemical changes *tend* to be spontaneous at 25 °C and 1 atm, based on their enthalpy changes. (Refer to Table 7.2 on page 288 of the text.)

    (a) $2PbO(s) \rightarrow 2Pb(s) + O_2(g)$

    (b) $NO(g) + SO_3(g) \rightarrow NO_2(g) + SO_2(g)$

    (c) $Fe_2O_3(s) + 3CO(g) \rightarrow 3CO_2(g) + 2Fe(s)$

                                                        Answer: _____

8. Give an example based on observations you have made in your daily life of a change that is endothermic and spontaneous.

    _____

New Terms

spontaneous change

20.3   Spontaneous changes tend to proceed from states of low probability to states of higher probability

Review

In this section we see that statistical probability is an important factor to consider in analyzing spontaneous events. In general, when a change is accompanied by an increase in the statistical probability of an energy distribution, the change has a tendency to occur spontaneously. Such a change increases the number of equivalent ways of distributing the energy, so a spontaneous process tends to disperse energy.

The thermodynamic quantity related to probability is the entropy, S. The more probable the state, the higher is the entropy, and the greater is the number of ways of distributing the energy in the system. Thus, an increase in entropy corresponds to a dispersal (spreading out) of energy over more possible arrangements. It also corresponds to an increase in the freedom of molecular motion — any change that increases the freedom of motion also increases the entropy. Mathematically, a change tends to occur spontaneously if $\Delta S > 0$ (ΔS is positive and $S_{final} > S_{initial}$).

The entropy change can often be predicted for a change. In general, the entropy of a system increases when

- the temperature of the system increases
- there is an increase in the volume of the system
- there is a change from solid → liquid, liquid → gas, or solid → gas

For a chemical reaction, ΔS will be positive when

- there is an increase in the number of moles of gas in the system
- there is an increase in the number of independent particles in the system.

Thinking It Through

Explain the thought processes involved in answering the following question.

1   The reaction $N_2O_4(g) \rightarrow N_2(g) + 2O_2(g)$ is exothermic. Is this reaction expected to be spontaneous? (Base your answer on the enthalpy and entropy changes involved.)

Self-Test

9.   If two containers holding different gases are connected together, the gases will gradually diffuse until the composition of the mixture is the same in both containers. Explain this in probability terms.

_____

_____

_____

_____

10.   What is the sign of ΔS for each of the following changes?

(a)   Solid iodine vaporizes.   _____

(b)   Waste oil is spilled into the ground.   _____

(c)   A gas is compressed into a tire.   _____

(d)   A liquid is cooled.   _____

(e)   Sugar dissolves in water to form a solution.

_____

11.   What is the sign of ΔS for each of the following reactions?

(a) $2HgO(s) \rightarrow 2Hg(g) + O_2(g)$ _____

(b) $CaO(s) + CO_2(g) \rightarrow CaCO_3(s)$ _____

(c) $H_2(g) + I_2(s) \rightarrow 2HI(g)$ _____

(d) $2C_2H_6(g) + 7O_2(g) \rightarrow 4CO_2(g) + 6H_2O(g)$ _____

New Terms

Write the definitions of the following terms, which were introduced in this section. If necessary, refer to the Glossary at the end of the text.

entropy   change in entropy, ΔS

20.4   All spontaneous processes increase the total entropy of the universe

Review

In regards to favoring spontaneity, the enthalpy change and entropy change sometimes work in the same direction, but at other times one favors spontaneity while the other does not. In the latter situations, the temperature becomes the controlling factor. All this is brought together by the second law of thermodynamics.

The *second law of thermodynamics* tells us that any spontaneous event is accompanied by an overall increase in the entropy of the universe. Earlier we saw that everything that happens is the net result of a spontaneous change of some sort, so every time something happens in the world there is an increase in entropy and some energy is dispersed and made unavailable for future use.

The Gibbs free energy is a composite function of the enthalpy and entropy. It allows us to gauge quantitatively how important these individual factors are in determining the spontaneity of an event.

At constant temperature and pressure,

$$\Delta G = \Delta H - T\Delta S$$

A change will be spontaneous when the enthalpy term (ΔH) and the entropy term ($T\Delta S$) combine to give a negative value for ΔG; that is, when the free energy decreases.

You should be able to analyze how (or if) temperature will affect the spontaneity of an event when you know the signs of ΔH and ΔS. Study the summary in Figure 20.9.

Thinking It Through

Explain the thought processes involved in answering the following question.

2   In the first Thinking It Through exercise, you learned that the reaction

$$N_2O_4(g) \ \rightarrow \ N_2(g) + 2O_2(g)$$

is exothermic. What is the expected sign of ΔG for this reaction and how do we expect the algebraic sign of ΔG to be affected by the temperature?

Self-Test

12.   For which of the following is ΔG *always* positive, regardless of the temperature?

(a) ΔH positive, ΔS negative

(b) ΔH negative, ΔS negative   _____

New Terms

Write the definitions of the following terms, which were introduced in this section. If necessary, refer to the Glossary at the end of the text.

second law of thermodynamics   exergonic

Gibbs free energy   endergonic

20.5   The third law of thermodynamics makes experimental measurement of absolute entropies possible

Review

At 0 K the entropy of any pure crystalline substance is zero. This is a statement of the *third law of thermodynamics*. Because the zero point on the entropy scale is known, the absolute amount of entropy that a substance possesses can be determined. (This can be compared with energy or enthalpy, where it is impossible to figure out exactly how much energy a substance has because it is impossible to know when a substance has zero energy.)

Standard entropies—entropies at 25 °C and 1 atm—can be used to calculate standard entropy changes by a Hess's Law type of calculation using Equation 20.6. Be sure to study Example 20.3.

Self-Test

13.   Calculate the standard entropy change, in J/K, for the following reactions.

(a) $2NaCl(s) \rightarrow 2Na(s) + Cl_2(g)$ _____

(b) $CO(g) + 2H_2(g) \rightarrow CH_3OH(l)$ _____

(c) $CH_4(g) + 2O_2(g) \rightarrow CO_2(g) + 2H_2O(g)$ _____

(d) $2KCl(s) + H_2SO_4(l) \rightarrow K_2SO_4(s) + 2HCl(g)$ _____

14.   If entropy changes were the *only* factors involved, which of the reactions in the preceding question would occur spontaneously at 25 °C and a pressure of 1 atm?

_____

New Terms

Write the definitions of the following terms, which were introduced in this section. If necessary, refer to the Glossary at the end of the text.

third law of thermodynamics   standard entropy change

standard entropy   standard entropy of formation

20.6   The standard free energy change, $\Delta G°$, is ΔG at standard conditions

Review

As with other thermodynamic quantities, the standard free energy change is the free energy change at 25 °C and 1 atm. We can compute it in two ways. One is from $\Delta H°$ and $\Delta S°$.

$$\Delta G° = \Delta H° - (298 \text{ K})\Delta S°$$

The other is from tabulated values of standard free energies of formation.

$$\Delta G° = (\text{sum } \Delta G_f° \text{ products}) - (\text{sum } \Delta G_f° \text{ reactants})$$

These are not difficult calculations. Study Examples 20.4 and 20.5 and work Practice Exercises 9 and 10; then try the Self-Test below.

Self-Test

15. Using Table 7.2 on page 288 and Table 20.1, calculate $\Delta G°$ (in kilojoules) for the reaction,

$$CaO(s) + H_2O(l) \rightarrow Ca(OH)_2(s)$$

_____

16. Use standard free energies of formation in Table 20.2 to calculate $\Delta G°$ (in kilojoules) for the following reactions:

 (a) $2C_2H_5OH(l) + 6O_2(g) \rightarrow 4CO_2(g) + 6H_2O(l)$

_____

 (b) $2HNO_3(l) + 2HCl(g) \rightarrow Cl_2(g) + 2NO_2(g) + 2H_2O(l)$

_____

New Terms

Write the definitions of the following terms, which were introduced in this section. If necessary, refer to the Glossary at the end of the text.

standard free energy change

standard free energy of formation

20.7 ΔG is the maximum work that can be done by a process

Review

The free energy change for a reaction is equal to the maximum amount of energy that, theoretically, can be recovered as useful work. In all real situations, however, somewhat less than this maximum is actually obtained. This is because the maximum work can only be produced if the process occurs reversibly. A reversible process takes an infinite length of time and consists of an infinite number of small changes in which the driving "force" is very nearly balanced by an opposing "force." No real process from which work is extracted occurs reversibly, so we always obtain less than the maximum. Nevertheless, ΔG gives us a goal to aim at, and allows us to gauge the efficiency of the way we are using a reaction to obtain work.

Example 20.1   Maximum Energy and Maximum Work

At 25 °C and 1 atm, what is the maximum amount of heat that could be extracted from the reaction of 1 mol CaO(s) with $H_2O(l)$?

$$CaO(s) + H_2O(l) \rightarrow Ca(OH)_2(s)$$

What is the maximum amount of work that could be obtained from this reaction?

Analysis:
The amount of heat obtained at constant pressure is the enthalpy change, so we need to calculate ΔH for the reaction. Because it occurs at 25 °C and 1 atm, this becomes the standard enthalpy change, $\Delta H°$.

To calculate the maximum work, we need to calculate ΔG for the reaction. Once again, because the conditions correspond to standard conditions, we need to calculate $\Delta G°$.

Solution:
If you worked Question 15 in the Study Guide, you already have all of the data that you need. For that question we found the following:

$$\Delta H° = -65.2 \text{ kJ}$$
$$\Delta S° = -34 \text{ J/K} \quad (T\Delta S = 10 \text{ kJ})$$
$$\Delta G° = -55 \text{ kJ (rounded)}$$

If the reaction is carried out so that no work is done, all the energy escapes as heat and the heat evolved is equal to $\Delta H°$. In other words, 65.2 kJ of heat is given off. If the reaction is carried out reversibly so that the maximum work is obtained, this work is equal to $\Delta G°$, or 55 kJ of work.

Notice that we get less energy as work than as heat. What happens to the rest of the energy? The answer is that it permits the entropy of the system to decrease. We know that for the reaction to be spontaneous, the entropy of the universe (system and surroundings taken together) must be positive, which means that overall there must be a net entropy increase and a net dispersal of energy. The 10 kJ that are *not* available for useful work amounts to the energy that must be dispersed into the surroundings to compensate for the lowering of the entropy of the system.

Is the Answer Reasonable?
We haven't really done any calculations, so the only thing to check is the reasoning in the analysis step, which seems to be sound.

Self-Test

17. Carbon monoxide is sometimes used as an industrial fuel.

   (a)   What is the maximum heat obtained at 25 °C and 1 atm by burning 1 mol of CO? The reaction is: $2CO(g) + O_2(g) \rightarrow 2CO_2(g)$

(b) What is the maximum work available at 25 °C and 1 atm from the combustion of 1 mol of $CO(g)$?

New Term

Write the definition of the following term, which was introduced in this section. If necessary, refer to the Glossary at the end of the text.

reversible process

20.8 ΔG is zero when a system is at equilibrium

Review

At equilibrium, the total free energy of the products equals the total free energy of the reactants and ΔG for the system is equal to zero.

$$\Delta G = 0 \quad \text{at equilibrium}$$

Because $\Delta G = 0$ at equilibrium, and because ΔG equals the maximum amount of work obtainable from a change, at equilibrium we can obtain no work from a system.

Equilibria in phase changes

For a phase change, equilibrium can occur at only one temperature. At this temperature, T,

$$T = \frac{\Delta H}{\Delta S}$$

Without much error, T can be calculated using $\Delta H°$ and $\Delta S°$ (which apply, strictly speaking, at 25 °C) because ΔH and ΔS do not change very much with temperature.

Equilibrium involving a phase change at 1 atm can only occur at one temperature. For example, pure liquid water at a pressure of 1 atm can be in equilibrium with ice *only* if its temperature is 0 °C. Liquid water can be in equilibrium with water vapor that has a pressure of 1 atm *only* when the temperature is 100 °C (the boiling point). At 25 °C and a pressure of 1 atm, (and with the absence of other gases such as air) water will exist entirely as a liquid. No equilibrium exists. (See Figure 20-1 on the next page.)

A typical free energy diagram for a phase change is illustrated in Figure 20.11. Notice that $G_{products} = G_{reactants}$ only at one temperature, which is the only temperature at which equilibrium can exist.

Equilibria in homogeneous chemical systems

Nearly all homogeneous chemical reactions are able to exist in a state of equilibrium at 25 °C, and the position of equilibrium is determined by the value of $\Delta G°$. Study the free energy diagrams in Figures 20.12 and 20.13. Notice that when $\Delta G°$ is positive, the position of equilibrium lies near the reactants. On the other hand, when $\Delta G°$ is negative, the position of equilibrium lies close to the products.

When $\Delta G°$ has a reasonably large negative value (–20 kJ, or so), the position of equilibrium lies far in the direction of the products, and when the reaction occurs it will appear to go essentially to completion. On the other hand, if $\Delta G°$ has a reasonably large positive value, hardly any products will be present at equilibrium. In other words, when the reactants are mixed, they will not *appear* to react at all. For all practical purposes, no reaction occurs when $\Delta G°$ is positive and has a value in excess of about 20 kJ. Therefore, we

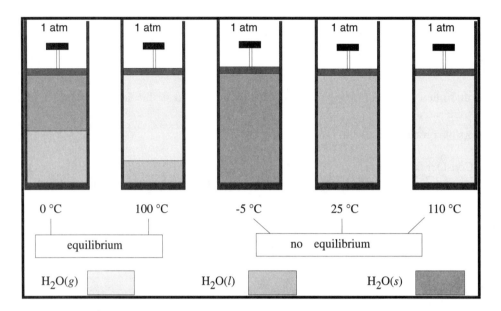

Figure 20-1 *At a pressure of 1 atm, equilibrium between any two phases of water can only occur at a single temperature. At temperatures of –5, 25, and 110 °C, only a single phase can exist.*

can use the sign and magnitude of $\Delta G°$ as an indicator of whether or not we expect to actually observe the formation of products in a reaction.

Summary

$\Delta G°$ large and negative	reaction goes very nearly to completion.
$\Delta G°$ large and positive	virtually no reaction occurs. The reaction does not appear to be spontaneous
$\Delta G°$ less than about ±20 kJ	reactants and products are both present in significant amounts at equilibrium.

In the text, we've used the symbol ΔG_T^o to stand for the equivalent of $\Delta G°$, but at a temperature other than 25 °C. Because ΔH and ΔS change little with temperature, we can approximate ΔG_T^o by the equation

$$\Delta G_T^o \approx \Delta H° - T\Delta S°$$

The sign and magnitude of ΔG_T^o can be used in the same way as for $\Delta G°$ in predicting the outcome of a reaction.

Keep in mind that even though $\Delta G°$ or ΔG_T^o may predict that a reaction should be "spontaneous," it tells us nothing about how rapid the reaction will be. The reaction of H_2 with O_2 has a very negative value of $\Delta G°$, so it is very "spontaneous." However, at room temperature the reaction is so slow that no reaction is observed. Thus, a change must not only be spontaneous, it must occur relatively fast for us to actually observe the change.

Thinking It Through

For the following, identify the information needed to solve the problem and show (or explain) what must be done with it.

3   Consider the reaction

$$2PCl_3(g) + O_2(g) \rightleftharpoons 2POCl_3(g)$$

Suppose an equilibrium mixture containing all three of the substances involved in this reaction is heated from 25 °C to 45 °C. How will the amount of PCl_3 in the mixture change?

Self-Test

18.   At the boiling point of water, liquid and vapor are in equilibrium at a pressure of 1 atm. Use $\Delta H°$ and $\Delta S°$ for the reaction,

$$H_2O(l) \rightleftharpoons H_2O(g)$$

to calculate the boiling point of water. How does the answer compare to the actual boiling point? Does this support the statement that ΔH and ΔS are nearly temperature independent?

_____

_____

19.   What would you expect to observe at 25 °C if 2 mol CO and 1 mol O_2 were mixed and allowed to react according to the equation:

$$2CO(g) + O_2(g) \rightarrow 2CO_2(g)?$$

_____

20.   What would you expect to observe at 25 °C if 2 mol of $N_2O(g)$ were mixed with 1 mol $O_2(g)$ in order to form NO(g) by the reaction: $2N_2O(g) + O_2(g) \rightarrow 4NO(g)?$

_____

21.   What should we expect to observe at 25 °C if we were to check a mixture of N_2O and O_2 for the presence of NO_2 in the reaction: $2N_2O(g) + 3O_2(g) \rightarrow 4NO_2(g)?$

_____

22.   Assuming that ΔH and ΔS are approximately independent of temperature, calculate the "standard" free energy change, ΔG_T^o, for the reaction in Question 15 at 1 atm and 50 °C

$$\Delta G_{323}^o = \text{_____}$$

Does this reaction proceed farther toward completion at this higher temperature? _____

New Terms

None

20.9   Calculating Equilibrium Constants from Thermodynamic Data

Review

For a given composition in a reaction mixture, the value of ΔG for the reaction is related to the reaction quotient by Equation 20.11.

$$\Delta G = \Delta G^\circ + RT \ln Q \qquad (20.11)$$

If you must apply this equation, remember that to calculate Q we use partial pressures (in atm) for gaseous reactions and molar concentrations if the reaction is in solution. Equation 20.11 can be used to determine where a reaction stands relative to equilibrium:

ΔG is negative	The reaction must proceed in the forward direction to reach equilibrium.
ΔG is zero	The reaction is at equilibrium.
ΔG is positive	The reaction must proceed in the reverse direction to reach equilibrium.

Example 20.11 illustrates how Equation 20.11 is applied.

Thermodynamic equilibrium constants

The principal purpose of this section is to show that the equilibrium constant is related to ΔG° for the reaction. You should learn Equation 20.12.

$$\Delta G^\circ = -RT \ln K \qquad (20.12)$$

In these equations, K is K_p for reactions involving gases. It is K_c for reactions in liquid solutions.

In using Equation 20.12, be sure to choose the value of R that matches the energy units of ΔG°. If ΔG° is in kilojoules, use $R = 8.314$ J mol^{-1} K^{-1} and be sure to change kilojoules to joules so the units cancel.

In this section we also see how we can estimate the value of the thermodynamic equilibrium constant at temperatures other than 25 °C. This is done by calculating the value of ΔG°_T from ΔH° and ΔS°. The appropriate equation is

$$\Delta G^\circ_T = \Delta H^\circ - T\Delta S^\circ$$

Once you've obtained ΔG°_T in this way, then you use Equation 20.12 to calculate the value of K.

Thinking It Through

For the following, identify the information needed to solve the problem and show (or explain) what must be done with it.

4   The reaction, $2C_4H_{10}(g) + 13O_2(g) \rightleftharpoons 8CO_2(g) + 10H_2O(g)$, has $\Delta G^\circ = -5406$ kJ at 25 °C. A certain mixture of these gases has the following partial pressures: for C_4H_{10}, 3×10^{-6} torr; for

O_2, 12.0 torr; for CO_2, 359 torr; for H_2O, 375 torr. Is this reaction mixture at equilibrium? If not, which way must the reaction proceed to reach equilibrium?

5　The reaction $2H_2(g) + O_2(g) \rightleftharpoons 2H_2O(g)$ has $K_c = 9.1 \times 10^{80}$ at 25 °C. What is the value of ΔG^{o}_{298} for this reaction?

Self-Test

23.　The reaction, $2N_2O(g) + 3O_2(g) \rightleftharpoons 4NO_2(g)$, has $\Delta G° = +0.17$ kJ. In a reaction mixture the gases involved in the reaction have the following partial pressures: N_2O, 220 torr; O_2, 120 torr; NO_2, 458 torr. Is this reaction mixture at equilibrium? If not, in which direction must the reaction proceed to reach equilibrium?

24.　The reaction, $NO(g) + NO_2(g) + H_2O(g) \rightleftharpoons 2HNO_2(g)$, has $K_p = 1.56$ atm^{-1} at 25 °C. What is $\Delta G°$ for this reaction expressed in kilojoules?

25.　At 25 °C, $K_p = 4.8 \times 10^{-31}$ for the reaction, $N_2(g) + O_2(g) \rightleftharpoons 2NO(g)$. What is $\Delta G°$ for this reaction expressed in kJ?

26.　The reaction, $2N_2O(g) + 3O_2(g) \rightleftharpoons 4NO_2(g)$, has $\Delta G° = +0.17$ kJ. What is K_p for this reaction?

27.　Use the data in Table 20.2 to compute the value of K_p for the reaction:

$$C_2H_2(g) + 2H_2(g) \rightleftharpoons C_2H_6(g)$$

28.　The reaction, $2C_4H_{10}(g) + 13O_2(g) \rightleftharpoons 8CO_2(g) + 10H_2O(g)$, has $\Delta G° = -5406$ kJ at 25 °C. What is the value of K_p for this reaction?

29.　The oxidation of sulfur dioxide to sulfur trioxide by molecular oxygen,

$$2SO_2(g) + O_2(g) \rightleftharpoons 2SO_3(g)$$

has a standard heat of reaction, $\Delta H° = -196.6$ kJ and a standard entropy of reaction $\Delta S° = -189.6$ J K^{-1}. What is the value of K_p for this reaction

(a)　at 25 °C?

(b)　at 500 °C?

New Terms

Write the definition of the following term, which was introduced in this section. If necessary, refer to the Glossary at the end of the text.

thermodynamic equilibrium constant

20.10   Bond Energies and Heats of Reaction

Review

In this section, you learn how values of ΔH_f° are used to calculate bond energies. The basis for these calculations is Hess's law and the fact that the enthalpy change is a state function; that is, the same enthalpy change takes place regardless of the path followed from the reactants to the products. Study Figure 20.14. The sum of the enthalpy changes corresponding to steps 1, 2, and 3 must be equal to ΔH_f° for the product.

In setting up an alternative path from the reactants (the elements in their standard states) to the product, we consider the following:

1   **The conversion of the elements on the reactant side into gaseous atoms.** The energy changes here are the standard heats of formation of the gaseous elements (Table 20.3). In the direction of the arrows in steps 1 and 2 in Figure 20.14, these changes are endothermic. We know this because it *always* takes energy to vaporize a solid element, and it *always* takes energy to break bonds to give atoms.

2   **The formation of all the bonds in the product molecule.** In discussing this energy change, we define the atomization energy, ΔH_{atom}, which is the energy needed to *break* all the bonds in the molecule. Whether we are forming bonds or breaking them, the amount of energy involved is the same. Only the sign of the ΔH is different, positive (endothermic) for bond breaking and negative (exothermic) for bond making. We obtain the atomization energy by adding up all the bond energies for the bonds in the molecule. (Table 20.4).

Once an alternative path is established, we can use it to calculate bond energies if the value of ΔH_f° is known, or we can calculate the value of ΔH_f° if all the bond energies are known. In the text, this is illustrated in the calculation of the heat of formation of methyl alcohol vapor. Study the steps in the calculations and then work the Self-Test below.

Self-Test

30.   Calculate the atomization energy of the molecule CH_3CN, in kJ/mol. The molecule has the structure

$$H-\overset{\displaystyle \overset{H}{|}}{\underset{\displaystyle \underset{H}{|}}{C}}-C\equiv N:$$

31. On a separate sheet of paper, construct a figure similar to that in Figure 20.14 for the formation of $CH_3CN(g)$ from its elements. Be sure to show both the direct and alternative paths.

32. Estimate the standard heat of formation of CH_3CN vapor in kJ mol^{-1} using data in Table 20.3 and your answers to Questions 30 and 31 above.

New Terms

Write the definitions of the following terms, which were introduced in this section. If necessary, refer to the Glossary at the end of the text.

bond energy   atomization energy

Solutions to Thinking It Through

1   Two factors determine spontaneity, the energy change and the entropy change. The reaction is exothermic, so the energy change is in favor of the reaction being spontaneous. Since four molecules are being formed from two, the entropy increases, so this factor is also in favor of spontaneity. Because both the energy and entropy changes favor spontaneity, we can anticipate that the reaction should be spontaneous.

2   The sign of ΔG is determined by the algebraic signs of ΔH and $T\Delta S$. The reaction is exothermic, so ΔH is negative. Three molecules are formed from one, so ΔS is positive, which means that $T\Delta S$ is positive. The sign of ΔG is determined by $\Delta G = (-) - (+)$.

   Since T is a positive quantity, the $T\Delta S$ term will be positive regardless of the temperature, so we conclude that ΔG will be negative regardless of the temperature.

3   The position of equilibrium is determined by the value of ΔG_T^o. For the reaction at 25 °C, we can use tabulated values of ΔG_f^o for the reactants and products to calculate ΔG_{298}^o. To obtain ΔG_T^o for the reaction at 45 °C, we use values of $\Delta H°$ and $\Delta S°$ for the reaction at 25°C and then apply the equation $\Delta G_T^o \approx \Delta H° - T\Delta S°$. To obtain $\Delta H°$, we can use tabulated values of ΔH_f^o and apply Hess's law. To obtain $\Delta S°$ we use tabulated value of $S°$ and perform a Hess's law-type of calculation.

   Once we know the values of ΔG_T^o at the two temperatures, we can determine which way the reaction will shift with temperature. If ΔG_T^o becomes more negative as T increases, then at the higher temperature there will be more product and less reactants.

   [If you wish to work through to an answer on this question, the necessary data is to be found in Appendix C. An alternative solution is presented here, too. We calculate ΔH_{298}^o and ΔS_{298}^o just as we would in the other solution. We use these values to calculate $\Delta G°$ at both temperatures and then compare. (This eliminates any inconsistencies in the data.)

$$\Delta H_{298}^o = [2(-1109.7 \text{ kJ})] - [2(-282.0 \text{ kJ})]$$

$$= -1644 \text{ kJ}$$

$$\Delta S_{298}^o = [2(324)] - [2(311.8) + 1(205)] \text{ (all J mol}^{-1}\text{ K}^{-1})$$

$$= -181 \text{ J mol}^{-1} \text{ K}^{-1}$$

$$= -0.181 \text{ kJ mol}^{-1} \text{ K}^{-1}$$

Now we calculate $\Delta G°$.

$$\Delta G^o_{298} = -1644 \text{ kJ} - (298 \text{ K})(-0.181 \text{ kJ mol}^{-1} \text{ K}^{-1})$$

$$= -1590 \text{ kJ}$$

$$\Delta G^o_{318} = -1644 \text{ kJ} - (318 \text{ K})(-0.181 \text{ kJ mol}^{-1} \text{ K}^{-1})$$

$$= -1586 \text{ kJ}$$

Because $\Delta G°$ is less negative at the higher temperature, the reaction does not proceed as far toward completion.]

4   We can use the equation $\Delta G = \Delta G° + RT \ln Q$ to calculate ΔG. If ΔG is zero, the reaction mixture is at equilibrium; if ΔG is negative the reaction must proceed in the forward direction to reach equilibrium; and if ΔG is positive, then the reaction must proceed in the reverse direction.

To calculate ΔG we need the value of the reaction quotient Q. This is the value of the mass action expression for the system, written using partial pressures.

$$\text{mass action expression} = \frac{P^8_{CO_2} P^{10}_{H_2O}}{P^2_{C_4H_{10}} P^{13}_{O_2}}$$

Substituting partial pressures expressed in atmospheres (obtained from partial pressures in torr by dividing by 760 torr/atm) gives the value of Q. We also need to use $T = 298$ K and $R = 8.314$ J mol^{-1} K^{-1}, and $\Delta G° = -5406 \times 10^3$ J. These quantities are substituted into the equation above to calculate ΔG. [$\Delta G = -5209$ kJ, so the reaction is spontaneous in the forward direction.]

5   To calculate $\Delta G°$, we need to have the value of K_p, not the value of K_c, because this is a reaction between gases. Therefore, we have to convert K_c to K_p. Rearranging equation 16.4 to solve for K_p gives

$$K_p = K_c (RT)^{-\Delta n_g}$$

We substitute $R = 0.0821$ L atm mol^{-1} K^{-1}, $T = 298$ K, $K_c = 9.1 \times 10^{80}$, $\Delta n_g = -1$, and then solve for K_p. Then we substitute K_p into the equation

$$\Delta G° = -RT \ln K_p$$

using $R = 8.314$ J mol^{-1} K^{-1} and $T = 298$ K. The answer will be in joules.

[The calculated value of $\Delta G°$ is -4.70×10^3 J, or -470 kJ.]

Answers to Self-Test Questions

1. $H = E + PV$
2. 507 J
3. $\Delta H > \Delta E$; at constant pressure there would be a volume decrease. Therefore, at constant pressure, the system absorbs some energy as work which is being done on it during the volume decrease. This extra energy can appear as extra heat given off during the reaction at constant pressure.
4. No volume change would occur at constant pressure, so ΔH and ΔE are the same.
5. $\Delta E = -55.5$ kJ
6. (a) spontaneous, (b) nonspontaneous, (c) spontaneous, (d) nonspontaneous

7. (a) nonspontaneous, $\Delta H° = +438.4$ kJ (b) nonspontaneous, $\Delta H° = +41.8$ kJ (c) spontaneous, $\Delta H° = -26.4$ kJ

8. The melting of ice on a warm day, or the evaporation of a puddle of water.

9. When the gas molecules in one container have the entire volume of the other available to it, a state of low probability exists until the gas expands into the other container as well.

10. (a) positive, (b) positive, (c) negative, (d) negative, (e) positive

11. (a) positive, (b) negative, (c) positive, (d) positive

12. (a) ΔG is positive at all temperatures.

13. (a) $+180.2$ J/K, (b) -332.3 J/K, (c) -5.2 J/K, (d) $+227.2$ J/K

14. (a) spontaneous, (b) nonspontaneous, (c) nonspontaneous (d) spontaneous

15. $\Delta G° = -55$ kJ

16. (a) $\Delta G° = -2651$ kJ, (b) $\Delta G° = -20.4$ kJ

17. (a) $\Delta H° = -283$ kJ/mol CO; heat evolved = 283 kJ, (b) $\Delta G° = -257.1$ kJ/mol of CO; max. work done = 257.1 kJ

18. $\Delta H° = 44.1$ kJ, $\Delta S° = 118.7$ J/K, $T_b = 372$ K = 99 °C. The actual boiling point is 100 °C = 373 K. The answer from $\Delta H°/\Delta S°$ is quite close, which supports the statement.

19. $\Delta G° = -514.2$ kJ; the reaction should go very nearly to completion.

20. $\Delta G° = +139.6$ kJ. No reaction should be observed.

21. $\Delta G° = +0.17$ kJ. Substantial amounts of both N_2O and NO_2 should be present at equilibrium.

22. $\Delta G_T^° = -54$ kJ at 50 °C (323 K). $\Delta G_T^°$ is less negative than $\Delta G°$, so the reaction should not proceed as far toward completion at the higher temperature.

23. Calculated $\Delta G = +15,022$ J. The mixture is not at equilibrium. The reverse reaction is spontaneous, so the reaction goes to the left to reach equilibrium.

24. $\Delta G° = +1.10$ kJ

25. $\Delta G° = +173$ kJ

26. $K_p = 0.93$

27. $K_p = 2.9 \times 10^{42}$

28. $K_p = 10^{948}$

29. (a) $K_p = 5.2 \times 10^2$ (b) $K_p = 3.5 \times 10^3$

30. $\Delta H_{atom} = 2478$ kJ/mol

31.

32. Estimated $\Delta H_f^°\ = +79$ kJ/mol (For comparison, the accepted value is $+95$ kJ/mol).

Tools you have learned

Consider removing this chart from the Study Guide so you can have it handy when tackling homework problems.

Tool	How it Works
$\Delta H = \Delta E + \Delta n_{gas} RT$	This is the tool you need to convert between ΔH and ΔE for a reaction
Predicting the sign of ΔS	Enables you to determine whether the entropy change favors spontaneity.
Sign of ΔG	Knowing the sign of ΔG lets you determine whether or not a change is spontaneous.
Standard entropies	We need these to calculate the value of $\Delta S°$ for a reaction.
$\Delta G° = \Delta H° - (298\ K)\Delta S°$	Use this equation to calculate $\Delta G°$ from $\Delta H°$ and $\Delta S°$ values.
Standard free energies of formation	We need these to calculate $\Delta G°$ for a reaction from tabulated values of $\Delta G_f°$.
Value of $\Delta G°$ for a reaction	The sign of $\Delta G°$ tells us whether or not a reaction can be observed. Only when $\Delta G°$ is negative do we expect to observe significant amounts of products.
$\Delta G_T° \approx \Delta H_{298}° - T\Delta S_{298}°$	Use this equation to calculate the value of $\Delta G°$ at temperatures other than 25 °C.
$\Delta G = \Delta G° + RT \ln Q$	Use this equation to determine whether a reaction is at equilibrium or to determine the direction a reaction must proceed to reach equilibrium.
$\Delta G° = -RT \ln k$	Use this equation to relate $\Delta G°$ to the equilibrium constant; K equals K_p for gaseous reactions and K_c for reactions in solution.

Summary of Important Equations

First law of thermodynamics

$$\Delta E = q - w$$

Definition of enthalpy

$$H = E + PV$$

Calculating work done *by* a system.

$$w = -P\Delta V$$

Converting between ΔE and ΔH

$$\Delta H = \Delta E + \Delta n\, RT$$

Remember, Δn is the change in the number of moles of *gas*.

Calculating $\Delta S°$

$$\Delta S° = (\text{ sum of } S° \text{ of products}) - (\text{ sum of } S° \text{ of reactants})$$

Gibbs free energy change

$$\Delta G = \Delta H - T\Delta S$$

Calculating $\Delta G°$ for a reaction from $\Delta G_f°$

$$\Delta G° = (\text{sum } \Delta G_f° \text{ of products}) - (\text{sum } \Delta G_f° \text{ of reactants})$$

Calculating $\Delta G°$ at a temperature other than 25°C

$$\Delta G_T^o = \Delta H_{298}^o - T\,\Delta S_{298}^o$$

Calculating ΔG from reaction mixture composition

$$\Delta G = \Delta G° + RT \ln Q$$

Relating $\Delta G°$ to the equilibrium constant

$$\Delta G° = -RT \ln K$$

Chapter 21

Electrochemistry

Some of the most useful and common practical applications of chemistry involve the use of or production of electricity. Chemical reactions that produce electrical power in batteries start our cars, run electronic calculators and portable radios, keep wristwatches running, and set proper exposures in cameras. Our lives are touched constantly by the ultimate fruits of electrolysis reactions, such as aluminum, bleach, halogenated organic molecules in plastics and insecticides, and soap. Besides all of these things, the relationship between electricity and chemical change has become an extremely useful tool in the laboratory for probing chemical systems of all kinds.

Learning Objectives

As you study this chapter, keep in mind the following goals:

1   To learn about the kinds of chemical reactions that can be studied electrically.

2   To learn how a spontaneous redox reaction can be set up to deliver electrical energy in a galvanic cell.

3   To learn to describe a galvanic cell using standard cell notation.

4   To see how the voltage, or potential, of a galvanic cell can be considered to arise as the difference between the potentials that each half-cell has for reduction.

5   To learn how reduction potentials are measured by comparison to a standard electrode called the hydrogen electrode.

6   To learn how to use standard reduction potentials to predict the spontaneous cell reaction, the cell potential, and whether or not a given reaction will proceed spontaneously.

7   To learn how to calculate ΔG° from a cell potential, and vice versa.

8   To learn how to calculate an equilibrium constant from the standard cell potential.

9   To learn how to calculate the effect on the cell potential of changing the concentrations of the ions in a galvanic cell.

10   To learn the chemistry of some common types of batteries and to look at possible future developments.

11   To learn what electrolysis is, to study how an electrolysis apparatus (electrolysis cell) is constructed, and to learn how to write equations for the reactions that take place in an electrolysis cell.

12   To learn how to use standard cell potentials to predict the products of electrolysis.

13   To learn how to compute the amount of chemical change caused by the flow of a given amount of electricity.

14   To study various practical applications of electrolysis.

21.1   Galvanic cells use redox reactions to generate electricity

Review

Electrical devices operate by the flow of electrons; that's what electricity is. Reactions that produce or consume electrical energy are called *electrochemical changes*. They are oxidation-reduction reactions, and their study constitutes the field of *electrochemistry*. As we discuss in this chapter, the applications of electrochemistry affect our daily lives as well as our activities in the laboratory.

When a redox reaction occurs, the energy released is normally lost to the environment as heat. By separating the half-reactions and making oxidation and reduction occur in different places, that is, in different *half-cells*, we can cause the electron transfer to take place by way of a wire through an external electrical circuit. In this way the energy of the reaction can be harnessed. The apparatus to accomplish this is called a *galvanic cell* or a *voltaic cell*. Study the discussion on page 914 that describes how a galvanic cell is put together.

The overall reaction in a galvanic cell is called the *cell reaction*, and is obtained by using the principles of the ion-electron method (Chapter 6) to combine the half-reaction that are taking place in the individual half-cells.

The electrodes in a galvanic cell are identified by the nature of the redox processes taking place. The electrode at which oxidation occurs is the *anode*; the electrode where reduction occurs is the *cathode*. In a galvanic cell, the anode carries a slight negative charge and the cathode a slight positive charge. During operation of the cell, electrons flow from the anode to the cathode through the external electrical circuit; in the solution, cations move toward the cathode and anions move toward the anode. Keep in mind that no electrons flow through the electrolyte solutions in the cell. The movement of electrons through the wires is *metallic conduction*; the transport of charge by the movement of ions through the solutions is called *electrolytic conduction*.

For the reactions to take place in the cell, the two half-cells must be connected electrolytically—for example, by a *salt bridge*. The salt-bridge permits ions to enter and leave the half-cells so that electrical neutrality can be maintained. In the absence of the salt bridge, no flow of electricity can occur.

Study the way we use *standard cell notation* to describe the nature of the electrodes and the electrolytes that make up the half-cells in a galvanic cell. Be sure that you understand Example 21.1 and can work Practice Exercises 1 and 2 before trying the Self-Test problems given next.

Self-Test

1.   The following reaction occurs spontaneously in a galvanic cell:

$$4H^+ + MnO_2 + Fe \rightarrow Fe^{2+} + Mn^{2+} + 2H_2O$$

   (a)   What half-reaction occurs in the cathode compartment?

   _____

   (b)   What electrical charge is carried by the iron electrode? _____

   (c)   Do electrons flow toward or away from the iron electrode? _____

   (d)   Using the iron half-cell as an example, explain how a salt bridge containing KNO_3 works.

   _____

2. Write the standard cell notation for the galvanic cell described in the preceding question.

_____

3. What are the anode and cathode half-reactions in the galvanic cell described by the notation

$$Zn(s) \mid Zn^{2+}(aq) \parallel Au^{3+}(aq) \mid Au(s)$$

_____

_____

New Terms

Write the definitions of the following terms, which were introduced in this section. If necessary, refer to the Glossary at the end of the text.

galvanic cell	voltaic cell	half-cell
anode	cathode	cell reaction
salt bridge	metallic conduction	electrolytic conduction
standard cell notation		

21.2 Cell potentials can be related to reduction potentials

Review

Central to the development of this section is the concept that each half-reaction has an intrinsic or innate tendency to proceed as a reduction. The magnitude of this tendency is given by the half-reaction's *reduction potential,* or the *standard reduction potential* if the concentrations of all the ions in the half-cell are 1 M, the partial pressure of any gas is 1 atm, and the temperature is 25 °C. When two half-cells compete for electrons, as they do in a galvanic cell, the one with the larger reduction potential proceeds as reduction and the other is forced to reverse and become an oxidation.

A functioning galvanic cell has a certain *cell potential.* This quantity is also called the *potential* or *electromotive force* produced by the cell and is measured in a unit called the *volt (V).* In a sense, it is a measure of the force by which the cell can push electrons through an external circuit. The standard cell potential, E^{o}_{cell}, is the cell potential under standard conditions (1 M concentrations of ions, 1 atm pressure for any gas, 25 °C). The equation by which E^{o}_{cell} is calculated from standard reduction potentials is

$$E^{o}_{cell} = \begin{pmatrix} \text{standard reduction} \\ \text{potential of the} \\ \text{substance reduced} \end{pmatrix} - \begin{pmatrix} \text{standard reduction} \\ \text{potential of the} \\ \text{substance oxidized} \end{pmatrix} \qquad (21.2)$$

This is a very important and useful equation, so be sure you've learned it.

The values of standard reduction potentials are compared to that of a reference electrode called the *standard hydrogen electrode,* which is assigned a potential of *exactly* 0 V.

$$2H^{+} (aq, 1.00\ M) + 2e^{-} \rightleftharpoons H_2(g, 1\ atm) \qquad E° = 0.00\ V$$

Be sure to study how reduction potentials are measured against the standard hydrogen electrode and review Example 21.2.

Self-Test

4. A lead half-cell was constructed using a lead electrode dipping into a 1.00 *M* Pb(NO$_3$)$_2$ solution. This was connected to a standard hydrogen electrode. A voltage of 0.13 V was measured for the cell when the positive terminal of the voltmeter was connected to the hydrogen electrode.

   (a) In the space below, sketch and label a diagram of the galvanic cell.

   (b) What substance is being reduced in the cell?   _____

   (c) What is $E^o_{Pb^{2+}}$ for the half-cell, Pb^{2+}(*aq*) + 2*e*$^-$ $\rightleftharpoons$ Pb(*s*)?

   _____

5. Referring to Table 19.1, to which electrode should the negative terminal of a voltmeter be connected in a cell constructed of the following half-cells?

$$Au^{3+} + 3e^- \rightleftharpoons Au \quad \text{and}$$

$$Zn^{2+} + 2e^- \rightleftharpoons Zn?$$

   _____

6. How is the volt defined in terms of SI units?_____

New Terms

Write the definitions of the following terms, which were introduced in this section. If necessary, refer to the Glossary at the end of the text.

cell potential, E_{cell} standard hydrogen electrode

potential   standard reduction potential

reduction potential   volt

standard cell potential, E^o_{cell}

21.3 Standard reduction potentials can predict spontaneous reactions

Review

As we noted above, for a given pair of half-reactions, the one having the higher (more positive) reduction potential occurs spontaneously as reduction; the other is reversed and occurs as oxidation. After setting up the half-reactions in this way, the cell reaction is obtained by adding the reduction and oxidation half-reactions in such a way that all electrons cancel. The procedure is the same as the one you used in the ion-electron method (Section 6.2). Factors are used to adjust the coefficients so that equal numbers of electrons are gained and lost. *Notice, however, that these factors are not used as multiplying factors when combining the half-reactions! The cell potential is obtained simply by subtracting one reduction potential from the other using Equation 20.2.* For a spontaneous cell reaction, this difference has a positive algebraic sign.

When asked whether or not a given overall reaction is spontaneous, first divide the reaction into its two half-reactions. Then find the reduction potential for each half reaction and compute the cell potential using Equation 20.2. If the result is positive, the reaction is spontaneous; if it is negative, however, the reaction is *not* spontaneous in the direction written. In fact, the reaction is spontaneous in the opposite direction.

Thinking It Through

For the following, identify the information needed to solve the problem and show (or explain) what must be done with it.

1.   What will be the spontaneous reaction if we add nickel and iron filings to a solution that contains $NiCl_2$ and $FeCl_2$?

Self-Test

7.   Suppose that the following half-reactions are used to prepare a cell.

$$ClO_3^-(aq) + 6H^+(aq) + 6e^- \rightleftharpoons Cl^-(aq) + 3H_2O \qquad E° = +1.45 \text{ V}$$
$$Hg_2HPO_4(s) + H^+(aq) + 2e^- \rightleftharpoons 2Hg(l) + H_2PO_4^-(aq) \qquad E° = +0.64 \text{ V}$$

    (a)   Determine the net spontaneous cell reaction.

    _____

    (b)   Determine the cell potential.   _____

8.   Without actually calculating E^o_{cell}, determine the spontaneous cell reaction involving the following half-reactions.

$$BrO_3^-(aq) + 6H^+(aq) + 6e^- \rightleftharpoons Br^-(aq) + 3H_2O \qquad E° = +1.44 \text{ V}$$
$$H_3AsO_4(aq) + 2H^+(aq) + 2e^- \rightleftharpoons HAsO_2(aq) + 2H_2O \qquad E° = +0.58 \text{ V}$$

    _____

9.   Will the following reaction occur spontaneously?

$$H_2SO_3(aq) + H_2O + Br_2(aq) \rightarrow SO_4^{2-}(aq) + 4H^+(aq) + 2Br^-(aq)$$

    _____

New Terms

None

21.4 Cell potentials are related to free energy changes

Review

The free energy change, ΔG, and the standard free energy change, ΔG°, for a reaction are related to the cell potential and standard cell potential, respectively, by the following equations.

$$\Delta G = -n \, \mathcal{F} \, E_{cell} \tag{21.5}$$

$$\Delta G^\circ = -n \, \mathcal{F} E^{\circ}_{cell} \tag{21.6}$$

where n is the number of electrons transferred and $\mathcal{F}$ is the Faraday constant (9.65×10^4 C/mol e^-). Since E_{cell} and E°_{cell} are in volts, and $1 \text{ V} = 1 \text{ J/C}$, ΔG and ΔG° are in units of joules.

The standard cell potential is also related to the equilibrium constant, K_c, for the reaction.

$$E^{\circ}_{cell} = \frac{RT}{n \, \mathcal{F}} \ln K_c \tag{21.7}$$

The value of R that must be used for this equation is $8.314 \text{ J mol}^{-1} \text{ K}^{-1}$. As noted above, $\mathcal{F}$ equals 9.65×10^4 C/mol e^-. Study Examples 21.8 and 21.9 plus Practice Exercises 9 and 10 in the text before trying the Self-Test questions below. Remember, for Equation 21.7, use natural logarithms and corresponding exponentials as you employ your pocket calculator.

Thinking It Through

For the following, identify the information needed to solve the problem and show (or explain) what must be done with it.

2   An equilibrium was set up for the reaction:

$$Ni(s) + Cd^{2+}(aq) \rightleftharpoons Cd(s) + Ni^{2+}(aq).$$

If the concentration of Ni^{2+} at equilibrium is 3.0×10^{-8} M, what will be the concentration of Cd^{2+}?

Self-Test

10.   Calculate ΔG° in kJ for the reaction in Question 7. _____

11.   The reaction, $2Al^{3+} + 3Cu \rightarrow 3Cu^{2+} + 2Al$, has $E^{\circ}_{cell} = -2.00$ V. What is ΔG° in kJ for this reaction?

_____

12. The reaction, $2AgBr + Pb \rightarrow PbBr_2 + 2Ag$, has $\Delta G^\circ = -237$ kJ. What is E^o_{cell} ?

      _____

13. What is the value of K_c for the reaction in Question 7?

      _____

14. What is the value of K_c for the reaction in Question 11?

      _____

15. The reaction, $PbI_2(s) + Zn(s) \rightleftharpoons Pb(s) + 2I^-(aq) + Zn^{2+}(aq)$, has an equilibrium constant, K_c = 2.2×10^{13} at 25 °C. What is the value of E^o_{cell} for this reaction?

      _____

New Terms

Write the definition of the following term, which was introduced in this section. If necessary, refer to the Glossary at the end of the text.

   Faraday constant, $\mathcal{F}$

21.5 Concentrations in a galvanic cell affect the cell potential

Review

The effect on the cell potential caused by nonstandard concentrations of solutes is given by the *Nernst equation.*

$$E_{cell} = E^o_{cell} - \frac{RT}{n\,\mathcal{F}} \ln Q \qquad (21.8)$$

where n is the number of electrons transferred and Q is the reaction quotient (the value of the mass action expression) for the reaction. Again, be careful about using natural logarithms with this equation.

   Be particularly careful to follow the correct procedures for writing Q, the mass action expression. The concentrations of pure solids and liquids do not appear in the mass action expression, and many electrochemical reactions are heterogeneous (one or more of the electrode materials are solids). The concentration of water also is omitted because it is essentially a constant, too. (If you need to review this, see Section 16.7 in the text.)

Example 21.1   **Writing the Nernst Equation for a Reaction**

What is the correct form for the Nernst equation for the following reaction at 25 °C for which E^o_{cell} = 0.38 V?

$$3PbSO_4(s) + 2Cr(s) \rightarrow 3Pb(s) + 3SO_4^{2-}(aq) + 2Cr^{3+}(aq)$$

Analysis:

We need to apply Equation 21.8 to this specific problem. Quantities that are specific to this reaction are the number of moles of electrons transferred, n, and the form of the mass action expression, Q. To find n, we need to determine the number of electrons gained or lost (they're the same, of course). In constructing the mass action expression, we have to be careful to omit the three solids.

Solution:

The oxidation of two chromium atoms to chromium(III) ions involves a transfer of $6e^-$, so $n = 6$ for this reaction. Therefore, because $R = 8.314$ J (mol $e^-)^{-1}$ K^{-1}, $T = 298.15$ K, and $\mathscr{F} = 9.65 \times 10^4$ C (mol $e^-)^{-1}$, we can substitute into Equation 21.8 as follows.

$$E_{cell} = 0.38 \text{ V} - \frac{8.314 \text{ J (mol } e^-)^{-1} \text{ K}^{-1} \times 298.15 \text{ K}}{6 \times 9.65 \times 10^4 \text{ C (mol } e^-)^{-1}} \times \ln Q$$

The large fraction reduces to 4.28×10^{-3} J C^{-1}, but the ratio of units, J C^{-1}, is the same as volts, V. In constructing Q, we omit the concentrations of the solids. This gives the answer we seek.

$$E_{cell} = 0.38 \text{ V} - 4.28 \times 10^{-3} \text{ V} \times \ln \left([SO_4^{2-}]^3 \, [Cr^{3+}]\right)$$

Is the Answer Reasonable?

To do a check here, the quickest thing is to check that we've placed the correct quantities into the fraction that precedes the natural log term. We can also recheck the arithmetic using our calculator to confirm that this part of the solution is correct. Then, we can check that we've omitted concentration terms for solids from the mass action expression (which we have) and that the concentration terms for the ions are both in the numerator (they are) and raised to the appropriate exponents (they are).

The measurement of cell potentials provides a means for determining unknown concentrations of ions in a half-cell, as illustrated by Example 21.11. Notice that in this calculation we first solve for Q (the reaction quotient). Then we substitute known concentrations and solve for the unknown value. Work Practice Exercises 12 and 13 before trying the exercises below.

Thinking It Through

For the following, identify the information needed to solve the problem and show (or explain) what must be done with it.

3   A galvanic cell was set up with a zinc electrode dipping into 100 mL of 1.00 M Zn^{2+} and an iron electrode dipping into 100 mL of 1.00 M Fe^{2+}. If the cell reaction

$$Zn(s) + Fe^{2+}(aq) \rightarrow Zn^{2+}(aq) + Fe(s)$$

delivers a constant current of 0.500 A, what will be the potential of the cell after 1500 min?

Self-Test

16.   Write the correct Nernst equation for the following reaction.

$$MnO_2(s) + 2H^+(aq) + H_3PO_2(aq) \rightarrow Mn^{2+}(aq) + H_3PO_3(aq) + H_2O \quad E^o_{cell} = 1.73 \text{ V}$$

17. Write the correct Nernst equation for the following (unbalanced) reaction. Use data in Table 21.1.

$$Cd(s) + Cr^{3+}(aq) \rightarrow Cd^{2+}(aq) + Cr(s)$$

18. What is the cell potential for the reaction in Question 16 if $[H_3PO_2] = 0.0010\ M$, $[Mn^{2+}] = 5.0 \times 10^{-4}\ M$, $[H_3PO_3] = 0.15\ M$, and the pH equals 5.0?

19. A chemist who wished to monitor the concentration of Cd^{2+} in the waste water leaving a chemical plant set up a galvanic cell consisting of a cadmium electrode that could be dipped into solutions suspected to contain Cd^{2+}, and a silver electrode that was immersed in a 0.100 M solution of $AgNO_3$. In a particular analysis, the potential of the cell was determined to be 1.282 V. The standard cell potential for the following reaction,

$$Cd + 2Ag^+ \rightarrow Cd^{2+} + 2Ag,$$

has been accurately measured to be 1.202 V. What was the Cd^{2+} concentration in the solution that was analyzed?

New Term

Write the definition of the following term, which was introduced in this section. If necessary, refer to the Glossary at the end of the text.

Nernst equation

21.6   Batteries are practical examples of galvanic cells

Review

This section describes the construction and the chemical reactions of a number of common batteries; see the New Terms, below. Be sure to learn the chemical reactions that take place at the electrodes and which substances serve as cathode and anode. You should also understand the advantages and disadvantages of the various cells.

Fuel cells offer increased thermodynamic efficiency in converting the energy of chemical reactions into work because they operate under conditions approaching reversibility. Another advantage is that the fuel can be fed to them continuously, so they don't need recharging.

Self-Test

20. What is the cathode reaction in the lead storage battery while it is being discharged?

21. What is the cathode reaction in the lead storage battery while it is being charged?

_____

22. A lead storage cell produces a potential of about 2 V. How can an automobile battery produce 12 V?

_____

23. What substance serves as the anode in the common dry cell?

_____

24. What substance serves as the anode in an alkaline battery?

_____

25. What is the anode reaction in a nicad battery when it is being discharged?

_____

26. Why can a hydrometer be used to test the state of charge of a lead storage battery?

_____

_____

27. Write the half-reaction that takes place at the cathode during the discharge of a silver oxide battery.

_____

28. Write equations for the cathode, anode, and net cell reaction in a nickel–metal hydride battery.

_____

_____

_____

29. Write equations for the cathode, anode, and net cell reaction for a lithium–manganese dioxide battery.

_____

_____

_____

30. What species is transported through the electrolyte between cathode and anode in a lithium ion battery?

_____

31. What function does graphite serve in a lithium ion battery?

_____

32. What chemical reaction occurs in a hydrogen-oxygen fuel cell?

_____

33.   Why is a fuel cell more efficient at producing usable energy than a conventional system that uses combustion of the fuel?

_____

34.   What reaction enables methanol to serve as a source of hydrogen for a hydrogen-oxygen fuel cell?

_____

New Terms

Write the definitions of the following terms, which were introduced in this section. If necessary, refer to the Glossary at the end of the text.

alkaline dry cell	nickel–cadmium battery	nicad battery
alkaline battery	lead storage battery	silver-oxide battery
Leclanché cell	zinc-carbon dry cell	lithium ion cell
nickel-metal hydride battery	intercalation	fuel cell
lithium-manganese dioxide battery		

21.7   Electrolysis uses electrical energy to cause chemical reactions

Review

When a nonspontaneous reaction is forced to occur by the passage of electricity, the process is called *electrolysis*. An *electrolysis cell (electrolytic cell)* consists of a pair of electrodes dipping into a chemical system in which there are mobile ions (formed by melting a salt or by dissolving an electrolyte in water). *When electricity flows, oxidation-reduction reactions are forced to occur at the electrodes.*

An important thing to learn in this section is that in *any* cell, regardless of whether it is using or producing electricity, the electrode at which oxidation occurs is called the *anode* and the electrode at which reduction occurs is called the *cathode*. Thus, we name an electrode according to the chemical reaction that occurs at it, not according to its charge. In an *electrolysis cell,* an external voltage source gives the anode a positive charge and the cathode a negative charge. These are the charges that they *must* have to force oxidation and reduction to occur. The positive charge of the anode pulls electrons from substances and causes them to be oxidized, and the negative charge of the cathode pushes electrons onto other substances and causes them to be reduced.

The equation for the *cell reaction* in an electrolysis apparatus is obtained by adding the individual oxidation and reduction half-reactions that occur at the electrodes. Remember to be sure that the electrons in the cell reaction cancel. As with galvanic cells, the procedure for this is the same as in the ion-electron method.

For reactions in aqueous solution, the redox of water is possible at the electrodes. You should know the following possible electrode reactions:

Anode (oxidation of H_2O)	$2H_2O(l) \rightarrow O_2(g) + 4H^+(aq) + 4e^-$
Cathode (reduction of H_2O)	$4H_2O(l) + 4e^- \rightarrow 2H_2(g) + 4OH^-(aq)$

Standard reduction potentials can often (but not always) be used to predict electrolysis reactions. When there are two competing reduction reactions at an electrode, the half-reaction with the most positive reduction potential will tend to occur. When there are competing oxidation reactions, the half-reaction with the least positive reduction potential will tend to occur. However, there are electrode peculiarities that can sometimes alter the expected outcome. Study Example 21.12.

Thinking It Through

For the following, identify the information needed to solve the problem and show (or explain) what must be done with it.

4   When a solution of NiF_2 is electrolyzed, metallic nickel is deposited on one electrode and O_2 is produced at the other. Which substances are oxidized and reduced, and at which electrodes?

Self-Test

35.   What is an *electrochemical change*?   _____

_____

36.   What is *electrochemistry*?   _____

_____

37.   At which electrode (anode or cathode) would these electrolysis half-reactions occur?

(a) $2I^-(aq) \rightarrow I_2(aq) + 2e^-$ _____

(b) $2Cr^{3+}(aq) + 7H_2O \rightarrow Cr_2O_7{}^{2-}(aq) + 14H^+(aq) + 6e^-$ _____

(c) $NO_3{}^-(aq) + 2H_2O + 3e^- \rightarrow NO(g) + 4OH^-(aq)$ _____

38.   Write the equation for the reduction of H_2O at the cathode of an electrolytic cell.

_____

39.   Write the equation for the oxidation of H_2O at the anode of an electrolytic cell.

_____

40.   When an aqueous solution of BaI_2 is electrolyzed, I_2 is formed at the anode and H_2 is formed at the cathode. What are the anode and cathode half-reactions?

anode: _____

cathode: _____

New Terms

Write the definitions of the following terms, which were introduced in this section. If necessary, refer to the Glossary at the end of the text.

electrolysis   electrolysis cell
electrolytic cell

21.8 Stoichiometry of electrochemical reactions involves electric current and time

Review

The amount of electricity consumed during electrolysis is proportional to the number of moles of electrons, (the number of *faradays*) passed through the electrolysis cell. Important relationships to remember are the following, where the *coulomb (C)* is the SI unit for amount of charge and the *ampere (A)* is the SI unit of electric current (coulombs per second or C s^{-1}).

$$1 \text{ mol } e^- = 9.65 \times 10^4 \text{ C}$$

$$1 \text{ C} = 1 \text{ A} \times \text{s}$$

$$1 \text{ } \mathcal{F} = 1 \text{ mol } e^- = 1 \text{ faraday}$$

The value of 9.65×10^4 C $\mathcal{F}^{-1}$ is called the *Faraday constant.* We use the relationships above, along with balanced *half-reactions* or a knowledge of the number of electrons transferred according to a balanced redox equation, to relate the amount of chemical change to amperes of electrical current and to time. Study Examples 21.13 through 21.15 and do Practice Exercises 15 to 18 before you work the Thinking It Through and Self-Test questions below.

Thinking It Through

For the following, identify the information needed to solve the problem and show (or explain) what must be done with it.

5   A solution of NaCl was electrolyzed for 30.0 min, producing Cl_2 at the anode and H_2 at the cathode. The resulting solution after electrolysis was titrated with 0.500 *M* HCl solution and required 22.3 mL of the acid to neutralize the solution. What was the current during the electrolysis?

Self-Test

41.   A current of 5.00 A flows for 25.0 minutes. How many moles of electrons does this deliver?

_____

42.   For how many seconds must a current of 6.00 A flow to deliver 0.225 mol of electrons?

_____

43.   What current must be supplied to deliver 0.0165 mol e^- in 155 s?

_____

44.   Calculate the number of moles of electrons that must pass through an electrolysis cell to produce 0.0150 mol $Cr_2O_7^{2-}$ by the reaction,

$$2Cr^{3+} + 7H_2O \rightarrow Cr_2O_7^{2-} + 14H^+ + 6e^-$$

_____

45. How many minutes are needed to make 0.0225 mol $Cr_2O_7^{2-}$ by the equation in Question 10 if the current is 4.00 A?

46. What current will produce 25.3 g Fe in 4.00 hours by reduction of Fe^{2+} in an aqueous solution?

New Terms

Write the definitions of the following terms, which were introduced in this section. If necessary, refer to the Glossary at the end of the text.

ampere (A)   coulomb (C)

faraday ($\mathscr{F}$)   Faraday constant

21.9   Electrolysis has many industrial applications

Review

This section describes *electroplating* and the methods of producing some important commercial metals and chemicals by electrolysis. In studying this section, you should learn the chemical reactions involved in the various electrolytic cells and processes—the *Hall–Héroult process,* the *Downs cell,* the *diaphragm cell,* and the *mercury cell*—and the reasons why the reactions are carried out as they are. When you feel you know the material, try the Self-Test.

Self-Test

47. What is the purpose of electroplating? _____

48. To which electrode do we connect the object to be electroplated?

49. What is the name and formula of the solvent for Al_2O_3 originally used in the Hall-Héroult process?

50. What is the net cell reaction in the Hall-Héroult process?

51. What is the major source of magnesium? _____ What salt of magnesium is used in the electrolysis reaction that produces the free metal?

52. What is the purpose of the special construction of the Downs cell?

53. Why is the electrolytic refining of copper so economical?

_____

54. Give the cathode, anode, and net cell reaction for the electrolysis of brine.

    anode reaction: _____

    cathode reaction: _____

    net reaction: _____

55. What products are formed if the brine solution is stirred while it is electrolyzed?

    _____

56. Name one advantage and one disadvantage of using a diaphragm cell in the electrolysis of brine.

    _____

57. What is an advantage and a disadvantage of using a mercury cell in the electrolysis of brine?

    _____

New Terms

Write the definitions of the following terms, which were introduced in this section. If necessary, refer to the Glossary at the end of the text.

diaphragm cell	electroplating	mercury cell
Downs cell	Hall–Héroult process	

Solutions to Thinking It Through

1   The mixture will contain $Ni(s)$, $Fe(s)$, $Ni^{2+}(aq)$, and $Fe^{2+}(aq)$. The two possible reduction half-reactions are

$$Ni^{2+}(aq) + 2e^- \rightarrow Ni(s)$$
$$Fe^{2+}(aq) + 2e^- \rightarrow Fe(s)$$

To determine the spontaneous reaction, we find the reduction potentials of Ni^{2+} and Fe^{2+} in Table 19.1. The one with the more positive reduction potential (Ni^{2+}) will be reduced, so we write its half-reaction as reduction. The other half-reaction (that for Fe^{2+}) will be reversed to occur as oxidation. We then add the two half-reactions, making sure the electrons cancel. The spontaneous reaction will be $Ni^{2+} + Fe(s) \rightarrow Ni(s) + Fe^{2+}$.

2   To answer the question, we need to have the equilibrium constant for the reaction. We can look up the reduction potentials of Ni^{2+} and Cd^{2+} in Table 19.1 and from them calculate the value of E^o_{cell} for the reaction *as written* ($E^o_{cell} = -0.15$ V). Then, we use this value of E^o_{cell} to calculate K_c with Equation 19.7 ($K_c = 8.6 \times 10^{-6}$). Once we have the value of K_c, we write the equilibrium law,

$$K_c = \frac{[\text{Ni}^{2+}]}{[\text{Cd}^{2+}]}$$

Then we substitute the known Ni^{2+} concentration and solve for the unknown Cd^{2+} concentration. The answer is $[\text{Cd}^{2+}] = 3.5 \times 10^{-3}$ M.

3   From the current and time, we can calculate the number of coulombs, and from that, the number of moles of electrons that flow in this time period. When one Zn exchanges electrons with one Fe^{2+}, two electrons are exchanged, so the number of moles of Zn that reacts equals the number of moles of e^- divided by 2. The value obtained is the amount by which the Zn^{2+} concentration increases and the amount by which the Fe^{2+} concentration decreases. We then compute the new concentrations, which we can use in the Nernst equation. To set up the Nernst equation, we have to compute the value of E^o_{cell} from tabulated reduction potentials of Zn^{2+} and Fe^{2+}. The value of n for the cell is 2, and the mass action expression is $[\text{Zn}^{2+}]/[\text{Fe}^{2+}]$. The value of Q is computed by substituting the new calculated values for $[\text{Zn}^{2+}]$ and $[\text{Fe}^{2+}]$.

$$E_{\text{cell}} = 0.38 \text{ V} - \frac{8.314 \text{ J } (\text{mol } e^-)^{-1} \text{ K}^{-1} \times 298.15 \text{ K}}{2 \times 9.65 \times 10^4 \text{ C } (\text{mol } e^-)^{-1}} \times \ln Q$$

$$E_{\text{cell}} = E^o_{\text{cell}} - 0.0128 \text{ V} \times \ln \frac{[\text{Zn}^{2+}]}{[\text{Fe}^{2+}]}$$

[After 1500 min, $[\text{Zn}^{2+}] = 1.23$ M and $[\text{Fe}^{2+}] = 0.767$ M.. The change in the cell potential is only 0.006 V. The cell potential becomes smaller by this amount.]

4   Oxidation and reduction can be identified by changes in oxidation number. In NiF_2, the nickel has an oxidation number of +2 and the metallic nickel deposited on the electrode has an oxidation number of zero, because it is now a free element.

$$\underset{+2}{\text{NiF}_2} \rightarrow \underset{0}{\text{Ni}}$$

A decrease in oxidation number from +2 to 0 corresponds to reduction, so the nickel is reduced and therefore must be deposited on the cathode.

The O_2 formed at the "other electrode" must come from H_2O, and the half-reaction that produces O_2 corresponds to oxidation, so the O_2 must be formed at the anode.

5   That the reaction produces H_2 at the cathode means that reduction of water must be taking place. This follows the following half-reaction, as you learned in the preceding section.

$$4\text{H}_2\text{O}(l) + 4e^- \rightarrow 2\text{H}_2(g) + 4\text{OH}^-(aq)$$

Therefore, the solution becomes basic as a result of the electrolysis. From the volume and concentration of the HCl solution, we can calculate the number of moles of HCl used. This is equal to the number of moles of OH^- that were formed in the solution (H^+ from the HCl and OH^- react in a 1-to-1 ratio). The half-reaction above tells us that the number of moles of OH^- equals the number of moles of e^- used in the electrolysis. Multiplying the moles of e^- by 9.65×10^4 C/mol e^- gives the number of coulombs used. Dividing the number of coulombs by the time in seconds (30.0 min $\times$ 60 s/min = 1800 s) gives the current in amperes. [The answer is 0.598 A]

Answers to Self-Test Questions

1. (a) $MnO_2 + 4H^+ + 2e^- \rightarrow Mn^{2+} + 2H_2O$
    (b) negative
    (c) away
    (d) NO_3^- ions flow into the iron half-cell compartment to compensate for the charge of the Fe^{2+} ions entering the solution.

2. $Fe(s) \mid Fe^{2+}(aq) \parallel Mn^{2+}(aq) \mid MnO_2(s)$

3. anode: $Zn(s) \rightarrow Zn^{2+}(aq) + 2e^-$; cathode: $Au^{3+}(aq) + 3e^- \rightarrow Au(s)$

4. (a)

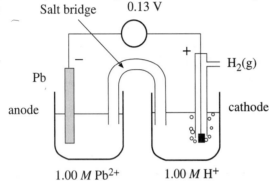

    (b) H^+ (c) –0.13 V

5. zinc

6. $1 \text{ V} = 1 \text{ J/C}$

7. (a) $ClO_3^-(aq) + 9H^+(aq) + 3Hg_2HPO_4(s) \rightarrow Cl^-(aq) + 3H_2O + 6Hg(l) + 3H_2PO_4^-(aq)$
    (b) $E° = 0.81$ V

8. $BrO_3^-(aq) + 3HAsO_2(aq) + 3H_2O \rightarrow Br^-(aq) + 3H_3AsO_4(aq)$

9. yes, $E_{cell}^o = +0.90$

10. $\Delta G° = -470$ kJ

11. $\Delta G° = +277$ kcal

12. $E° = 1.23$ V

13. $K_c = 2.3 \times 10^{82}$

14. $K_c = 2 \times 10^{-203}$

15. $E_{cell}^o = 0.395$ V

16. $E_{cell} = 1.73 \text{ V} - (0.0129 \text{ V}) \times \ln\left(\dfrac{[Mn^{2+}] [H_3PO_3]}{[H^+]^2 [H_2PO_2]}\right)$

17. $E_{cell} = -0.34 \text{ V} - (0.00429 \text{ V}) \times \ln\left(\dfrac{[Cd^{2+}]^3}{[Cr^{3+}]^2}\right)$

18. $E_{cell} = 1.47$ V

19. 2.0×10^{-5} M

20. $PbO_2(s) + 4H^+(aq) + SO_4^{2-}(aq) + 2e^- \rightarrow PbSO_4(s) + 2H_2O$

21. $PbSO_4(s) + 2e^- \rightarrow Pb(s) + SO_4^{2-}(aq)$

22. Six cells are connected in series, so their voltages add.

23. Zinc

24. Zinc

25. $Cd(s) + 2OH^-(aq) \rightarrow Cd(OH)_2(s) + 2e^-$

26. During discharge, H_2SO_4 is used up, and the density of the electrolyte changes (decreases).

27. $Ag_2O(s) + H_2O + 2e^- \rightarrow 2Ag(s) + 2OH^-(aq)$

28. anode: $MH(s) + OH^-(aq) \rightarrow M(s) + H_2O + e^-$
    cathode: $NiO()H)(s) + H_2O + e^- \rightarrow Ni(OH)_2 + OH^-(aq)$
    cell reaction: $MH(s) + NiO(OH)(s) \rightarrow Ni(OH)_2(s) + M(s)$

29. anode: $Li \rightarrow Li^+ + e^-$
    cathode: $Mn^{IV}O_2 + Li^+ + e^- \rightarrow Mn^{III}O_2(Li^+)$
    cell reaction: $Li + Mn^{IV}O_2 \rightarrow Mn^{III}O_2(Li^+)$

30. Li^+ ions

31. When the battery is charged, graphite accepts Li^+ ions which slip between layers of carbon atoms.

32. $2H_2 + O_2 \rightarrow H_2O$

33. The reactions at the electrodes occur under more nearly thermodynamically reversible conditions.

34. $CH_3OH(g) + H_2O(g) \xrightarrow{\text{catalyst}} CO_2(g) + 3H_2(g)$

35. Electrochemical changes produce or are caused by electricity.

36. Electrochemistry is the study of electrochemical changes.

37. (a) anode
    (b) anode
    (c) cathode

38. $2H_2O + 2e^- \rightarrow H_2(g) + 2OH^-(aq)$

39. $2H_2O \rightarrow 4H^+(aq) + O_2(g) + 4e^-$

40. anode: $2I^-(aq) \rightarrow I_2(aq) + 2e^-$
    cathode: $2H_2O + 2e^- \rightarrow H_2(g) + 2OH^-(aq)$

41. 7.77×10^{-2} mol e^-

42. 3.62×10^3 s

43. 10.3 A

44. 0.0900 mol e^-

45. 36.2 minutes

46. 6.08 A

47. To beautify and protect metals.

48. cathode

49. cryolite, Na_3AlF_6

50. $4Al^{3+}(l) + 6O^{2-}(l) \rightarrow 4Al(l) + 3O_2(g)$

51. the ocean; $MgCl_2$

52. To keep the Cl_2 and Na apart so they don't reform NaCl.

53. The anode mud contains precious metals whose value helps pay for the electricity that's used.

54. cathode: $2e^- + 2H_2O \rightarrow H_2(g) + 2OH^-(aq)$
    anode: $2Cl^-(aq) \rightarrow Cl_2(g) + 2e^-$
    net: $2Cl^-(aq) + 2H_2O \rightarrow H_2(g) + Cl_2(g) + 2OH^-aq)$

55. Cl^- is gradually changed to OCl^-.

56. Advantage: No OCl^- is formed by reaction of Cl_2 with OH^-.
    Disadvantage: NaOH solution is contaminated by small amounts
    of unreacted NaCl.

57. Advantage: Very pure NaOH is produced.
    Disadvantage: There is a potential for mercury pollution.

Tools you have learned

Consider removing this chart from the Study Guide so you can have it handy when tackling homework problems.

Tool	How it Works
Standard reduction potentials	In a galvanic cell, the difference between two reduction potentials equals the standard cell potential. Comparing reduction potentials lets us predict the electrode reactions in electrolysis.
Standard cell potentials	By calculating E^o_{cell} we can predict the spontaneity of a redox reaction. We can use E^o_{cell} to calculate $\Delta G°$ and equilibrium constants. They are also needed in the Nernst equation to relate concentrations of species in galvanic cells to the cell potential.
$\Delta G° = -n\,\mathcal{F}\,E°$	This equation lets us calculate standard free energy changes from cell potentials, and vice versa.
$E^o_{cell} = \dfrac{RT}{n\mathcal{F}} \ln K_c$	This equation lets us calculate equilibrium constants from cell potentials.
Nernst equation $$E_{cell} = E^o_{cell} - \dfrac{RT}{n\mathcal{F}} \ln Q$$	This equation lets us calculate the cell potential from E^o_{cell} and concentration data; we can also calculate the concentration of a species in solution from E^o_{cell} and a measured value for E_{cell}.
Faraday constant $1\,\mathcal{F} = 9.65 \times 10^4$ C/mol $e-$	Besides being a constant in the equations above, it allows us to relate coulombs (obtained from the product of current and time) to moles of chemical change in electrochemical reactions.

Summary of Important Equations

Stoichiometric relationships in electrolysis

$$1 \text{ C} = 1 \text{ A} \times 1 \text{ s}$$

$$1 \text{ mol } e^- = 1 \text{ } \mathscr{F} = 9.65 \times 10^4 \text{ C}$$

Using reduction potentials to calculate cell potentials

$$E^o_{\text{cell}} = \begin{pmatrix} \text{standard reduction} \\ \text{potential of the} \\ \text{substance reduced} \end{pmatrix} - \begin{pmatrix} \text{standard reduction} \\ \text{potential of the} \\ \text{substance oxidized} \end{pmatrix}$$

Cell potential and free energy change

$$\Delta G = -n \text{ } \mathscr{F} \text{ } E_{\text{cell}}$$
$$\Delta G^\circ = -n \text{ } \mathscr{F} \text{ } E^o_{\text{cell}}$$

Cell potential and K_{c}

$$E^o_{\text{cell}} = \frac{RT}{n \text{ } \mathscr{F}} \ln K_{\text{c}}$$

Nernst equation

$$E_{\text{cell}} = E^o_{\text{cell}} - \frac{RT}{n \text{ } \mathscr{F}} \ln Q$$

Chapter 22

Nuclear Reactions and Their Role in Chemistry

The unstable nuclei of many naturally occurring and synthetic isotopes present both risks and opportunities. To understand them, we have to learn what radiations are, their energies and penetrating abilities, how to detect and measure them, and both the dangers and benefits they make possible.

Learning Objectives

1   To learn the circumstances that allow us to use the law of conservation of mass and the law of conservation of energy as independent laws; and to learn the combined law of conservation of mass-energy and when it must be used.

2   To learn how nuclear binding energy is calculated and how the nuclear binding energy per nucleon is a measure of nuclear stability.

3   To learn how the strong force and the electrostatic force are involved in holding nucleons together; to learn the modes of radioactive decay and the associated radiations; and to learn how to balance nuclear equations.

4   To study factors that are associated with the stability of nuclei, such as the odd-even rule, the existence of "magic numbers," and how the ratio of neutrons to protons correlates with the stability of a nucleus.

5   To learn how the bombardment of the atoms of specific isotopes by various particles causes transmutations.

6   To learn how radiations are detected and how radioactive materials or their radiations are quantitatively described; to learn about the background radiation; and to see how protections against radiations can be achieved.

7   To see how radiations from radionuclides can be used in tracer analysis; in neutron activation analysis; and in dating ancient artifacts or geological strata.

8   To learn how fission occurs in a nuclear chain reaction; to see how fission gives energy and how this energy can be used to make electricity; and to study various aspects of safety—radioactive wastes and their storage, and loss-of-coolant emergencies.

22.1   Mass and energy are conserved in *all* their forms

Review

Radionuclides, isotopes that are *radioactive*, emit streams of particles or of electromagnetic radiation. The study of the associated energies required a rethinking of the concepts of both matter and energy, which carried Einstein to the development of a relationship that drew a distinction between the *rest mass* of a particle and its mass in motion. To develop the relationship between mass and energy, Einstein made a combined law, the *law of conservation of mass-energy*, and proposed what is now called the *Einstein equation*, $\Delta E = \Delta m_0 c^2$. This equation is essential to a discussion of nuclear stability because it lets us calculate nuclear binding energies (Section 22.2). It also lets us understand the huge energy yields from small quantities of "fuel" in nuclear fission (Section 22.8).

Chemists can ignore the distinction between mass and energy in all situations involving the stoichiometry of chemical reactions. A calculation in this Section using the Einstein equation illustrates how extremely small the error is when we do this for enthalpy changes in chemical reactions.

Self-Test

1. The enthalpy of combustion of acetylene is -1.30×10^3 kJ/mol. When 1.00 mol of acetylene burns, how much mass (in nanograms) changes to energy?

_____

New Terms

Write the definitions of the following terms, which were introduced in this section. If necessary, refer to the Glossary at the end of the text.

Einstein equation	radioactive	radioactivity
law of conservation of mass-energy	radionuclide	

22.2 The energy required to break a nucleus into separate nucleons is called the nuclear binding energy

Review

The energy that leaves the system when nucleons come together to form a nucleus would be the energy required to break up the nucleus. This is why the energy that leaves the system is called the *nuclear binding energy*. The greater this binding energy is, the more stable is the nucleus. In another sense, the nuclear binding energy is the energy the nucleus does not have because some of the mass of the nucleons changed to energy and left the system as the nucleus formed. Without this energy, the nucleus is more stable than it could have been with this energy.

When binding energies per nucleon are plotted against atomic number (Figure 22.1 in the text), the curve rises rapidly from the least stable nuclei to reach a peak in the vicinity of the isotopes of atomic number 26 (iron). Then the curve drops slowly as the highest atomic numbers are approached. In other words, on strictly the grounds of net energy changes, remembering that nature tends to favor events that are exothermic, the fusion of small nuclei into larger ones should release energy. And nuclear *fusion* does this. Likewise, the breaking up of very large nuclei into those of intermediate atomic numbers, should also give an overall gain in nuclear stability and the release of energy. Nuclear *fission* does this. Many nuclei change in the direction of greater stability by less drastic events. They emit radiations that transport energy out of their nuclei.

Self-Test

2. Why do we call the nuclear binding energy the energy that a nucleus does not have?

_____

3. How do we explain the fact that the total mass of the nucleons in helium-4 is less than the actual mass of its nucleus?

    _____

    _____

    _____

4. In the curve of Figure 22.1 in the text, what does the maximum point correspond to, a point of *high stability* or a point of *low stability* for a nucleus at or near it?

    _____

New Term

Write the definition of the following term, which was introduced in this section. If necessary, refer to the Glossary at the end of the text.

binding energy   nuclear fission

mass defect   nuclear fusion

22.3   Radioactivity is an emission of particles and/or electromagnetic radiation by unstable atomic nuclei

Review

Within a nucleus, the *electrostatic force* causes protons to repel each other, which lessens nuclear stability. But the *nuclear strong force,* which causes nucleons to attract each other, acts to overcome the electrostatic force. The electrostatic force (of repulsion), however, is able to act over longer distances than the strong force, so if a nucleus does not have enough neutrons to "dilute" the electrostatic force, the nucleus is unstable. A common consequence of such instability is *radioactive decay.* Various *radionuclides* decay by one of the following modes.

Decay Mode	Change in the Nucleus	Change in Mass no.	Change in At. no.
alpha emission	loss of $_2^4$ He (and usually also $_0^0$ γ)	-4	-4
beta emission	loss of $_{-1}^0$ e (and usually also $_0^0$ γ)	none	none
gamma emission	loss of $_0^0$ γ (1 MeV range)	none	none
positron emission	loss of $_1^0$ e (then an annihilation collision produces gamma radiation)	none	none
neutron emission	loss of $_0^1$ n	-1	-1
electron capture	change of a proton into a neutron	none	none

The energy of a radiation is usually described by some multiple of the *electron-volt* (eV), and the relative instability of a radionuclide is described by its half-life.

When we write *nuclear equations,* the sums of the mass numbers on each side of the arrow must be equal as well as the sums of the atomic numbers on each side.

The most penetrating radiations are those with neither mass nor charge (gamma and X rays) or with mass but no charge (neutrons).

Several radionuclides of high mass number do not achieve stable nuclei by one nuclear change. Additional changes occur as a *radioactive disintegration series* is descended to a stable isotope.

Self-Test

5. Consider the natures of the electrostatic force and the strong force in an atomic nucleus.

    (a) Which acts between both protons and neutrons? _____

    (b) Which acts only between protons? _____

    (c) Which destabilizes nuclei? _____

    (d) Which acts over the shorter distance? _____

    (e) Which is a force of attraction? _____

6. What is present in a nucleus, besides the strong force, that helps to lessen repulsions between protons?

    _____

7. Write the nuclear equations for the decay of a hypothetical isotope, $^{279}_{111}X$, by each process. (Use Z as the atomic symbol for any new nuclide that forms from each process.)

    (a) by beta and gamma emission _____

    (b) by alpha and gamma emission _____

    (c) by positron emission _____

    (d) by neutron emission _____

    (e) by electron capture and X-ray emission _____

8. State what kind of particle or photon is *emitted* when

    (a) a neutron changes to a proton. _____

    (b) an electron capture takes place. _____

    (c) the radionuclide's atomic number increases by 1. _____

    (d) the atomic number decreases by 2. _____

    (e) 2 photons of gamma radiation are produced following decay. _____

    (f) the mass number decreases by 1. _____

(g)   the atomic number decreases by 1.   _____

(h)   the mass number decreases by 4.   _____

(i)   no change to a different element occurs.   _____

9.   Gamma rays have energies on the order of

(a) 0.1 MeV   (b) 1.0 MeV   (c) 10 MeV   (d) 1.0 keV   _____

10.   The radiation with the best ability to penetrate lead is

(a)   alpha radiation   (c)   gamma radiation
(b)   beta radiation   (d)   positron radiation

_____

11.   Annihilation radiation photons result from the collision of an electron with

(a)   a positron   (c)   a neutron
(b)   another electron   (d)   a proton   _____

12.   The net effect of electron capture is the conversion of

(a)   an electron into a proton
(b)   a proton into a neutron
(c)   a neutron into a proton
(d)   a positron into an electron   _____

New Terms

Write the definitions of the following terms, which were introduced in this section. If necessary, refer to the Glossary at the end of the text.

alpha particle	nuclear equation	alpha radiation
positron	beta particle	beta radiation
antimatter	radioactive	radioactive decay
radioactive disintegration series	electron capture	radioactivity
radionuclide	electron-volt (eV)	gamma radiation
X ray	neutron emission	

22.4   Stable isotopes fall within the "band of stability" on a plot based on numbers of protons and neutrons

Review

The *odd-even rule* says that nuclear instability is prevalent among nuclides having odd numbers for either the mass number or the atomic number, and particularly when both are odd and when both make the isotope lie outside the *band of stability*. When one or both numbers is a *magic number* (2, 8, 20, 28, 50, 82 or 126), the nuclide is more stable than those nearby in the band of stability. Among the elements below atomic number 83, radionuclides with too high a neutron/proton ratio tend to be beta emitters. Those with too low a value of this ratio tend to emit positrons. Radionuclides with atomic numbers above 83 are most often alpha emitters.

Self-Test

13. The most stable isotope of the following four isotopes (where we use *hypothetical* atomic symbols) is

   (a) $^{15}_{8} X$ (b) $^{131}_{53} Y$ (c) $^{16}_{8} Z$ (d) $^{32}_{15} A$ _____

14. At which atomic number is the nuclide most likely to be both stable and have a neutron to proton ratio very nearly equal to 1?

   (a) 10   (b) 40   (c) 80   (d) 106   _____

15. An isotope of atomic number 65 and mass number 140

   (a)   lies below the band of stability.
   (b)   lies within the band of stability.
   (c)   lies above the band of stability.
   (d)   has one of the magic numbers.   _____

16. If a radionuclide lies above and outside the band of stability, then the ejection of what particle will move it closer to this band?

   (a)   beta particle
   (b)   gamma ray photon
   (c)   positron
   (d)   a photon of gamma emission   _____

New Terms

Write the definitions of the following terms, which were introduced in this section. If necessary, refer to the Glossary at the end of the text.

   band of stability   odd-even rule

   magic numbers

22.5   Transmutation is the change of one isotope into another

Review

When *transmutation* is caused by the bombardment of nuclei with high energy particles (e.g., $^{4}_{2}$ He, $^{1}_{1}$ p, or $^{2}_{1}$ d), generally a *compound nucleus* first forms. It then sheds its excess energy by emitting a different particle or gamma radiation. Exactly what mode of decay is taken by the compound nucleus depends only on the energy it acquired by the initial bombardment and particle capture, not on the kind of particle captured. Hundreds of isotopes, nearly all of them radioactive, and including all of the *transuranium elements* have been made this way.

Self-Test

17. To make a compound nucleus of $^{27}_{13}$Al from each of the following bombarding particles, what must be the target isotope? Give its symbol.

    (a) proton   _____

    (b) deuteron   _____

    (c) alpha particle   _____

18. What particle or photon must the compound nucleus, $^{27}_{13}$Al*, eject to change into each of the following nuclides? Give the name and symbol.

    (a) $^{23}_{11}$Na   _____

    (b) $^{26}_{12}$Mg   _____

    (c) $^{27}_{13}$Al   _____

19. What is the general name for elements 93-114?

    _____

    For all the elements 93 and up?   _____

New Terms

Write the definitions of the following terms, which were introduced in this section. If necessary, refer to the Glossary at the end of the text.

    compound nucleus   transuranium elements

    transmutation

22.6   How is radiation measured?

Review

The various kinds of atomic radiation are sometimes called *ionizing radiation* because they create ions (and free radicals) in their wakes, or they make phosphors scintillate (give off bursts of light). This property accounts both for the hazards of radiation and for the ease of detection. You should be able to describe in general terms how the Geiger counter, a scintillation counter, and a film dosimeter work.

In learning the units for various measurements discussed in this section, notice that the *becquerel* (Bq) is the SI version of the *curie* (Ci), and that both describe the *activity* of a radioactive source, not the energy of its radiations. The becquerel and the curie are thus extensive quantities—they depend on the mass of the source (as well as the half-lives of the radionuclides present).

The activity is the number of disintegrations per second and is related to the first order rate constant (also called the *decay constant*) for the process

$$activity = kN$$

The half-live is related to the decay constant by

$$t_{1/2} = \frac{\ln 2}{k}$$

Be sure to study Example 22.2 to see how we can calculate the activity of a sample of an isotope from its half-life.

The *gray* (Gy) is the SI version of the *rad*, and both refer to the *energy absorbed* by a quantity of matter because of the radiation it receives, not to the activity of the source and not even solely to the actual energy associated with the radiation. Thus the rad and the gray are also extensive quantities. They depend on the duration of the exposure (as well as on the energy, usually given in some multiple of the *electron volt*, of the radiation itself).

The *rem* is always some fraction (sometimes a very large fraction) of a rad (or a gray). The exact fraction depends on the kind of radiation, because the damage to tissue varies with this factor even when different kinds of radiation have identical energies and duration of exposure. The rem is a unit for describing potential harm to humans, because it concerns effects on tissue. The SI equivalent of the rem is the *sievert* (*Sv*); 1 rem = 10^{-2} Sv.

One of the reasons radiation is dangerous to living creatures is its ability to disrupt bonds to give highly reactive species called free radicals. These are molecular fragments that have one or more unpaired electrons. The damage caused by free radicals is large because of the ability of these reactive particles to set off undesirable reactions within living cells.

Because of several radionuclides in the earth's crust as well as cosmic radiation, we are bathed constantly in a low level of *background radiation* averaging close to 360 mrem per year for each person in the United States—more depending on an individual's use of medical radiation. In all applications, workers can protect themselves to a considerable extent by using dense shielding materials (e.g., lead) and by getting at some distance from the source. For every doubling of the distance, the radiation intensity drops by a factor of four, according to the *inverse square law*.

Self-Test

20. What is the name of a radiation detector that

    (a) uses photographic film or plates?   _____

    (b) contains a phosphor?   _____

    (c) lets radiation generate ions in a gas at low pressure?   _____

21. One rd = _____ J/g

22. One Gy = 1 _____ (supply units)

23. One Bq = _____ (supply units)

24. One Ci = _____ Bq

25. 1 Gy = _____ rd

26. How are even very low doses in rems dangerous to humans?

    _____

27.   What is the relationship of the rem to the rad, in general terms?

_____

28.   "Rad" stands for _____

29.   "Rem" stands for _____

30.   The exposure we all experience per year to background radiation is about

    (a) 3 rem   (b) 36 rem   (c) 360 mrem   (d) 1600 µCi   _____

31.   If the intensity of radiation is 100 units at a distance of 1.50 m from a source, how far away must one move to reduce the exposure by 1.00 unit?

    (a) 0.0015 m   (b) 15 m   (c) 12.2 m   (d) 1500 m   _____

32.   The half-life of cobalt-60 is 5.27 yr. Calculate the activity of a 35 mg sample of this isotope in becquerels.

        _____

New Terms

Write the definitions of the following terms, which were introduced in this section. If necessary, refer to the Glossary at the end of the text.

background radiation	gray (Gy)	rad (rd)
becquerel (Bq)	ionizing radiations	rem
curie (Ci)	inverse square law	sievert (Sv)
free radical		

22.7   Radionuclides have many medical and analytical applications

Review

In nearly all applications, gamma emitters are the best kinds of radionuclides. This radiation is the most penetrating of all, and therefore it is the easiest to detect, and it lets the scientist use very small quantities of the radionuclide. All of its radiation serves the purpose because little if any is blocked.

    In *tracer analysis,* the ability of a bodily fluid to enter a particular tissue can be traced if the fluid contains a small concentration of a radionuclide (e.g., $^{99m}_{43}$ Tc as TcO_4^-).

    When a sample is bombarded by neutrons in *neutron activation analysis*, its various nuclei capture neutrons and become compound nuclei that then emit gamma radiation. Which frequency of gamma radiation comes out is determined by the kinds of atoms in the sample, and the intensities at these frequencies give measures of the concentrations of the atoms.

    For *radiological dating,* pairs of radionuclides have to be identified and their relative concentrations in the sample measured. For dating very ancient rock formations, members of the pair might belong to the same radioactive disintegration series—for example, uranium-238 and lead-206—with the lighter one assumed to be produced solely by the decay of the heavier one at a rate of decay that has held constant over the millennia. For dating organic remains, the relative concentrations of carbon-14 and carbon-12 are used. As long as the living thing (e.g., a tree) lives, its level of carbon-14 is presumed to be constant. Once it dies, it no longer takes in carbon-14 and now the decay of this radionuclide at a known rate means that the age of any object made from the living thing (e.g., a wooden article) can be measured.

Self-Test

33. What method involving the use of radionuclides would be used in each situation?

(a) Determine the existence and concentration of lead as an impurity in the fingernails of children who have eaten chips of lead-based paints.

_____

(b) Measure the age of the Laurentian shield of bedrock in a southern Ontario province in Canada.

_____

(c) Measure how well blood circulates through a region of the lower leg suspected of having an early stage of gangrene with the hope of doing no more serious an amputation than absolutely necessary.

_____

34. How does carbon-14 originate in the upper atmosphere? Write a nuclear equation.

_____

35. In what chemical form is carbon-14 taken in by plants? Write a chemical formula.

_____

36. For the carbon-14 method to work without any correction factors, what would have to be true about the ratio of carbon-14 to carbon-12 in all living things both today and in the past?

_____

37. When carbon-14 dating methods were used on a sample of wood taken from a door post of an ancient archaeological site, it was found to have a specific activity of 382 Bq/g. How old was this wood sample according to calculations uncorrected for the factors that are known to cause some errors in the method? (Calculate to two significant figures.)

_____

New Terms

Write the definitions of the following terms, which were introduced in this section. If necessary, refer to the Glossary at the end of the text.

neutron activation analysis

radiological dating

tracer analysis

22.8   Nuclear fission is the breakup of a nucleus into two fragments of comparable size after capture of a slow neutron

Review

The capture of a neutron by a uranium-235 nucleus produces an unstable, compound nucleus that breaks apart—undergoes *fission*. The products are isotopes of intermediate atomic number, neutrons, and energy. If the neutrons can be slowed enough by moderators (e.g., graphic or heavy water or ordinary water), some may be captured by unchanged uranium-235 nuclei and so launch a chain reaction. (Plutonium-239 is also a fissile isotope.) The difference in binding energy between uranium-235 and the product isotopes is released largely as heat that can change water to steam and so drive electrical turbines.

To operate a reactor safely, control rods can be used to capture enough neutrons to make the multiplication factor equal 1. Now the reactor is said to be critical, because exactly as many neutrons are left at the end of the fission cycle as started the cycle.

Since the concentration of fissile isotope in the fuel elements of a nuclear reactor is small, a critical mass of the isotope is not possible and an atomic bomb explosion cannot occur. Should the coolant be lost, then the reactor could melt through its containment vessel. And heat could cause a steam explosion that would rupture the vessel (as occurred at Chernobyl, Russia).

Radioactive wastes include gases, liquids, and solids. Iodine-131, cesium-137, and strontium-90 are particular problems when they get into the environment because the blood can carry them throughout the body where their radiations cause harm. Long-lived solid wastes must be kept out of human touch for centuries.

Self-Test

38.   Why can nuclei capture neutrons much more easily than protons? _____

_____

39.   What is nuclear fission? _____

_____

40.   The isotopes initially formed from nuclear fusion have neutron-to-proton ratios that are too *high* or too *low*? _____

How do they adjust these ratios? _____

_____

41.   Which is generally higher, the sum of the binding energies of the nuclei produced by fission or the binding energy of the uranium-235 nucleus?

_____

42.   What makes the fission of uranium-235 self-sustaining? _____

_____

43. What is meant by "pressurized" in the pressurized water reactor?

_____

44. What is the function of each of the following in a pressurized water nuclear reactor?

(a) The moderator

_____

(b) The cladding

_____

(c) The primary coolant loop

_____

(d) The secondary coolant loop

_____

(e) The control rods

_____

45. Why do each of the following radionuclide wastes pose human health problems?

(a) Iodine-131

_____

(b) Cesium-137

_____

(c) Strontium-90

_____

New Terms

Write the definitions of the following terms, which were introduced in this section. If necessary, refer to the Glossary at the end of the text.

    fissile isotope   fission   nuclear chain reaction

Answers to Self-Test Questions

1. 14.4 ng
2. Because it is the energy *lost* from the system when some mass changed to energy as the nucleons formed into a nucleus.
3. Some mass of the nucleons changed to energy when the nucleus formed.
4. High stability
5. (a) strong force (b) electrostatic force (c) electrostatic force (d) strong force (d) strong force

6. neutrons

7. (a) $^{279}_{111}X \rightarrow ^{0}_{-1}e + ^{279}_{112}Z + ^{0}_{0}\gamma$

    (b) $^{279}_{111}X \rightarrow ^{4}_{2}He + ^{275}_{109}Z + ^{0}_{0}\gamma$

    (c) $^{279}_{111}X \rightarrow ^{0}_{1}e + ^{279}_{110}Z$

    (d) $^{279}_{111}X \rightarrow ^{1}_{0}n + ^{278}_{111}Z$

    (e) $^{279}_{111}X + ^{0}_{-1}e \rightarrow ^{279}_{110}Z + X\ rays$

8. (a) beta particle (b) X ray photon (c) beta particle (d) alpha particle (e) positron (f) neutron (g) positron (h) alpha particle (i) neutron (or gamma ray photon, only)

9. b

10. c

11. a

12. b

13. c

14. a

15. a

16. a

17. (a) $^{26}_{12}Mg$ (b) $^{25}_{12}Mg$ (c) $^{23}_{11}Na$

18. (a) alpha particle (b) proton $^{1}_{1}p$ (c) gamma ray photon $^{0}_{0}\gamma$

19. actinide elements; transuranium elements

20. (a) dosimeter (b) scintillation counter (c) Geiger counter

21. 10^{-5}

22. J/kg

23. 1 disintegration/s

24. 3.7×10^{10}

25. 100

26. They generate unstable ions and radicals that initiate other chemical changes of danger to the individual.

27. The rem is a fraction of a rad, the fraction depending on how damaging a particular rad dose is in tissue.

28. radiation absorbed dose

29. radiation equivalent for man

30. c

31. b

32. 1.5×10^{12} Bq

33. (a) neutron activation analysis (b) radiological dating (c) tracer analysis

34. $^{1}_{0}n + ^{14}_{7}N \rightarrow ^{15}_{7}N^{*} \rightarrow ^{14}_{6}C + ^{1}_{1}p$

35. CO_2

36. a constant ratio of carbon-14 to carbon-12

37. 7.2×10^{3} years

38. Neutrons are electrically neutral and so are not repelled by nuclei as they approach.

39. The spontaneous breaking of an unstable nucleus roughly in half.

40. Too high. They eject neutrons.

41. The sum of the binding energies of the products.

42. It produces more neutrons than needed to cause further fission events.

43. The water in the primary coolant loop is under such high pressure that even at high temperatures it is in the liquid state.

44. (a) Convert high energy neutrons into slower (thermal) neutrons.
    (b) Hold both the fuel in place and retain radioactive wastes.
    (c) Remove heat from the cladding elements as it is produced by fission.
    (d) Remove heat from the primary coolant loop and let this heat generate steam under pressure to drive the turbines.
    (e) Manage the flux of neutrons in the core so that the reactor will be critical during operation and go subcritical at shutdown.

45. (a) Concentrates in the thyroid gland where its radiation could cause thyroid cancer or other loss of thyroid function.
    (b) Transported by the blood wherever sodium goes.
    (c) Is attracted to bone tissue.

Summary of Important Equations

Einstein equation

$$\Delta E = \Delta m_0 c^2$$

Inverse square law (or radiation protection)

$$\frac{I_1}{I_2} = \frac{d_2^2}{d_1^2}$$

Chapter **23**

Metallurgy and the Properties of Metals and Their Complexes

In this chapter we will examine some of the properties of metals, including how they are recovered from their compounds, the degree of covalent bonding in their compounds, and the colors of metal compounds. We will also look more deeply into the nature of metal complexes of the type discussed in Chapter 17, particularly those formed by the transition metals. You will learn how complexes are named, about their structures, and about the bonding in these substances. You will also learn about the important role they play in biological systems.

Learning Objectives

As you study this chapter, keep in mind the following objectives:

1   To learn where metals occur in nature.

2   To learn what factors determine the thermal stability of metal compounds.

3   To learn how ores of metals are concentrated and how metals are obtained from them.

4   To learn how aluminum ore is purified and how magnesium is separated from seawater.

5   To learn the reactions that take place in a blast furnace during the reduction of iron ore.

6   To learn the kinds of chemical reactions that are used to extract metals from their compounds.

7   To learn how the degree of covalent character in a metal-nonmetal bond can be related to the charge and size of the metal ion and to the size and charge of the nonmetal anion.

8   To learn how the degree of covalent character of a metal-nonmetal bond affects the melting point of the compound.

9   To learn how the degree of covalent character of a metal-nonmetal bond affects the color of the compound.

10   To learn about the kinds of substances that serve as ligands in complex ions.

11   To learn how to correctly write the formulas of complex ions.

12   To learn how the stability of a complex ion is affected by the formation of chelate rings.

13   To learn how to apply the basic IUPAC rules of nomenclature for metal complexes.

14   To learn how the structures of complexes depend on the number of ligands that surround the metal ion.

15   To learn how metal complexes are often able to exist in different structural forms called isomers.

16   To learn how a theory called crystal field theory explains the relative stabilities of oxidation states, colors, and magnetic properties of complexes of the transition metals.

23.1   Metals are prepared from compounds by reduction

Review

Metals whose salts are water-soluble are often found in the sea. The sea is the major source of sodium and magnesium. Many other metals are found in compounds in the earth. Metals whose compounds are very unstable often occur in nature in the free state.

Recovery of metals from compounds

In compounds, metals usually occur in positive oxidation states. Therefore, recovery of the metal from a compound involves reduction.

Unstable compounds can be caused to undergo thermal decomposition. The temperature required depends on the heat of formation of the compound. If ΔH_f° is positive or only slightly negative, thermal decomposition proceeds at a relatively low temperature. Be sure to study Examples 23.1 and 23.2.

Metals of moderate activity have compounds that can be obtained by chemical reduction of their oxides. Study the reactions of carbon and oxygen with the oxides of tin and lead. If the metal is itself very easily oxidized, electrolysis of one of its compounds is required to obtain the free metal. Halide salts are used if possible because they have relatively low melting points.

Self-Test

1.   Write an equation for the reduction of copper(II) oxide by hydrogen.

_____

2.   Estimate the temperature in degrees Celsius above which the thermal decomposition of lead(II) oxide would occur to an appreciable extent. For $PbO(s)$, $\Delta H_f^\circ = -217$ kJ mol^{-1}, $S^\circ = 68.6$ J mol^{-1} K^{-1}. For $Pb(s)$, $S^\circ = 64.8$ J mol^{-1} K^{-1}.

_____

3.   Magnesium is obtained electrolytically by reduction of molten $MgCl_2$. Write an equation for the electrolysis reaction.

_____

New Terms

None

23.2   Metallurgy is the science and technology of metals

Review

Obtaining metals for practical applications can be divided into a number of steps: (1) mining, (2) pretreatment of ores, (3) reduction of metal compounds, and (4) purification.

Sources of metals are the sea and the earth. You should learn which metals are extracted from the sea and how they are obtained. When metal ores are dug from the ground, they must be enriched; dirt and other useless gangue has to be removed to make recovery of the metal economical. Washing with water and flotation are two physical methods of enriching ores.

Some ores, especially sulfide ores, must be roasted in air to convert metal sulfides to metal oxides, which are more easily reduced. You should study the reactions for the chemical purification of bauxite, the ore of aluminum.

Once an ore has been enriched and pretreated (if necessary), it is reduced. The strength of the reducing agent needed depends on the chemical activity of the metal. The most active metals, such as sodium, magnesium, and aluminum, must be obtained by electrolysis. Many metals can be obtained by reduction of the oxide with carbon. Study especially the reactions that take place in the blast furnace.

After the free metal is obtained from its ore, it usually must be purified further before it can be used commercially. Today the basic oxygen process is used to convert pig iron from the blast furnace to steel. Study the reactions that take place during the operation of the basic oxygen furnace.

Self-Test

4. Which two metals are extracted in quantity from the ocean?

   _____

5. Give the chemical equations for the separation of magnesium from sea water and its conversion to magnesium chloride.

   _____

   _____

6. What kinds of ores are treated by the flotation process?

   _____

7. Give the chemical equation for the roasting of a lead sulfide ore.

   _____

8. What property of aluminum oxide is made use of in the purification of bauxite?

   _____

   Give the chemical equations for the reactions used to purify the bauxite ore.

   _____

9. What is the active reducing agent in the blast furnace? _____

   Give the *overall* equation for the reduction of hematite, Fe_2O_3, using this reducing agent.

   _____

10. Write equations that illustrate the reactions that $CaCO_3$ undergoes when it is added to the charge in a blast furnace in the reduction of iron ore.

    _____

    _____

11. Why has the basic oxygen process become the principal steel-making process?

_____

_____

New Terms

Write the definitions of the following terms, which were introduced in this section. If necessary, refer to the Glossary at the end of the text.

metallurgy	coke
flotation	blast furnace
gangue	slag
roasting	basic oxygen process

23.3   Metal compounds exhibit varying degrees of covalent bonding

Review

You learned in Chapter 9 that differences in electronegativity can be used to judge the degree of ionic character in covalent bonds. In this section we learn another way to judge the degree of covalent character in a bond that is based on the concept that a cation will tend to polarize an anion and, by drawing some electron density toward the cation, create some partial covalent character to the ionic bond. The ability of a cation to polarize an anion increases as the charge on the cation increases and the size of the cation decreases. The *ionic potential*, ϕ, which is defined as the ratio of cation charge to size, serves to incorporate both factors. As the ionic potential increases, the metal cation becomes increasingly effective at polarizing the anion.

Anion polarization also is affected by the charge and size of the anion itself. For a given charge, the larger the anion, the more easily polarized it is. For a given size, the larger the negative charge, the more easily polarized is the anion.

The relative degrees of covalent character in metal-nonmetal bonds can be judged by comparing the melting points of the compounds. In general, the more covalent the compound, the lower the melting point. The colors of metal compounds are also affected by the degree of covalent character to the bonds. Compounds with colorless ions can absorb photons that cause electrons to be transferred from the anion to the cation. This produces a "charge transfer absorption" band — a band of absorbed wavelengths that usually lies in the UV region of the spectrum. Light so absorbed does not impart color to the compound because our eyes are not sensitive to wavelengths in this spectral region. However, as the metal-nonmetal bond becomes more covalent, less energy is needed to cause the charge transfer process and the absorption band shifts somewhat into the visible part of the spectrum at the violet end. Removal of violet light from white light causes the reflected light to appear somewhat yellow. The more covalent the compound becomes, the more visible light is absorbed and the more deeply colored the compound appears.

Self-Test

12. Which ion has the larger ionic potential, Fe^{2+} or Fe^{3+}?   _____

13. Which compound is more covalent, FeO or FeS?   _____

14. Which compound has the lower melting point, CaS or $CaCl_2$? (S^{2-} and Cl^- have nearly the same ionic radii.)

_____

15. Which compound is likely to be the more deeply colored, $SnCl_2$ or SnI_2?

_____

16. Which is the more covalent compound, MgI_2 or BaI_2?   _____

17. Which is the more covalent compound, $MgCl_2$ or $AlCl_3$?   _____

New Terms

Write the definitions of the following terms, which were introduced in this section. If necessary, refer to the Glossary at the end of the text.

ionic potential

charge transfer absorption band

23.4   Complex ions are formed by many metals

Review

Complex ions were discussed in Chapter 19 where the focus was on their equilibria and how their formation affects the solubilities of salts. Here we look at these substances in greater detail to find out the kinds of substances that form complexes with metals and the kinds of structures that they form.

Compounds that contain *complex ions* (often referred to simply as *complexes)* are sometimes called *coordination compounds* because coordinate covalent bonds generally bind the pieces that make up the complex ion. One piece is an electron pair acceptor (a metal ion at the center of the complex) and the other is an electron pair donor (Lewis base) that we call a *ligand.* Many ligands are anions (e.g., Cl^-, S^{2-}, CN^-, OH^-, SCN^-, $S_2O_3^{2-}$, and NO_2^-). Other ligands are electrically neutral, such as NH_3 and H_2O. All these are *monodentate ligands* because they "bite" or are joined to the metal ion by one coordinate covalent bond. *Bidentate ligands,* such as $NH_2CH_2CH_2NH_2$ or the oxalate ion ($C_2O_4^{2-}$) contain two donor atoms and are held by two coordinate covalent bonds to the metal ion. EDTA is an example of a *polydentate ligand.* Study some of the practical uses of EDTA.

In the formula of a complex, the *acceptor atom* is always written first followed by the ligands. Anionic ligands are listed first in alphabetical order, followed by neutral ligands, also in alphabetical order. When several ligands are held by the same acceptor atom, as in $[CrCl(H_2O)_5]^{2+}$, the net charge, which is the algebraic sum of the charges on the acceptor and the ligands, is placed outside the square brackets.

Complexes containing bidentate ligands, which form chelate ring structures, are significantly more stable than complexes in which the metal is surrounded by the same number of donor atoms provided by monodentate ligands. This phenomenon is called the *chelate effect.* The reason is because a bidentate ligand is less likely than a monodentate ligand to become completely detached from a metal ion.

Self-Test

18. The zinc ion forms a complex with four OH^- ions.

(a)   What is the formula of this complex ion?   _____

(b)   Could the complex ion be isolated as a potassium salt or a chloride salt? Write the formula.

_____

19. Fe^{3+} and SCN^- form a complex ion with a net charge of 3–. Write the formula of this complex ion.

_____

20. Ni^{2+} and NH_3 form a complex ion with a net charge of 2+. The ratio of donor atom to ligand is 1 to 6. Write the formula for the complex.

_____

21.   A complex is formed from Co^{2+}, one ethylenediamine, 2 water molecules, and 2 chloride ions. Write the formula for the complex.

_____

22.   Sketch chelate rings formed by

(a)   ethylenediamine

(b)   oxalate ion

23.   Which complex would you expect to be more stable, $[Cr(en)_3]^{3+}$ or $[Cr(NH_3)_6]^{3+}$?

_____

New Terms

Write the definitions of the following terms, which were introduced in this section. If necessary, refer to the Glossary at the end of the text.

complex ion	acceptor ion	bidentate ligand
complex	donor atom	polydentate ligand
ligand	monodentate ligand	chelate effect

23.5   The nomenclature of metal complexes follows an extension of the rules developed earlier

Review

In this section you learn the IUPAC rules for naming metal complexes. In summary, they are as follows:

Name the cation first, followed by the anion.

Name the ligands first (in alphabetical order according to the *ligand name,* not according to any number prefixes such as di or tri), followed by the metal. Write the oxidation number of the metal in parentheses using Roman numerals after the name of the metal.

Use prefixes to specify numbers of ligands (except *bis, tris,* etc. are used when there might be confusion if *di, tri,* etc. are used).

The names of anionic ligands end in *-o.*

A neutral ligand has the same name as the molecule (except H_2O is aqua and NH_3 is ammine).

Negatively charged complexes always end in the suffix *-ate,* which is appended to the name of the metal. When the suffix -ate is used, we also use the Latin stem for the metal name when the metal has a symbol not derived from the English name. (The exception is mercury, e.g., mercurate).

Self-Test

24. Name the following complexes:

    (a) $[Cr(NH_3)_6]^{2+}$

    _____

    (b) $[Cu(OH)_4]^{2-}$

    _____

    (c) $[MnCl_4(NH_3)_2]^{2-}$

    _____

    (d) $K_3[Co(CN)_6]$

    _____

    (e) $[Ni(NH_3)_2(H_2O)_4]Cl_2$

    _____

25. Write formulas for the following complexes:

    (a) dichlorobis(ethylenediamine)cobalt(III) nitrate

    _____

    (b) ammonium tetrachloro(ethylenediamine)chromate(III)

    _____

    (c) tetrabromostannate(II) ion

    _____

    (d) diaquatetrahydroxoaluminate(III) ion

    _____

    (e) diiodoargentate(I) ion

    _____

New Terms

None

23.6   Coordination number and structure are often related

Review

Coordination number is the number of ligand donor atoms that are bonded to the metal in a complex. Common coordination numbers in complexes include 2, 4, and 6. For coordination number 2, the complex usually has a linear geometry. For coordination number 4, both tetrahedral and square planar structures are observed. For coordination number 6, almost all complexes are octahedral. Such octahedral complexes are formed with monodentate ligands as well as polydentate ligands. Study Figures 23.13 and 23.14. Be sure you can sketch an octahedral complex; review Figure 23.15.

Self-Test

26. What property do metal ions that form tetrahedral complexes usually have?

_____

27. Sketch the structure of the complex $[Co(NH_3)_6]^{3+}$.

28. What is the coordination number of the metal ion in each of the following?

 (a) $[Cu(NH_3)_4]^{2+}$ _____

 (b) $[Co(NH_3)_4Cl_2]^+$ _____

 (c) $[Ni(en)_3]^{2+}$ _____

New Term

Write the definition of the following term, which was introduced in this section. If necessary, refer to the Glossary at the end of the text.

 coordination number

23.7   Isomers of coordination complexes are compounds with the same formula but different structures

Review

When two or more different compounds have the same chemical formula, they are said to be *isomers* of each other. For coordination compounds, there are several ways for this to occur, but the most important one for you to learn about is the type of isomerism called *stereoisomerism*. This occurs when two or more compounds have the same formula, but differ in the way their atoms are arranged in space.

Geometric isomers of square planar complexes such as $[Pt(NH_3)_2Cl_2]$, or octahedral complexes such as $[Cr(H_2O)_4Cl_2]^+$, exist in *cis* and *trans* forms. In the *cis* isomer, the chloride ions are next to each other on *the same side* of the metal ion. In the *trans* isomer, the chloride ions are opposite each other. In general, *cis* and *trans* isomers can exist for square planar complexes with the general formula Ma_2b_2 (where M is a metal ion and *a* and *b* are monodentate ligands). *Cis* and *trans* isomers also exist for octahedral complexes with the general formula Ma_2b_4, and for octahedral complexes with the general formula MA_2b_2 (where *A* is a bidentate ligand and *b* is a monodentate ligand).

Chiral isomers occur when two structures differ only in that one is the nonsuperimposable mirror image of the other—that is, when the mirror image of one isomer looks exactly like the other isomer, but the two isomers themselves do not match exactly when one is placed over the other. Chiral isomers exist for complexes with the general formula MA_3 and $cis–MA_2b_2$ (where *A* stands for a bidentate ligand and *b* stands for a monodentate ligand). Study Figures 23.18 and 23.19. Chiral isomers are also known as *optical isomers* because the two isomers affect polarized light in opposite ways. (See Facets of Chemistry 23.1.)

Self-Test

29.   Is *cis-trans* isomerism possible for tetrahedral complexes? Explain.

_____

_____

30.   How many different isomers exist for the complex $[Co(en)_2Br_2]^+$?

_____

31.   Is the *trans* isomer of $[Ni(C_2O_4)_2(CN)_2]^{4-}$ chiral? (Note: $C_2O_4{}^{2-}$ is oxalate ion, a bidentate ligand.)

_____

New Terms

Write the definitions of the following terms, which were introduced in this section. If necessary, refer to the Glossary at the end of the text.

isomerism	cis-isomer	superimposability
stereoisomerism	trans-isomer	enantiomers
geometric isomerism	chirality	optical isomers

23.8 Bonding in transition metal complexes involves *d* orbitals

Review

The crystal field theory is used to explain the properties of complexes in which the metal ion has a partially filled *d* subshell. To understand the theory, it is necessary to know the shapes and directional properties of the *d* orbitals. Study Figure 23.21.

In an octahedral complex, we can imagine the ligands to lie along the *x*, *y*, and *z* axes of a Cartesian coordinate system with the metal ion in the center (at the origin). The negative charges of the ligands (either anions or the negative ends of ligand dipoles) point directly at the metal ion's $d_{x^2-y^2}$ and d_{z^2} orbitals, but they point between the d_{xy}, d_{xz}, and d_{yz} orbitals. Because of this, the ligands repel electrons in the $d_{x^2-y^2}$ and d_{z^2} orbitals more than they repel electrons in the other three. This raises the energies of the $d_{x^2-y^2}$ and d_{z^2} orbitals above the energies of the d_{xy}, d_{xz}, and d_{yz} orbitals. The net result is an energy level diagram like that shown in Figure 23.23. The energy difference between the two energy levels is called the *crystal field splitting* and is given the symbol Δ.

The magnitude of Δ depends on the nature of the ligands attached to the metal ion, the oxidation state of the metal, and the period in which the metal occurs. In general, as the oxidation state increases, other things being equal, the size of Δ becomes larger. Going down a group, Δ becomes larger, too.

In this section, the usefulness of the crystal field theory is illustrated by considering three phenomena—the stabilities of certain oxidation states of metal ions in complexes, the origin of the colors of complexes, and the magnetic properties of complexes.

For chromium(II) ion, you see that the removal of a high-energy electron, along with an increase in the magnitude of Δ that accompanies the increase in oxidation state, helps to make the oxidation of $[Cr(H_2O)_6]^{2+}$ to $[Cr(H_2O)_6]^{3+}$ energetically favorable. Stated in another way, in water, chromium(III) ion is the more stable oxidation state, because chromium(II) ion is so easily oxidized to it.

According to crystal field theory, the color of a complex arises from the absorption of a photon that has an energy equal to Δ. For transition metal complexes, this photon has a frequency that places it in the visible region of the spectrum. The color observed for the complex is the color of the light that *isn't* absorbed.

Some ligands always produce a large Δ, regardless of the metal ion, and some ligands always produce a small Δ. The list of ligands arranged in order of their ability to produce a large Δ is called the spectrochemical series. Your teacher will tell you whether you should memorize this series of ligands.

The amount of energy needed to cause two electrons to become paired in the same orbital is called the *pairing energy*, to which we have given the symbol *P*. For certain numbers of *d* electrons, there is a choice as to how the electrons are to be distributed among the higher and lower *d*-orbital energy levels. Pairing an electron with another in a low-energy *d* orbital costs energy equal to the pairing energy, but it saves an energy equal to Δ. On the other hand, placing the electron in the higher energy orbital costs an energy equal to Δ, but saves energy equal to the pairing energy. Which energy distribution prevails depends on how the magnitudes of Δ and *P* compare. When $\Delta > P$, pairing of electrons in the lower energy level is preferred and a *low spin* complex is formed; when $\Delta < P$, then spreading the electrons out as much as possible is preferred and a *high spin* complex is formed.

Self-Test

32. Which of the *d* orbitals point directly along the *x, y,* and *z* axes?

33. Which complex has the larger Δ, $[CrCl_6]^{4-}$ or $[CrCl_6]^{3-}$?

_____

34. Which complex has the larger Δ, $Ni(CN)_4{}^{2-}$ or $Pt(CN)_4{}^{2-}$?

_____

35. Cyanide ion produces a very large crystal field splitting. Should it be easy or difficult to oxidize $[Co(CN)_6]^{4-}$ to $[Co(CN)_6]^{3-}$? Explain your answer in terms of the populations of the d orbitals of the metal ion.

36. Which complex would be expected to absorb light of longer wavelength?

    (a) $[Ti(H_2O)_6]^{3+}$ or $[Ti(H_2O)_6]^{2+}$ _____

    (b) $[Ni(H_2O)_6]^{2+}$ or $[Ni(CN)_6]^{4-}$ _____

37. Which complex has a larger Δ, one that absorbs red light or one that absorbs blue light?

    _____ _____

38. How many unpaired electrons would you expect to find in each of the following?

    (a) $[CrI_6]^{4-}$ _____   (c) $[Fe(H_2O)_6]^{3+}$ _____

    (b) $[Cr(CN)_6]^{4-}$ _____   (d) $[Fe(CN)_6]^{3-}$ _____

New Terms

Write the definitions of the following terms, which were introduced in this section. If necessary, refer to the Glossary at the end of the text.

crystal field theory	crystal field splitting	spectrochemical series
low-spin complex	high-spin complex	pairing energy

23.9   Metal ions serve critical functions in biological systems

Review

Most metal ions in living systems serve their function when they are held in complex ions of various kinds. The porphyrin ring system is found in hemoglobin and myoglobin, which contain Fe^{2+} that binds O_2 molecules as a ligand. A similar ring system containing Co^{2+} is found in vitamin B_{12}.

Self-Test

39. Name two biologically important complexes that contain the porphyrin structure.

_____   _____

40. How many donor atoms does the porphyrin ring contain? _____

New Terms

None

Answers to Self-Test Questions

1. $CuO(s) + H_2(g) \xrightarrow{\text{heat}} Cu(s) + H_2O(g)$
2. 1930 °C (rounded from 1925 °C)
3. $MgCl_2(l) \xrightarrow{\text{electrolysis}} Mg(l) + Cl_2(g)$
4. Magnesium and sodium
5. $Mg^{2+}(aq) + 2OH^-(aq) \rightarrow Mg(OH)_2(s)$
   $Mg(OH)_2(s) + 2H^+(aq) + 2Cl^-(aq) \rightarrow Mh^{2+}(aq) + 2Cl^-(aq) + 2H_2O$
6. Sulfide ores of copper and lead
7. $2PbS + 3O_2 \rightarrow 2PbO + 2SO_2$
8. Amphoteric behavior; $Al_2O_3(s) + 2OH^- \rightarrow 2AlO_2^- + H_2O$
   $AlO_2^- + H^+ + H_2O \rightarrow Al(OH)_3(s)$
   $2Al(OH)_3 \xrightarrow{\text{heat}} Al_2O_3 + 3H_2O$
9. Carbon monoxide; $Fe_2O_3 + 3CO \rightarrow 2Fe + 3CO_2$
10. $CaCO_3 \rightarrow CaO + CO_2$
    $CaO + SiO_2 \rightarrow CaSiO_3$ (slag)
11. It is fast and produces good-quality steel.
12. Fe^{3+}
13. FeS
14. CaS
15. SnI_2
16. MgI_2
17. $AlCl_3$
18. (a) $[Zn(OH)_4]^{2-}$
    (b) It could be isolated as a potassium salt, $K_2[Zn(OH)_2$.
19. $[Fe(SCN)_6]^{3-}$
20. $[Ni(NH_3)_6]^{2+}$
21. $[CoCl_2(H_2O)_2(en)]$
22. (a) (b)

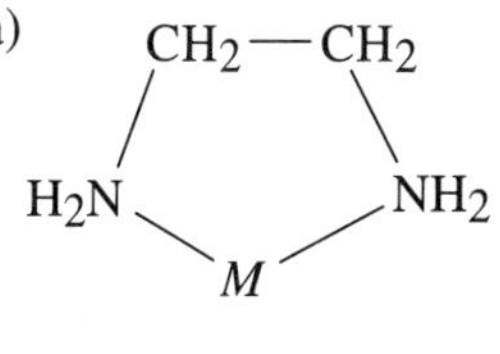

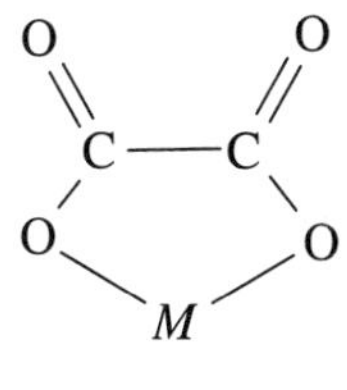

23. $[Cr(en)_3]^{3+}$ (chelate effect)
24. (a) hexaamminechromium(II) ion
    (b) tetrahydroxocuprate(I) ion
    (c) diamminetetrachloromanganate(II) ion
    (d) potassium hexacyanocobaltate(III)
    (e) diamminetetraaquanickel(II) chloride
25. (a) $[CoCl_2(en)_2]NO_3$
    (b) $NH_4[CrCl_4(en)]$
    (c) $[SnBr_4]^{2-}$
    (d) $[Al(OH)_4(H_2O)_2]^-$
    (e) $[AgI_2]^-$
26. They usually have filled or empty d subshells.
27.

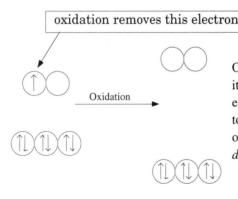

28. (a) 4, (b) 6, (c) 6
29. No. For complexes of the type Ma_2b_2, only one tetrahedral structure can be constructed.
30. Three. A trans isomer and a pair of chiral cis isomers that are enantiomers.
31. No. The complex and its mirror image are superimposable.
32. $d_{x^2-y^2}$ and d_{z^2}
33. $[CrCl_6]^{3-}$, because it has chromium in the higher oxidation state.
34. $[Pt(CN)_4]^{2-}$, because Pt is below Ni in its group.
35.

oxidation removes this electron

Oxidation

Oxidation should be easy because it involves removing a high-energy electron and it also leads to a lowering of the energy of the orbitals that hold the remaining d electrons.

$[Co(CN)_6]^{4-}$ $[Co(CN)_6]^{3-}$

36. (a) $[Ti(H_2O)_6]^{3+}$ (b) $[Ni(CN)_6]^{4-}$
37. The one that absorbs the higher energy blue light.
38. (a) 4   (b) 2   (c) 5   (d) 1
39. hemoglobin, myoglobin
40. four

Chapter 24

Some Chemical Properties of the Nonmetals and Metalloids

In this chapter we examine some of the properties of the nonmetallic elements. Our goal is to learn some of the general ways the free elements can be prepared as well as some of the general methods that are used to make certain of their compounds. We will also look at structural similarities that exist between the oxoacids and oxoanions of the nonmetallic element. As in Chapter 23, one of our goals is to see how principles of thermodynamics, kinetics, and chemical bonding can help us understand similarities and differences among a variety of substances. Although the specific factual information is important to learn, pay particular attention to the generalizations that are discussed, because these tie together the chemistry of a variety of elements and make it easier to remember specific details.

Learning Objectives

As you study this chapter, keep in mind the following objectives:

1   To learn how nonmetallic elements are obtained from their compounds.

2   To become familiar with the molecular structures of the nonmetallic elements.

3   To learn how the ability of nonmetals to form π bonds to each other affects the structures of the nonmetals.

4   To learn the methods that can be used to prepare hydrogen compounds of nonmetals.

5   To learn how the tendency of nonmetals to bind to other like atoms varies in the periodic table.

6   To learn how oxygen compounds of the nonmetallic elements can be prepared.

7   To learn the structures of some of the oxides of the nonmetals and metalloids.

8   To learn the structures of the oxoacids and oxoanions of the nonmetals.

9   To learn about the factors that determine the formulas of the halogen compounds of nonmetallic elements.

10   To learn how the structure of the nonmetal halide and the electronic structure of the central atom affects the ability of the compound to react with water.

24.1   Obtaining Metalloids and Nonmetals in their Free States

Review

Metalloids tend to exist in positive oxidation states in their compounds, so they are prepared by chemical reduction of their compounds (usually their oxides). This is accomplished using hydrogen or carbon as the reducing agent.

Some nonmetals are found in the free state. These include the noble gases, oxygen, nitrogen, sulfur and carbon. Except for the noble gases, the nonmetals are also found in compounds. Study the kinds of compounds in which sulfur and carbon are found.

No broad generalizations can be made about the recovery of nonmetals from their compounds because of the variety of oxidation states, both positive and negative, in which they are found. When the nonmetal is in a negative oxidation state, it can be obtained by oxidation. Study the ways in which the halogens can be made from compounds containing the halide ions. If the nonmetal is in a positive oxidation state, it is recovered from compounds by reduction.

Self-Test

1.   Which kind of process, oxidation or reduction, would be needed to recover

   (a)   Sulfur from Cu_2S _____

   (b)   Phosphorus from $Ca_3(PO_4)_2$ _____

   (c)   Arsenic from As_2O_3 _____

   (d)   Antimony from Na_3Sb _____

2.   Write a chemical equation for the reduction of PoO_2 by

   (a)   hydrogen   _____

   (b)   carbon   _____

3.   Complete and balance the following equations. If no reaction occurs, write N.R.

   (a) $Cl_2(g)$ + $NaI(aq) \rightarrow$ _____

   (b) $Br_2(aq)$ + $NaF(aq) \rightarrow$ _____

   (c) $Br_2(aq)$ + $NaI(aq) \rightarrow$ _____

4.   Why can't F_2 be prepared by electrolysis of aqueous NaF?

   _____

New Terms

   None

24.2   The free elements have structures of varying complexity

Review

Except for the noble gases, the nonmetals exist in nature in covalently bonded molecular structures. One of the principal factors that determines the complexity of these structures is the ability of the nonmetal to form strong π bonds to other atoms of the same kind. Elements in Period 2 are small and the sideways overlap of their p orbitals is quite effective, so they form strong π bonds to other like atoms. As we descend a group, however, the strengths of the π bonds between the atoms decreases substantially, so the atoms prefer to form single bonds instead.

You should study the molecular structures formed by the various nonmetals and metalloids. Review the structures of the allotropes of carbon in Chapter 13 (page 588). Study the structures of the S_8 and P_4 molecules, as well as the kind of structures formed by silicon and boron.

Self-Test

5. Why does bromine, a period 4 element, form diatomic Br_2 instead of a more complex molecular structure?

6. Draw the resonance structures of ozone.

7. How do the rhombic and monoclinic allotropes of sulfur differ?

8. On a separate sheet of paper, sketch the structures of (a) the S_8 molecule, and (b) the P_4 molecule.

9. What structural unit characterizes crystals of elemental boron?

10. Describe the structure of elemental silicon. (How many bonds does each Si atom form; how are they arranged? How does the structure compare to that of an allotrope of carbon?)

11. In crystals of solid polonium (Po), how many bonds would you expect each Po atom to form to other Po atoms?

New Terms

Write the definitions of the following terms, which were introduced in this section. If necessary, refer to the Glossary at the end of the text.

allotrope	ozone	allotropy
monoclinic sulfur	orthorhombic sulfur	white phosphorus
red phosphorus	black phosphorus	

24.3   Hydrogen forms compounds with most nonmetals and metalloids

Review

Binary hydrogen compounds are often called hydrides, regardless of whether the hydrogen is in a positive or negative oxidation state. You should be able to write the formulas and predict the structures of the simple hydrides of the nonmetals and metalloids.

Boron hydrides are unusual because they contain three center bonds in which a hydrogen atom binds to two boron atoms simultaneously to give a hydrogen "bridge." Study Figure 24.8.

Preparation of Hydrides

Two methods for preparing nonmetal hydrides are described. One involves the direct combination of the elements, which is only feasible when the hydride has a negative free energy of formation. If ΔG_f° is positive, the equilibrium constant for the formation of the hydride is very small and direct combination is not a practical method for preparing the compound.

The second method involves adding protons to the anion of the nonmetal. The general equation is

$$X^{n-} + nHA \rightarrow H_nX + A^-$$

The strength of the Brønsted acid, HA, needed to carry out the reaction varies with the location of X in the periodic table. The acid strength of H_nX increases from left to right across a period and from top to bottom in a group. Conversely, the basicity of the anion, X^{n-}, increases from right to left in a period and from bottom to top in a group. The stronger X^{n-} is as a base, the *weaker* the proton donor must be to prepare the hydride, so the generalization is as follows: The strength of the acid required to protonate X^{n-} increases as the position of X varies from left to right in a period and from top to bottom in a group.

Any of the anions with charges equal to or larger than 2– are so basic that they react with water. The Group VIA nonmetal anions give the corresponding HX^- ions. For example, with O^{2-}, addition of water produces OH^-

$$O^{2-} + H_2O \rightarrow 2OH^-$$

Anions with charges of 3– or larger react completely to form the hydride.

$$Mg_3P_2 + 6H_2O \rightarrow 3Mg(OH)_2 + 2PH_3$$

Self-Test

12. Write equations for the reaction of the elements to form

    (a) ammonia   _____

    (b) hydrogen chloride _____

13. Write an equation for the formation of HCl from KCl.

    _____

14. Write an equation for the reaction of aqueous Na_2S with $HCl(aq)$.

    _____

15. How many three center bonds are there in the diborane molecule?   _____

16. Why can't concentrated H_2SO_4 be used to make HI from NaI?

    _____

    What acid is used instead of H_2SO_4?   _____

17. Which is the more basic anion?

    (a) S^{2-} or Si^{4-} _____

    (b) As^{3-} or Br^- _____

    (c) F^- or I^- _____

New Terms

Write the definitions of the following terms, which were introduced in this section. If necessary, refer to the Glossary at the end of the text.

hydride   hydrolysis   silanes

24.4   Catenation between Nonmetal Atoms

Review

Catenation is the ability of atoms of the same element to link together to form chains of atoms. The maximum chain length varies from element to element, with the maximum ability being demonstrated by carbon. Many compounds in which catenation takes place are hydrides, as illustrated in the text. Sulfur is an element that demonstrates catenation in a variety of its elemental forms and compounds. Examples include the polysulfide ions, S_x^{2-}, thiosulfate ion, and polythionate ions (study their structures).

Self-Test

18. On a separate sheet of paper, draw structural formulas for the following.

    (a) Si_3H_8 (b) tetrathionate ion   (c) thiosulfate ion   (d) azide ion.

New Term

Write the definition of the following term, which was introduced in this section. If necessary, refer to the Glossary at the end of the text.

catenation

24.5   Oxygen combines with almost all nonmetals and metalloids

Review

Oxides of nonmetals form acidic solutions in water if they are able to react with the solvent. You should be able to write the equations for the reactions of CO_2, SO_3, and NO_2 with water.

In this section we focus on two principal aspects of the oxides of nonmetallic elements. One is their preparation. The other is their structures.

Preparation of hydrides

Among the general methods of preparing oxides of nonmetal are the following: (1) direct combination of the elements, (2) oxidation of a lower oxide, (3) reaction of a nonmetal hydride with oxygen, and (4) other redox reactions.

1   Direct combination of oxygen with a nonmetal is only feasible when the heat of formation of the oxide is negative. Formation of nitrogen oxides in general are endothermic, so they cannot be prepared in practical amounts by reaction of O_2 with N_2. Among the nonmetals that will react with molecular oxygen are carbon, sulfur, and phosphorus.

2   Often, O_2 will combine with a nonmetal oxide in which the central nonmetal has less than its maximum oxidation state. For example, SO_2 (with S in the +4 oxidation state) will react with O_2 to form SO_3 (in which S is in the +6 state). The reaction is slow, however, and is accomplished commercially using a catalyst.

3   Nonmetal hydrides react with oxygen to form the nonmetal oxide and water.

4   Nonmetal oxides are often among the products of redox reactions, as illustrated by the reactions of HNO_3 with copper described in the text.

Molecular structures of the oxides

Study the structures of the oxides of carbon, nitrogen, and sulfur discussed in the text. You should be able to sketch the structures of the oxides of phosphorus, P_4O_6 and P_4O_{10}. In particular, note how their structures are related to that of white phosphorus, P_4. Study the structure of quartz. Note the tetrahedral arrangement of oxygen atoms around silicon. You should be able to explain why quartz crystals are chiral.

Self-Test

19. Complete and balance the following equations. If no reaction occurs, write N.R.

(a) $S + O_2 \rightarrow$ _____

(b) $C + O_2 \rightarrow$ _____

(c) $P_4 + O_2 \rightarrow$ _____

(d) $N_2 + O_2 \rightarrow$ _____

20. Complete and balance the following equations.

(a) $NO + O_2 \rightarrow$ _____

(b) $CO + O_2 \rightarrow$ _____

(c) $P_4O_6 + O_2 \rightarrow$ _____

21. Complete and balance the following equations.

(a) $CH_4 + O_2 \rightarrow$ _____

(b) $PH_3 + O_2 \rightarrow$ _____
     (Hint: the product contains phosphorus in the +5 oxidation state.)

(c) $H_2S + O_2 \rightarrow$ _____
     (Hint: the product contains sulfur in the +4 oxidation state.)

22. Hot concentrated sulfuric acid is a fairly potent oxidizing agent. It reacts with metallic copper to give Cu^{2+} and SO_2 among the products. Write a balanced net ionic equation for the reaction.

_____

23. On a separate sheet of paper, sketch the structures of P_4O_6 and P_4O_{10}.

24. What equilibrium exists between $NO_2(g)$ and $N_2O_4(g)$? What are the colors of these two gases?

_____

25. Quartz crystals are chiral, but molten quartz is not. Why?

_____

_____

New Terms

None

24.6   Nonmetals for a variety of oxoacids and oxoanions

Review

Oxoacids have the general formula $XO_m(OH)_n$. Examples discussed are phosphoric acid, sulfuric acid, and nitric acid. These compounds and their ions are divided into two categories: simple monomeric species, which contain just one atom of the nonmetal; and polymeric species, which are formed (at least in principle) by the linking of monomeric acids or anions through the formation of oxygen bridges.

Oxoacids of elements in Period 2

For the period 2 elements, the oxoacids are HOF (unstable), HNO_3 and HNO_2, H_2CO_3, and H_3BO_3 [which is better written $B(OH)_3$]. Boric acid is not a Brønsted acid like the others; instead it functions as a Lewis acid. Study the diagrams that illustrate how it behaves as an acid.

Oxoacids of elements in Periods 3 and below

The halogens undergo disproportionation in water to give hypohalous acids, HOX, and the binary acid, HX (which is fully ionized). The hypohalate anions, OX^-, have a tendency to disproportionate further. Study how kinetic stability, rather than thermodynamic stability, affects these reactions.

Study the reactions of the oxoacids of sulfur. In particular, note that attempts to isolate bisulfite salts lead to the formation of disulfite ion, $S_2O_5^{2-}$. Also, note the formulas of the oxoacids of selenium and tellurium.

Phosphorus forms two principal monomeric oxoacids, phosphorous acid (H_3PO_3) and phosphoric acid (H_3PO_4). Study their structures. Note that H_3PO_3 is only a diprotic acid. Arsenic forms two oxoacids with similar formulas, but they are both triprotic acids.

No simple oxoacid of silicon can be isolated, although the anion SiO_4^{4-} is observed and is the anion of the hypothetical orthosilicic acid, H_4SiO_4.

Self-Test

26. Why is HOF so unstable?

_____

27. Why does boric acid behave as a Lewis acid?

_____

28. Draw Lewis structures for HNO_3 and HNO_2.

_____

29. If fluorine were to form HFO_4, its Lewis structure might be drawn as shown below.

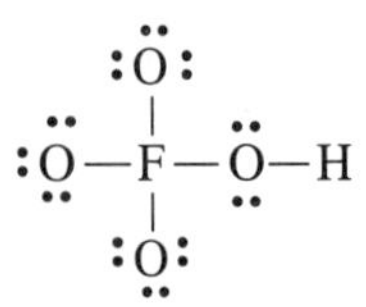

I 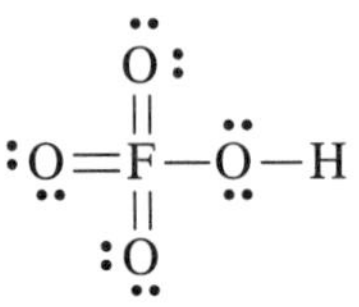   II

Explain why each of these structures is unlikely (or impossible) for this acid. Does is seem likely that this acid could be prepared?

_____

_____

30. Hypophosphorous acid, H_3PO_2, has the structure shown below. How many hydrogens per acid molecule can be released from the molecule?

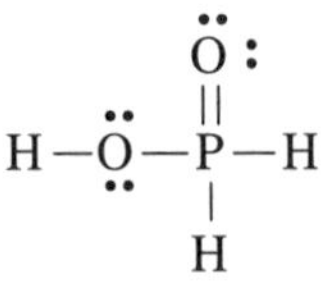

Answer _____

31. Write a balanced equation for the disproportionation of BrO^-.

_____

New Terms

Write the definitions of the following terms, which were introduced in this section. If necessary, refer to the Glossary at the end of the text.

disproportionation

monomer

24.7   Halogen compounds are formed by most nonmetals and metalloids

Review

The halogens form diatomic molecules because they need to form only one bond to complete their octets. Substitution of one halogen for another is possible, so a series of interhalogen compounds can be formed. Simple halogen compounds of the other nonmetallic elements is determined by the number of electrons these elements need to reach an octet. Elements of Period 3 and below can expand their valence shells and bind to additional halogen atoms, two at a time. (For each pair of electrons made available to form bonds, two halogens become attached.) The maximum number of halogens that can be bonded to a nonmetal is equal to the number of valence electrons of the element. However, the size of the nonmetal also influences the number of halogen atoms that can be attached (in general, the larger the nonmetal, the larger the number of halogen atoms can be attached).

Nonmetal-halogen bonds are susceptible to attack by water, which is used in the preparation of silicone polymers. However, the SF_6 and CCl_4 molecules are not able to react with water because of the tight packing of halogens around the central atom, which leaves no room for an attacking water molecule to reach the central atom. NCl_3 is attacked by water, as described in the text. Study the suggested reaction mechanism. NF_3 is not attacked by water because of the weak basicity of the electron pair on the nitrogen.

Self-Test

32. Use VSEPR theory to predict the shapes of the following.

    (a) PCl_3 _____

    (b) PCl_5 _____

    (c) ClF_3 _____

    (d) ClF_5 _____

    (e) SF_4 _____

    (f) SF_6 _____

33. Would you expect ClF_5 to be easily attacked by water? Explain.

    _____

    _____

34. Write a balanced equation for the reaction of BrF_5 with water.

    _____

New Terms

None

Answers to Self-Test Questions

1. (a) oxidation, (b) reduction, (c) reduction, (d) oxidation
2. (a) $PoO_2 + 2H_2 \rightarrow Po + 2H_2O$
    (b) $PoO_2 + C \rightarrow Po + CO_2$
3. (a) $Cl_2(g) + 2NaI(aq) \rightarrow I_2(aq) + 2NaCl(aq)$
    (b) N.R.
    (c) $Br_2(aq) + 2NaI(aq) \rightarrow I_2(aq) + 2NaBr(aq)$
4. Fluoride ion is more difficult to oxidize than water, so O_2 is formed instead of F_2.
5. Each Br atom needs to form only one bond to another atom to complete its octet.
6. $:\overset{..}{O}=\overset{..}{O}-\overset{..}{\underset{..}{O}}:$ $:\overset{..}{\underset{..}{O}}-\overset{..}{O}=\overset{..}{O}:$
7. They both contain S_8 molecules, but the molecules are stacked differently in their crystals.
8. See Figures 24.4 and 24.5.
9. Icosahedral B_{12} units linked through some of their vertices.
10. Each Si atom forms four bonds arranged tetrahedrally. The structure is like that of diamond.
11. Two bonds. Being a Group VIA element, each Po atom needs two electrons to complete its octet.
12. (a) $N_2(g) + 3H_2(g) \rightarrow 2NH_3(g)$
    (b) $Cl_2(g) + H_2(g) \rightarrow 2HCl(g)$
13. $KCl(s) + H_2SO_4(l) \rightarrow HCl(g) + KHSO_4(s)$
14. $Na_2S(aq) + 2HCl(aq) \rightarrow H_2S\ (g) + 2NaCl(aq)$
15. Two
16. Concentrated H_2SO_4 is able to oxidize I^- to I_2. The proton donor used to prepare HI is H_3PO_4.
17. (a) Si, (b) As^{3-}, (c) F^-
18. (a)

$$
\begin{array}{c}
\quad\ \ H \quad\ \ H \quad\ \ H \\
\quad\ \ | \qquad | \qquad | \\
H-Si-Si-Si-H \\
\quad\ \ | \qquad | \qquad | \\
\quad\ \ H \quad\ \ H \quad\ \ H
\end{array}
$$

(b)
$$
\left[:\overset{..}{\underset{..}{O}}-\overset{\overset{..}{O}:}{\underset{:\underset{..}{O}:}{S}}-\overset{..}{\underset{..}{S}}-\overset{..}{\underset{..}{S}}-\overset{\overset{..}{O}:}{\underset{:\underset{..}{O}:}{S}}-\overset{..}{\underset{..}{O}}: \right]^{2-}
$$

(c)
$$
\left[:\overset{..}{\underset{..}{S}}-\overset{\overset{..}{O}:}{\underset{:\underset{..}{O}:}{S}}-\overset{..}{\underset{..}{O}}: \right]^{2-}
$$

(d)
$$
\left[:\overset{..}{N}=N=\overset{..}{N}: \right]^{-}
$$

19. (a) $S + O_2 \rightarrow SO_2$
    (b) $C + O_2 \rightarrow CO_2$
    (c) $P_4 + 5O_2 \rightarrow P_4O_{10}$ or $P_4 + 3O_2 \rightarrow P_4P_6$
    (d) $N_2 + O_2 \rightarrow$ N.R. (at room temperature)
20. (a) $2NO + O_2 \rightarrow 2NO_2$
    (b) $2CO + O_2 \rightarrow 2CO_2$
    (c) $P_4O_6 + 2O_2 \rightarrow P_4O_{10}$
21. (a) $CH_4 + 2O_2 \rightarrow CO_2 + 2H_2O$
    (b) $4PH_3 + 8O_2 \rightarrow P_4O_{10} + 6H_2O$
    (c) $2H_2S + 3O_2 \rightarrow 2H_2O + 2SO_2$
22. $Cu + 4H^+ + SO_4^{2-} \rightarrow Cu^{2+} + SO_2 + H_2O$

23. See Figure 24.12

24. $2NO_2(g) \rightleftharpoons N_2O_4(g)$; NO_2 is red-brown, N_2O_4 is colorless.

25. When quartz is melted, the chiral helices of the crystals break down to produce a jumbled, nonchiral liquid.

26. Because of the high electronegativity of fluorine, oxygen is forced to have a zero oxidation state.

27. Boron has less than an octet in $B(OH)_3$, so it easily binds to the oxygen of water. This polarizes the H_2O molecule and allows it to release H^+ relatively easily.

28.

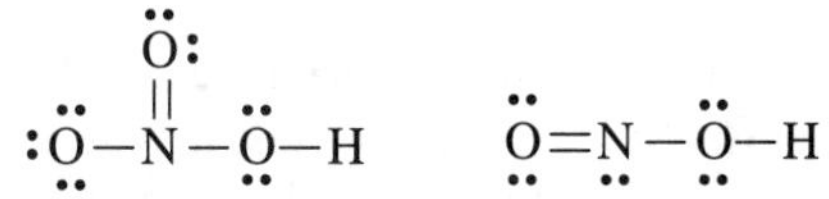

$$\overset{\overset{\displaystyle \ddot{O}:}{\|}}{:\ddot{O}-N-\ddot{O}-H} \qquad \ddot{O}=N-\ddot{O}-H$$

29. In structure I, the formal charge on fluorine is +3 and each of the lone oxygens has a formal charge of -1. The high degree of charge separation and the fact that fluorine, the most electronegative element, has a positive formal charge, would make this a very "high energy" structure (and therefore very unstable). The second structure, with no formal charges, is impossible because fluorine cannot have more than seven electrons in its valence shell. It is not likely that this acid can be prepared.

30. One. It is a monoprotic acid because it has only one –OH group.

31. $3BrO^- \rightarrow BrO_3^- + 2Br^-$

32. (a) trigonal pyramidal, (b) trigonal bipyramidal, (c) T-shaped, (d) square pyramidal
    (e) distorted tetrahedral (see-saw shaped), (f) octahedral

33. Yes. The molecule has a square pyramidal shape, which means the Cl is exposed and can be attacked by water.

34. Bromine is in the +5 oxidation state, which is the oxidation state of Br in $HBrO_3$. The reaction would be:
    $BrF_5 + 3H_2O \rightarrow HBrO_3 + 5HF$

Chapter 25

Organic Compounds and Biochemicals

There are probably ten times as many organic compounds as any other kind, so this chapter can only serve as a very brief introduction to a large field of chemistry. Yet the study of organic compounds assumes many of the features of the study of any other field. You learn in this chapter about classifying organic compounds into families defined by functional groups. You see how such groups confer common chemical properties to all members of the same family, but that the nonfunctional, hydrocarbon groups contribute much to physical properties.

All of living processes in nature have a molecular basis, and *biochemistry* describes them. The complex molecules of biochemistry have functional groups like those in simpler substances and so they have similar chemical properties. Thus, this chapter is also meant to introduce you to the major kinds of biochemicals.

Learning Objectives

1   To learn about the major structural features of organic molecules and the importance of functional groups to the study of organic chemistry.

2   To learn the principles of formal (IUPAC) nomenclature—characteristic family name endings, rules for identifying "parents," rules for numbering parent chains or rings, and the names of hydrocarbon groups (alkyl groups).

3   To learn to predict physical states and general physical properties from structure, properties such as solubilities in water or nonpolar solvents.

4   To learn to "read" a molecular structure to tell if a substance is likely to undergo addition, substitution, oxidation, reduction (hydrogenation), or hydrolysis reactions.

5   To learn *how* certain reactions occur:

    (a)   how the carbon–carbon double bond undergoes an addition reaction;

    (b)   how an alcohol undergoes dehydration.

6   To learn to predict the products that form when

    (a)   alcohols are oxidized,

    (b)   alcohols undergo dehydration,

    (c)   alcohols and carboxylic acids form esters,

    (d)   ammonia or amines and carboxylic acids form amides,

    (e)   amines neutralize strong acids,

    (f)   aldehydes and ketones are hydrogenated,

    (g)   esters or amides are hydrolyzed or saponified,

    (h)   the benzene ring undergoes halogenation, nitration, or sulfonation.

7   To be able to look at a structural formula of a biochemical and assign it to its appropriate family.

8   To learn the structure of glucose, both its cyclic and its open-chain forms and learn how glucose is related to the nutritionally important disaccharides and polysaccharides.

9   To learn how to write the structure of a triacylglycerol molecule such as typically found among animal fats or vegetable oils.

10   Given the structure of a triacylglycerol, to write an equation for its digestion.

11   To distinguish between the terms "protein" and "polypeptide."

12   To be able to write the sequence of atoms joined in the "backbone" of a polypeptide and explain how polypeptides can be alike in backbones but still different.

13   To describe in general terms the causes of unique overall shapes for proteins and explain how such shapes are important at the molecular level of life.

14   To explain what enzymes are.

15   To learn the names, symbols, and functions of the substances directly involved in the chemistry of heredity.

16   To describe what is meant by the "genetic code."

17   To describe the technology of genetic engineering and the purposes served by it.

25.1   Organic chemistry is the study of carbon compounds

Review

The study of *organic chemistry* is helped by the concept of *functional groups*, the main idea in this Section. Organic chemistry is organized around families defined by such groups, and we introduce many of the most important families here. Table 25.1 lists the principal functional groups we discuss in this chapter. We suggest you take a look at them now, but you will probably find this table more useful for review later.

Substances whose molecules have oxygen or nitrogen atoms in them tend to be more polar and so more soluble in water than others of the same size. We note that some groups bear similarities in structure and therefore in chemistry to simple inorganic species. Amines are ammonia-like, for example, and alcohols are water-like.

Besides functional groups, organic molecules contain hydrocarbon-like, *nonfunctional groups*. Be sure to understand the value of using one symbol, R, for all such groups, no matter how many carbons are present. The fact that nonfunctional groups do not change chemically in most (if not all) of the reactions of a family is what makes such a simplifying symbol possible. R groups have few chemical reactions because they are essentially nonpolar and therefore unattractive to ionic or polar reactants.

The carbon atoms making up an R group can be in *straight chains, branched chains,* and *rings.* When rings incorporate an atom other than carbon, the ring is said to be *heterocyclic.*

Self-Test

1. Circle what is most likely the functional group in

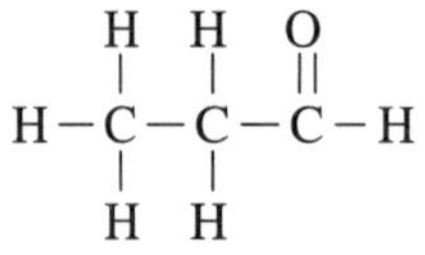

$$
\begin{array}{ccc}
H & H & O \\
| & | & \| \\
H-C-C-C-H \\
| & | \\
H & H
\end{array}
$$

2. Which molecule would be more polar, CH_3—CH_3 or CH_3—NH_2?

   _____

3. What forms in the following reaction?

   CH_3—$NH_2 + H$—$Br \rightarrow$ _____

4. What forms in the following reaction?

   R—$NH_2 + H$—$Cl \rightarrow$ _____

5. What significance does the functional group concept have for the study of organic chemistry?

   _____

   _____

6. Study the members of the following pairs of compounds and decide if they represent *isomers*, or are *identical*, or *neither*.

   (a) $CH_3OCH_2CH_3$ and $CH_3CH_2CH_2OH$ _____

   (b) 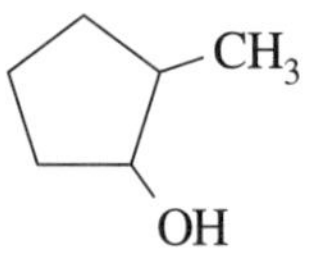   and   _____

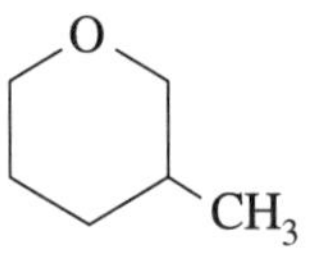

   (c) $CH_3CH_2CH_2OH$ and $HOCH_2CH_2 CH_3$ _____

7. Which compound in question 7 is a heterocyclic compound?

   _____

New Terms

Write the definitions of the following terms, which were introduced in this section. If necessary, refer to the Glossary at the end of the text.

branched chain	organic chemistry
functional group	ring compound
heterocyclic compound	straight chain compound

25.2   Hydrocarbons consist of only C and H atoms

Review

When the molecules of a compound have no double or triple bonds, it is a *saturated compound.* Otherwise, it is *unsaturated.* (Compounds with double or triple bonds have the capacity to take up additional hydrogen or other atoms, much as an unsaturated solution has the capacity to take up additional solute. That's why such compounds are said to be unsaturated.) The molecules of *hydrocarbons* are made solely of C and H. The saturated hydrocarbons have only single bonds and are called *alkanes* (or *cycloalkanes*). Whether straight-chain, branched-chain, or ring, the saturated hydrocarbons have very few chemical properties. All hydrocarbons are hydrophobic; they do not dissolve in water.

Nomenclature

Here's how to sort out the *IUPAC rules* for all families. *Regardless of the family,* the IUPAC rules all begin with the idea of a "parent" unit—a molecular portion which is named and whose carbon skeleton is numbered. Each family has its own rule for identifying the parent. For the alkanes, the "parent" is the longest continuous sequence of carbons. In other words, a parent alkane is always, by itself, a straight-chain alkane. That's why it's important that you learn the names of the straight-chain alkanes through $C_{10}H_{22}$ (Table 25.2 in the text).

The next key idea in IUPAC nomenclature is that each family has a characteristic *name ending.* This is the same for all members of a given family. Thus "-ane" is the name ending for all alkanes and cycloalkanes, and "-ol" is the ending of all IUPAC names of alcohols.

Each family has its own IUPAC rule for selecting the *parent chain* to which substituents are attached.

Each family also has its own IUPAC rule for *numbering the parent chain;* a rule that tells us which position must be numbered 1. For the alkanes, the rule is to number from whichever end of the parent chain is nearest the first branch.

Hydrocarbon groups, called *alkyl groups,* have their own names. You should learn the names and structures for those having one to three carbons—the *methyl, ethyl, propyl,* and *isopropyl* groups.

The locations of the alkyl groups on the parent chain are identified by the numbers given to the carbons of the parent chain when the chain is properly numbered.

When you assemble all the parts of a name into one whole, remember the following rules.

1   Each alkyl group (or other substituent) must be associated with a chain location number.

2   Multiplier prefixes, like di- and tri-, sometimes have to be added to the names of alkyl groups.

3   The names of the alkyl groups are organized into the final name alphabetically, ignoring the multiplier prefixes (di-, tri-, etc.). Thus "trimethyl" would precede "propyl" because "m" precedes "p" in the alphabet.

4   Two numbers in a name are always separated by a comma; e.g., 2,2-dimethylpropane.

5   A hyphen is always used to separate a number from a word-part of a name; e.g., 2,3-dimethylbutane.

These are illustrated in Example 25.1.

Alkane chemistry

At room temperature almost nothing attacks alkanes. They do burn, and fluorine reacts violently with them. At higher temperatures, alkanes react with chlorine and bromine, and alkane molecules can be *cracked.*

Alkenes

The parent of an *alkene* must be the longest chain *that includes the double bond.* The parent is numbered from whichever end gets to the double bond first, regardless of the location of alkyl branches, and the numbering proceeds through the double bond.

Many alkenes exhibit *geometric isomerism* (cis-trans isomerism) and have molecules that differ only in their geometry at the double bond. Notice that this isomerism in alkenes is possible only when neither end of the double bond holds identical groups.

Alkene chemistry

The carbon–carbon double bond has π electrons, making the double bond somewhat electron-rich and attractive to protons and other electron-poor species. So alkenes undergo *addition reactions* in which the reactant (or a catalyst) is able to donate H^+ to the double bond. Thus alkenes react with hydrogen chloride and with water, provided there is an acid catalyst. When gaseous molecules of HCl add to an alkene, H^+ from HCl goes to one end of the double bond using the pi electrons to make a new C–H bond. This leaves the other end, now electron poor, to accept the chloride ion. Overall, we have

$$
\underset{/}{\overset{\backslash}{C}} = \underset{\backslash}{\overset{/}{C}} \quad + \quad H-Cl \quad \longrightarrow \quad H - \overset{|}{\underset{|}{C}} - \overset{|}{\underset{|}{C}} - Cl
$$

Compare this addition with the addition of water and note that again one end of the double bond gets an H and the other end gets the rest of the adding molecule, OH in this case. Now the product is an alcohol.

$$
\underset{/}{\overset{\backslash}{C}} = \underset{\backslash}{\overset{/}{C}} \quad + \quad H-O^H \quad \xrightarrow{\text{acid catalyst}} \quad H - \overset{|}{\underset{|}{C}} - \overset{|}{\underset{|}{C}} - O^H
$$

Two halogens, Cl_2 and Br_2, also add to carbon–carbon double bonds. One Cl or Br goes to one end and the other Cl or Br goes to the other end of the double bond. H_2 adds similarly, but special catalysts are needed.

As a study goal, practice writing the structures of the products of the following alkenes with hydrogen chloride, water (in the presence of an acid catalyst), Cl_2, Br_2, and H_2 (assuming the special conditions).

$$
H_2C=CH_2 \qquad CH_3CH=CHCH_3
$$

Aromatic hydrocarbons

The typical Lewis structure for an *aromatic hydrocarbon* such as benzene would suggest that it is an unsaturated hydrocarbon.

However, aromatic hydrocarbons do not give addition reactions because the benzene ring strongly resists anything that breaks up its unique pi electron network. Instead, *substitution reactions* occur, which leave the benzene ring system intact. Practice writing the structures of what forms when benzene reacts with concentrated sulfuric acid, nitric acid, and the two halogens, Cl_2 and Br_2, when an iron(III) halide catalyst is present.

Self-Test

8. Write the IUPAC names of the following.

(a) $CH_3CH_2CH_2CH_2CH_3$

_____

(b) 
$$CH_3\overset{\overset{\displaystyle CH_3}{|}}{C}HCH_2CH_2CH_2CH_3$$

_____

(c)
$$CH_3\overset{\overset{\displaystyle CH_3}{|}}{\underset{\underset{\displaystyle CH_3}{|}}{C}}CH_2CH_2CH_3$$

_____

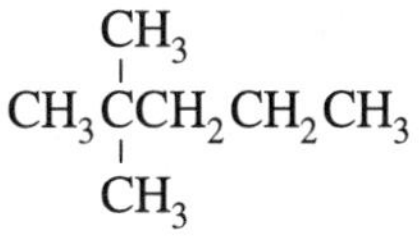

(d) 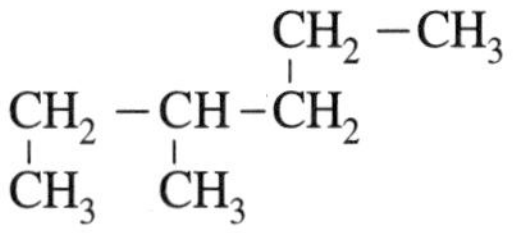
$$\begin{array}{l} \overset{\overset{\displaystyle CH_2-CH_3}{|}}{CH_2-CH-CH_2} \\ \underset{\displaystyle CH_3}{|}\underset{\displaystyle CH_3}{|} \end{array}$$

_____

(e) 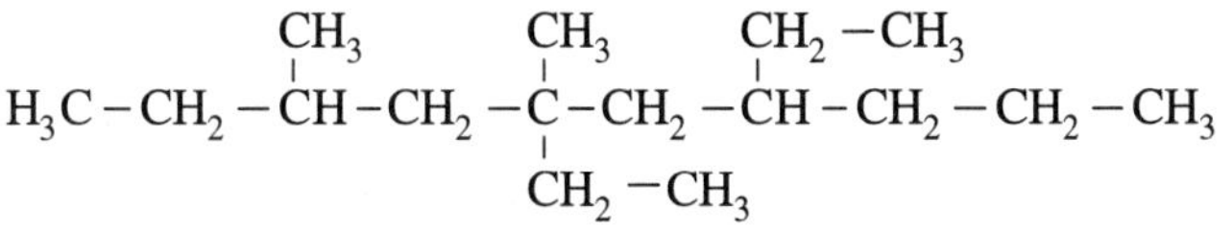
$$H_3C-CH_2-\overset{\overset{\displaystyle CH_3}{|}}{C}H-CH_2-\overset{\overset{\displaystyle CH_3}{|}}{\underset{\underset{\displaystyle CH_2-CH_3}{|}}{C}}-CH_2-\overset{\overset{\displaystyle CH_2-CH_3}{|}}{C}H-CH_2-CH_2-CH_3$$

_____

9. Write the structure of 2,3,3-trimethyl-4-ethylheptane.

10. Write the IUPAC names and structures of the isomers of C_5H_{12}.

11. Write the IUPAC name of the following compound.

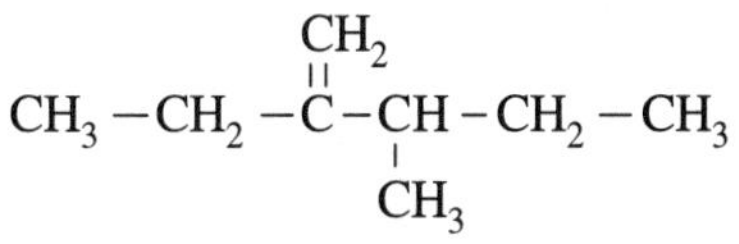

$$CH_3-CH_2-\overset{\overset{\displaystyle CH_2}{||}}{C}-\overset{\overset{\displaystyle }{\underset{\underset{\displaystyle CH_3}{|}}{C}}}H-CH_2-CH_3$$

_____

12. Write the structures of the *cis* and *trans* isomers of 4-methyl-2-pentene.

_____

13. Write the condensed structural formulas for the products of the following reactions.

    (a)   The addition of hydrogen to propene   _____

    (b)   The addition of water to 3-hexene   _____

    (c)   The addition of chlorine to 1-pentene   _____

    (d)   The reaction of bromine in the presence of $FeBr_3$ with benzene

                                     _____

14. Compare and contrast the chemical behavior toward hot concentrated sulfuric acid of cyclohexane, cyclohexene, and benzene.

    (a)   Cyclohexane   _____

    (b)   Cyclohexene   _____

    (c)   Benzene   _____

New Terms

Write the definitions of the following terms, which were introduced in this section. If necessary, refer to the Glossary at the end of the text.

addition reaction	geometric isomerism	hydrocarbons
alkanes	geometric isomers	saturated organic compounds
alkenes	alkynes	substitution reactions
alkyl groups	aromatic compounds	unsaturated organic compounds

25.3   Alcohols and ethers are organic derivatives of water

Review

To be an *alcohol*, the molecule must have the OH group (or it can be written HO) attached to a *saturated* carbon, one with four *single* bonds. (A molecule can have two or more such groups. In fact most organic compounds have more than one functional group, either alike or different.)

A molecule is an *ether* if it has the ether group, an O attached to two hydrocarbon groups, like alkyl groups, cycloalkyl groups, or benzene rings. The ether group has few chemical reactions, none of which we study in this text.

When you see an alcohol group, you think "This is a water-like group that confers on the molecule the following properties."

1   The OH group helps to make the substance more soluble in water and have a higher boiling point than the corresponding hydrocarbon.

2   The molecules of the substance can be oxidized provided that the carbon holding the OH group also holds H.

3   Alcohols of the RCH_2OH type are oxidized, first to aldehydes and then to carboxylic acids.

4   Alcohols of the R_2CHOH type are oxidized to ketones.

5   Alcohols of the R_3COH type are not oxidized by simple loss of H_2.

6   Alcohols give *elimination reactions,* those in which the pieces of a water molecule depart from adjacent carbons and leave behind a carbon–carbon double bond.

7   Alcohols give *substitution reactions,* those in which the OH group is substituted for by Cl, Br, or I, using reactions with the corresponding concentrated HX solution (where X = Cl, Br, or I).

For the sake of completeness, anticipating the next section, add a fifth property.

8   Alcohols react with carboxylic acids to give esters.

Self-Test

15. Write the structure of 2,3-dimethyl-1-butanol.

_____

16. Write the IUPAC name of the following compound.

$$\overset{\overset{\displaystyle CH_3}{|}}{CH_3CH}CH_2\overset{\overset{\displaystyle OH}{|}}{CH}CH_3$$ _____

17. Write the structures of the products that could be made by the oxidation of each compound. If no oxidation can occur, state so.

(a)   2-propanol

_____

(b)   1-propanol

_____

(c)   2-methyl-2-butanol

_____

18. Examine the following structures and then answer the following questions.

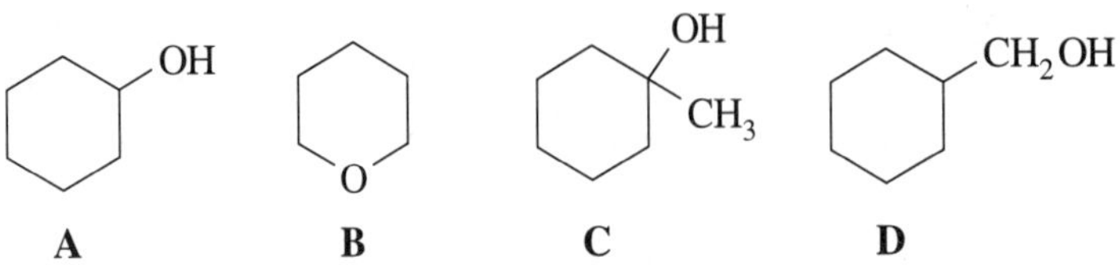

  A   B   C   D

 (a) Which compound(s) cannot be oxidized?   _____

 (b) Which compound(s) cannot be dehydrated? _____

 (c) Which compound(s) can be oxidized to a ketone? _____

 (d) Which compound(s) can be oxidized to an aldehyde? _____

 (e) Which compound(s) can be oxidized to a carboxylic acid

    (with the same number of carbons)?   _____

 (f) Which compound(s) can be dehydrated (to alkenes)? _____

19. Write the structure of the organic product in each situation.

 (a)

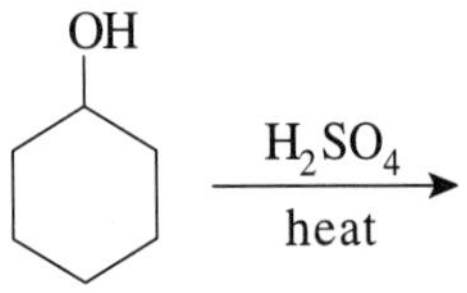

 (b)

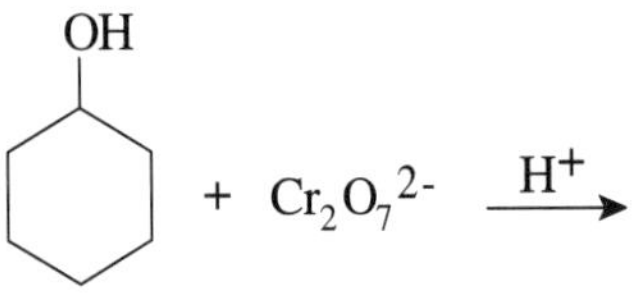

New Terms

Write the definitions of the following terms, which were introduced in this section. If necessary, refer to the Glossary at the end of the text.

 alcohol   elimination reaction   ether

 dehydration   substitution reaction

25.4 Amines are organic derivatives of ammonia

Review

Amines are alkyl derivatives of ammonia and so, like ammonia, are proton acceptors or Brønsted bases. Molecules of amines can also accept and donate hydrogen bonds. These are the essential characteristics of amines, but to anticipate a later section, amines can be converted to amides.

The protonated forms of amines are like the ammonium ion in that they are proton donors or Brønsted acids. They can neutralize a strong base, such as OH^-. Because they are cations and interact strongly with water molecules, protonated amines are more soluble in water than the neutral amine molecules.

Self-Test

20.   What forms when methylamine neutralizes hydrochloric acid? Write the structure.

21.   If $CH_3CH_2NH_3{}^+Cl^-$ reacts with aqueous sodium hydroxide, what forms? Write the structure.

New Term

Write the definition of the following term, which was introduced in this section. If necessary, refer to the Glossary at the end of the text.

amine

25.5   Organic compounds with carbonyl groups include aldehydes, ketones, and carboxylic acids

Review

The first task is to learn to recognize by name the functional groups that involve the *carbonyl group* when you see them in a complex structure. This is like being able to recognize that a wiggly blue line on a map represents a river, because functional groups are the "map signs" of organic structures. Certain chemical and physical properties are associated with each functional group "map sign."

When you learn functional groups, be sure not to tie their structures to particular positions on a page. The *sequences* of atoms are what count, not whether they are written left-to-right or top-to-bottom. For example, all of the following structures are *esters* because each has the "carbonyl-oxygen-carbon" sequence of atoms that defines the ester group.

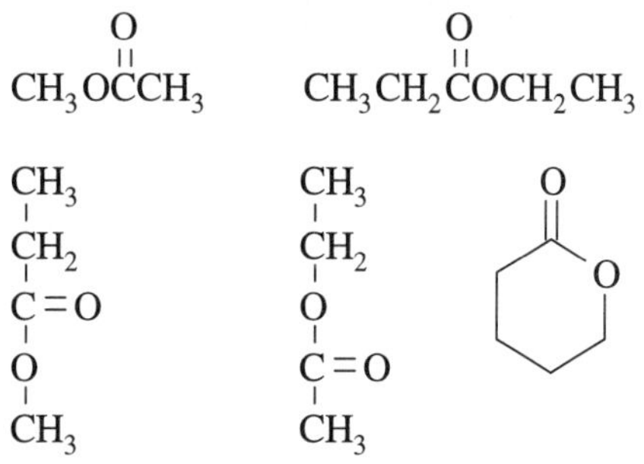

Some important esters, e.g., those in vegetable oils or animal fats, have three ester groups per molecule.

It's important to get *aldehydes* and *ketones* straight, because they differ so much in ease of oxidation. In the *aldehyde group,*

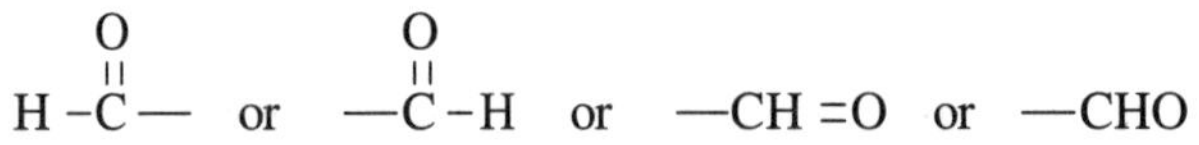

$$H-\overset{\overset{\displaystyle O}{\|}}{C}-\quad \text{or}\quad -\overset{\overset{\displaystyle O}{\|}}{C}-H\quad \text{or}\quad -CH=O\quad \text{or}\quad -CHO$$

Representations of the aldehyde group

the carbonyl group holds at least one hydrogen atom. In addition it holds either another hydrogen atom or it is joined to a carbon atom. The carbon can be saturated or unsaturated. In ketones the carbonyl is flanked on both sides by carbon atoms.

$$-\overset{\overset{\displaystyle O}{\|}}{C}-\qquad\qquad CH_3-\overset{\overset{\displaystyle O}{\|}}{C}-CH_3$$

The ketone group.   Propanone,
(also called a keto group)   a ketone.

Carboxylic acids all have the "carbonyl-oxygen-hydrogen" sequence.

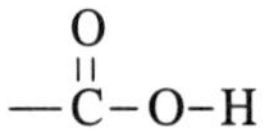

$$-\overset{\overset{\displaystyle O}{\|}}{C}-O-H$$

Thus all of the following structures are of carboxylic acids.

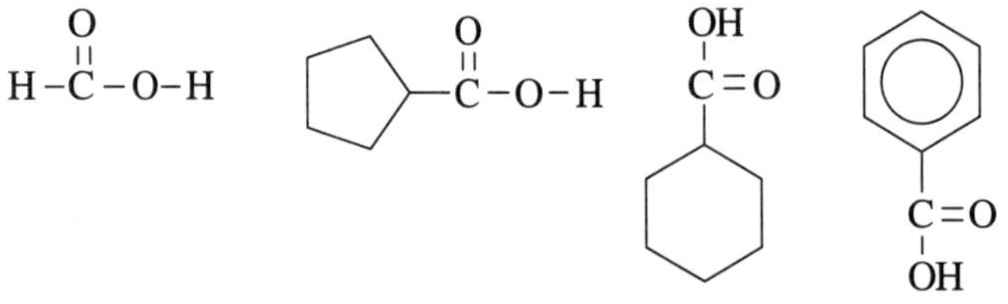

You will often see the carboxylic acid group abbreviated as CO_2H or COOH, and sometimes it's written "backward" as HO_2C or HOOC.

Be sure to catch the structural difference between an *amine* and an *amide*. In the *amides*, there is *always* a carbonyl group directly attached to the nitrogen atom. The difference is important because the properties are so different. Amines are basic compounds; amides are not. Amides are broken apart by water; amines are not. Both groups are polar and both can participate in hydrogen bonds.

Amides: $CH_3-\overset{\overset{\displaystyle O}{\|}}{C}-NH_2$ $CH_3-\overset{\overset{\displaystyle O}{\|}}{C}-NH-CH_3$ $CH_3-\overset{\overset{\displaystyle O}{\|}}{C}-\overset{\overset{\displaystyle CH_3}{|}}{N}-CH_2-CH_3$

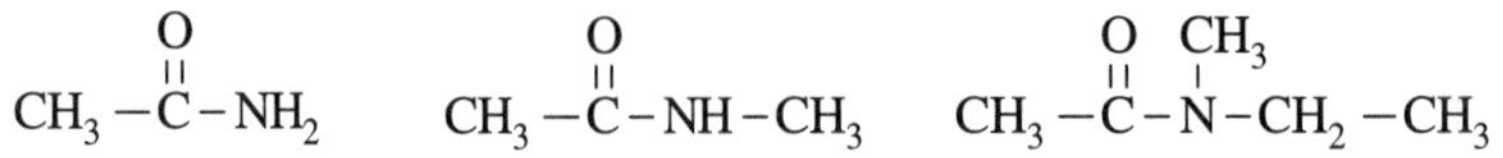

Amines: $CH_3CH_2NH_2$ $CH_3CH_2NHCH_3$ $CH_3CH_2\overset{\overset{\displaystyle CH_3}{|}}{N}CH_2CH_3$

As for the chemical properties of carbonyl compounds, concentrate on the following characteristics.

Concerning *aldehydes*, you should learn the following.

1   The aldehyde group is one of the most easily oxidized of all functional groups, being changed by oxidation to the carboxyl group.

2   The aldehyde group adds hydrogen, catalytically, to give alcohols of the RCH_2OH type.

♦ ♦ ♦

Concerning *ketones*, there are only two properties that we studied.

1   The keto group strongly resists oxidation.

2   The keto group adds hydrogen, catalytically, to give alcohols of the R_2CHOH type.

♦ ♦ ♦

With respect to *carboxylic acids,* learn the following.

1   Carboxylic acids are weak acids (toward water), but they readily neutralize strong base, like OH^-, and form the corresponding carboxylate ions, RCO_2^-.

2   Carboxylic acids react with alcohols (acid-catalysis) to give esters.

3   The carboxyl group can be changed to the amide group by a reaction with either ammonia or an amine.

4   The carboxyl group strongly resists oxidation.

5   The carboxylate group, CO_2^-, is a good Brønsted base; when it accepts H^+, it becomes the (weakly acidic) carboxyl group, CO_2H, again.

♦ ♦ ♦

Esters undergo the following reactions.

1   The ester group is hydrolyzed (acid catalysis) to give the carboxylic acid and the alcohol from which the ester is made.

2   The ester group is also hydrolyzed by the action of aqueous alkali, like NaOH, to give the parent alcohol, only now the carboxylate ion of the parent carboxylic acid, not the free acid, forms. (This reaction is often called the *saponification* of an ester.)

♦ ♦ ♦

Amides have the following properties.

1   The nitrogen of an amide is not a proton acceptor (not a Brønsted base), unlike the nitrogen of an amine (or ammonia).

2   Amides react with water to give the parent carboxylic acid and amine (or ammonia).

Self-Test

Questions 22-28 refer to the following structures. Some questions draw on knowledge from the preceding sections.

$$CH_3CH_2\overset{\displaystyle O}{\overset{\|}{C}}-O-CH_3 \qquad HOCH_3 \qquad H-\overset{\displaystyle O}{\overset{\|}{C}}-CH_3 \qquad \qquad HO-\overset{\displaystyle O}{\overset{\|}{C}}CH_2CH_3$$

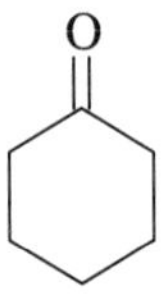

1   2   3   5

4

$$CH_3CH_2OH \qquad CH_3-\overset{\displaystyle O}{\overset{\|}{C}}-O-CH_2CH_3 \qquad CH_3\overset{\displaystyle O}{\overset{\|}{C}}-OH \qquad HO-CH_2-\overset{\displaystyle O}{\overset{\|}{C}}-CH_3$$

6   7   8   9

$$CH_3-O-\overset{\displaystyle O}{\overset{\|}{C}}-CH_2-O-CH_3$$
10

O-CH₃

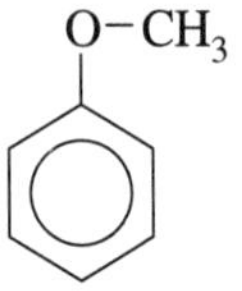

11

22. Which structures contain each of the following groups?

    (a) ester group   _____

    (b) ether group   _____

    (c) alcohol group   _____

    (d) carboxylic acid group   _____

    (e) aldehyde group   _____

    (f) ketone group   _____

23. Which compound is the most easily oxidized?   _____

24. Which compound(s) will rapidly neutralize sodium hydroxide at room temperature?

    _____

25. Two of the compounds shown will react to give compound 1. Which are they? _____

26. Compound 7 will react with water to give two of the compounds. Which two are they?

    _____

27. Which compound has a benzene ring?   _____

28. Which compound will react with hydrogen to give compound 6?   _____

Questions 29-34 refer to the structures below. Some of the questions require knowledge of material given earlier.

$$NH_2-CH_2-\overset{\overset{\displaystyle O}{\|}}{C}-CH_3 \qquad NH_2-CH_2CH_3 \qquad CH_3\overset{\overset{\displaystyle O}{\|}}{C}-NH_2 \qquad NH_3 \quad CH_3NH_2$$

$$\qquad\qquad 1 \qquad\qquad\qquad 2 \qquad\qquad 3 \qquad\qquad 4 \qquad 5$$

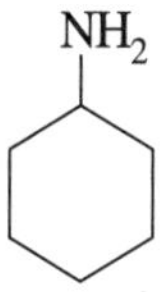

$$\underset{6}{\overset{NH_2}{\bighexagon}} \qquad CH_3\overset{\overset{\displaystyle O}{\|}}{C}-OH \quad NH_2CH_3 \quad CH_3NHCH_3 \quad CH_3NH\overset{\overset{\displaystyle O}{\|}}{C}CH_3 \quad CH_3NH_3{}^+$$

$$\qquad\qquad 7 \qquad\qquad 8 \qquad\quad 9 \qquad\qquad 10 \qquad\quad 11$$

29. The named groups given next are found in which structures?

    (a) the amine group   _____

    (b) an amino ketone   _____

    (c) the amide group   _____

30. Which two structures are of the same compound?   _____

31. The hydrolysis of structure 10 gives which two compounds?   _____

32. Which structures represent compounds that neutralize aqueous acids rapidly at room temperature?

    _____

33. Which structure has a carboxyl group?   _____

34. Which structure results when an amine neutralizes an acid?   _____

New Terms

Write the definitions of the following terms, which were introduced in this section. If necessary, refer to the Glossary at the end of the text.

aldehyde	carboxylic acid
amide	ester
carbonyl group	ketone

25.6   Most biochemicals are organic compounds

Review

This very short section is meant only to be a very brief overview of *biochemistry* as well as to introduce you to the general kinds of biochemicals and how each serves in providing a living system with materials, energy, and information.

Self-Test

35. What is studied in the field of biochemistry?

_____

36. What two kinds of substances are the chief sources of chemical energy for living systems?

_____

37. The biochemicals most closely involved in providing information for living things are in what family?

_____

38. The general name for the catalysts in cells is what? _____

New Term

Write the definition of the following term. If necessary, refer to the Glossary at the end of the text.

biochemistry

25.7   Carbohydrates include sugars, starch, and cellulose

Review

The simplest *carbohydrates*—the *monosaccharides*—involve alcohol plus aldehyde or ketone groups, so monosaccharides partake of the properties of these functional groups. Monosaccharide molecules are normally in cyclic forms that are in equilibrium with open-chain forms, and only in the open-chain forms are the aldehyde or keto groups present.

The *disaccharides* and *polysaccharides* give the monosaccharides when they react with water. These systems are made from the cyclic forms of the monosaccharides, being strung together by means of oxygen bridges. Water reacts at these bridges when disaccharides and polysaccharides are digested. We can write equations for the hydrolysis (digestion) of di- and polysaccharides as follows, and these equations will help you remember the important relationships.

Disaccharides:

$$\text{lactose} + H_2O \xrightarrow[\text{(hydrolysis)}]{\text{digestion}} \text{galactose} + \text{glucose}$$

$$\text{sucrose} + H_2O \xrightarrow[\text{(hydrolysis)}]{\text{digestion}} \text{glucose} + \text{fructose}$$

Polysaccharides:

$$\text{starch} + nH_2O \xrightarrow[\text{(hydrolysis)}]{\text{digestion}} n\text{glucose}$$

$$\text{cellulose} + nH_2O \xrightarrow[\text{(hydrolysis)}]{\text{digestion}} n\text{glucose}$$

Starch is actually a mixture of two polysaccharides, amylose and amylopectin. Cellulose, the chief constituent of the cell walls of plants, is not digestible in humans, but its acid-catalyzed hydrolysis also gives glucose as the only product.

Self-Test

39. The functional group generally absent from carbohydrates is

    (a) alcohol   (c) aldehyde

    (b) alkene   (d) ketone   _____

40. Animals store glucose units as

    (a) sucrose   (c) starch

    (b) glycogen   (d) cellulose   _____

41. Sucrose digestion leads to

    (a) glucose and fructose   (c) glucose only

    (b) galactose and glucose   (d) malt sugar   _____

42. The sugar in milk is

    (a) glucose   (c) lactose

    (b) maltose   (d) sucrose   _____

43. A carbohydrate that makes up most of cotton is

    (a) cellulose   (c) lactose

    (b) maltose   (d) starch   _____

44. The digestion of lactose gives

    (a) glucose   (c) fructose and glucose

    (b) table sugar   (d) galactose and glucose   _____

45. Because of the many OH groups in glucose molecules, glucose is

    (a) insoluble in water   (c) soluble in water

    (b) nonpolar   (d) hypotonic   _____

46. The digestion of starch is an example of

    (a) oxidation   (c) neutralization

    (b) reduction   (d) hydrolysis   _____

New Terms

Write the definitions of the following terms, which were introduced in this section. If necessary, refer to the Glossary at the end of the text.

carbohydrates   monosaccharides

disaccharides   polysaccharides

25.8   Lipids

Review

The ester group is the key functional group in the *triacylglycerols*, members of the family of *lipids* that react with water during digestion to give long-chain carboxylic acids—fatty acids—and glycerol. The fatty acids often carry one or more alkene double bonds. It is not the presence of an ester group that defines the larger family of the lipids, however. To be a lipid, all that a natural product has to be is mostly hydrocarbon-like so that it tends to be far more soluble in nonpolar solvents than in water. Thus cholesterol, which has no ester group, is a lipid.

Self-Test

47.   Which of the following compounds could *not* be obtained by the digestion of triacylglycerol?

$$
\text{(a)}\quad CH_3CH_2CH_2CH_2CH_2CH_2CH_2CH_2CH_2CH_2CH_2\overset{\displaystyle O}{\overset{\|}{C}}OH
$$

$$
\text{(b)}\quad CH_3CH_2CH_2CH_2CH_2CH_2CH_2CH_2CH_2CH_2CH_2CH_2\overset{\displaystyle O}{\overset{\|}{C}}OCH_3
$$

$$
\text{(c)}\quad HOCH_2\overset{\displaystyle OH}{\underset{|}{C}}HCH_2OH
$$

$$
\text{(d)}\quad CH_3CH_2CH_2CH_2CH_2CH_2CH_2CH_2CH =CHCH_2CH_2CH_2CH_2CH_2CH_2CH_2\overset{\displaystyle O}{\overset{\|}{C}}OH
$$

_____

48.   Because the vegetable oils have several alkene groups per molecule, they are called

   (a)   polyunsaturated   (c)   polymers
   (b)   polyenes   (d)   polypeptides   _____

49.   When triacylglycerols are digested, the reaction is the hydrolysis of

   (a)   glycerol   (c)   alkene groups
   (b)   fatty acids   (d)   ester groups   _____

50.   The hydrocarbon-like portions of a phosphoglyceride are

   (a)   cationic   (c)   hydrophilic
   (b)   hydrophobic   (d)   anionic   _____

51.   The surfaces of the lipid bilayer are dominated by

   (a)   hydrophilic groups   (c)   cholesterol
   (b)   polyunsaturation   (d)   nonpolar tails   _____

52. One of the services performed by proteins embedded in lipid bilayer membranes is

    (a) digestive enzyme function   (c) hormone synthesis

    (b) enzyme manufacture   (d) ion channels   _____

New Terms

Write the definitions of the following terms, which were introduced in this section. If necessary, refer to the Glossary at the end of the text.

triacylglycerols

lipids

25.9   Proteins are almost entirely polymers of amino acids

Review

All *proteins* consist of molecules of one or more *polypeptides*, and many proteins also include another organic molecule or a metal ion. Each polypeptide is a polymer of several (usually hundreds of) *α-amino acids*. The specific amino acids used, the number of times each is employed, and the order in which they are joined are the three factors that give each polypeptide its uniqueness.

When amino acids link together, water splits out from the carboxyl group of one amino acid and the α-amino group of the next one to give a *peptide bond*. (In a sense, a polypeptide is a condensation polymer similar in certain respects to nylon, which you studied in Chapter 13.) Polypeptides coil over much of their lengths into helices that are stabilized by hydrogen bonds. These helices usually undergo further kinking and folding. Thus each polypeptide has its own unique shape as well as unique amino acid sequence, and if this shape is lost, the protein no longer can function biologically.

Enzymes are proteins that catalyze reactions in cells. A ***lock-and-key*** mechanism enables an enzyme molecule to "recognize" and fit to only those substrate molecules meant for it.

Self-Test

53. The side chain in serine is —CH_2OH. Therefore, serine's structure is

$$(a) \quad {}^+NH_3\overset{\overset{\displaystyle O}{\|}}{\underset{\underset{\displaystyle OCH_2OH}{|}}{C}HCO^-}$$

$$(c) \quad HOCH_2\overset{+}{N}H_2\overset{\overset{\displaystyle O}{\|}}{\underset{\underset{\displaystyle OH}{|}}{C}HCO^-}$$

$$(b) \quad {}^+NH_3\overset{\overset{\displaystyle O}{\|}}{\underset{\underset{\displaystyle CH_2OH}{|}}{C}HCO^-}$$

$$(d) \quad {}^+NH_3CH_2\overset{\overset{\displaystyle O}{\|}}{C}OCH_2OH$$

_____

54.   Which arrow points to the peptide bond?

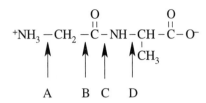

(a) A   (b) B   (c) C   (d) D   _____

55.   What is the side chain in the following compound?

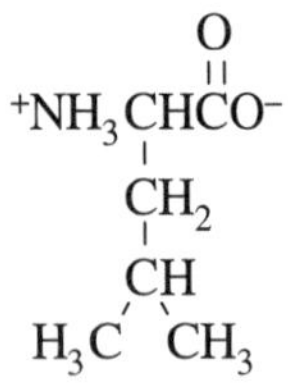

$$^+NH_3CHCO^-$$
$$| \atop CH_2$$
$$| \atop CH$$
$$H_3C \quad CH_3$$

(a) $^+NH_3 -$ (c) an alkyl group

(b) $-\overset{O}{\overset{||}{C}}O^-$ (d) $^+NH_3\overset{O}{\overset{||}{CHCO}}^-$ _____

56.   Which structure best represents the nature of the main chain or "backbone" in polypeptides?

(a) $^+NH_3CH_2\overset{O}{\overset{||}{C}}NHCH_2\overset{O}{\overset{||}{C}}NHCH_2\overset{O}{\overset{||}{C}}$ — etc.   (c) $^+NH_3CH_2\overset{O}{\overset{||}{C}}CHC\overset{O}{\overset{||}{C}}HC$ — etc.
                                                          $-NH \quad NH-$

(b) $^+NH_3CH_2\overset{O}{\overset{||}{C}}O\overset{O}{\overset{||}{C}}CH_2NHNHCH_2\overset{O}{\overset{||}{C}}$ — etc.   (d) $^+NH_3CH_2\overset{O}{\overset{||}{C}}ONHCH_2\overset{O}{\overset{||}{C}}ONHCH_2\overset{O}{\overset{||}{C}}O$ — etc.

_____

57.   One of the possible dipeptides that can form from alanine (side chain = CH_3) and cysteine (side chain = $-CH_2SH$) is

(a) $^+NH_3\overset{O}{\overset{||}{CHC}}NH\overset{O}{\overset{||}{CHCO}}^-$ (c) $^+NH_3CH_2CH_2\overset{O}{\overset{||}{C}}NH\overset{O}{\overset{||}{CHCO}}^-$
        $| \atop SH \quad | \atop CH_3$ $| \atop CH_2SH$

(b) $^+NH_3\overset{O}{\overset{||}{CHC}}SCH_2\overset{O}{\overset{||}{CHCO}}^-$ (d) $^+NH_3\overset{O}{\overset{||}{CHC}}NH\overset{O}{\overset{||}{CHCO}}^-$
        $| \atop CH_3 \quad | \atop NH_2$ $| \atop CH_2SH \quad | \atop CH_3$

_____

58. In protein chemistry, "native form" refers to what?

    (a)   a building unit of a protein

    (b)   the location of the hydrophobic groups in a protein molecule

    (c)   the shape of a protein molecule after denaturation

    (d)   the final shape of a protein molecule   _____

59. Enzymes are

    (a)   catalysts   (b) B-vitamins   (c) substrates   (d) denatured proteins   _____

New Terms

Write the definitions of the following terms, which were introduced in this section. If necessary, refer to the Glossary at the end of the text.

α-amino acid

enzyme

peptide bond

polypeptide

protein

25.10   Nucleic acids carry our genetic information

Review

DNA either directs the synthesis of more of itself—*replication*—or it directs the apparatus for making polypeptide molecules having particular sequences of their amino acid side chains. DNA is one of the two kinds of nucleic acids, and the backbone of DNA is an alternating sequence of deoxyribose-phosphate units, each one bearing a side-chain amine or *base*. The kind, number, sequence, and hydrogen-bonding abilities of these bases—there are four of them—determine the properties of the *gene* units of individual *DNA double helices,* according to the Crick-Watson theory. The amines come as matched base pairs, with adenine (A) pairing by hydrogen bonds to thymine (T) and guanine (G) pairing to cytosine (C).

The molecules of *RNA*, of which there are several types, consist of alternating sequences of ribose-phosphate units, each ribose bearing a side-chain base. The bases are the same as those in DNA except that uracil (U) replaces thymine (T). Base pairing occurs between A and U as well as between A and T.

Just prior to cell division, each of the two strands in a DNA double helix uses the pairing requirements of its side-chain bases to guide *replication.*

Between cell divisions, the sequence of bases in DNA directs the synthesis of heterogeneous nuclear RNA, abbreviated hnRNA, in which the bases are complementary to those of the parent DNA. This process—*transcription*—transfers the genetic message to RNA. Both the *intron* and *exon* segments of DNA are transcribed into hnRNA, but only the exon units of DNA make up parts of a gene. The base sequences of hnRNA that came from introns are next deleted, and the sequences that came from the exons are joined to give, after this processing, a molecule of messenger RNA, or mRNA. Each adjacent series of three bases on mRNA is a *codon.* It will eventually direct a particular amino acid unit into place during the synthesis of a polypeptide. Thus the *genetic code* is the match-up between codons on mRNA and the amino acids used to make polypeptides.

The mRNA next becomes associated with a cluster of ribosomal RNA molecules (rRNA) and proteins at a particle called a ribosome. Here the synthesis of a polypeptide is directed by the mRNA. Transfer RNA molecules, or tRNA, bear amino acid units to this synthesis site. An adjacent series of three bases, called an *anticodon*, located on tRNA can fit by hydrogen bonds only to its matching codon on mRNA. This overall process—*translation*—uses the unique, transcribed genetic message (on mRNA) to give a unique polypeptide structure.

Genetic defects can arise at any stage, but those that are inherited occur as incomplete or faulty sequences of bases on DNA.

Viruses consist of nucleic acids—some viruses with DNA and others with RNA—combined with proteins.

In *genetic engineering,* DNA material corresponding to some desired protein is inserted into a cell where it then proceeds to make the protein for which it is coded. When a cell's DNA—it might be a cell of some bacterium or a yeast—is altered by new DNA, the resulting DNA is called *recombinant DNA*. The technique has been used to manufacture human insulin.

Self-Test

In questions 64 to 74, fill in the blanks with the best technical terms.

60. The one-gene—one-enzyme relationship involves a master code that consists of a distinctive sequence of _____ along a backbone in mRNA. Each of these groups consists of three consecutive_____ joined to _____ units on the mRNA backbone.

61. The force of attraction responsible for the pairing of amines in the _____ double helix is the _____ and (use the code letters) _____ pairs

    with T and _____ pairs with G.

62. The product of replication is another _____

63. The sequences of bases in DNA that together make up a whole gene are called _____

    and the sequences that separate these from each other and are not associated with genes are

    called _____.

64. The type of RNA made at the direction of DNA is   _____

65. The type of RNA whose base sequence is complementary just to exons is called

    _____

66. The overall series of events from DNA to the RNA that directs polypeptide synthesis is called

    _____

67. After translation has occurred, the product is a _____

68. The RNA that carries amino acid units is called _____

69. A _____ is a particle made of nucleic acid and proteins that is capable of causing an infection.

70. When a _____ in a bacterial cell is made to accept a DNA molecule unrelated to the normal inventory of the cell, it then carries a DNA referred to as _____ , and when such a bacteria is made to manufacture some desired proteins, the overall operation is called_____

New Terms

Write the definitions of the following terms, which were introduced in this section. If necessary, refer to the Glossary at the end of the text.

anticodon	nucleic acid	codon
intron	DNA	recombinant DNA
DNA double helix	replication	exon
RNA	genetic code	transcription
genetic engineering	translation	

Answers to Self-Test Questions

1.

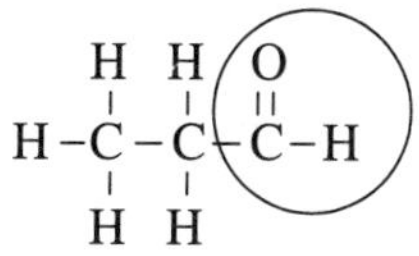

2. CH_3—NH_2
3. CH_3—$NH_3^+ + Br^-$
4. R—$NH_3^+ + Cl^-$
5. It greatly simplifies it. There are very few kinds of functional groups among the millions of organic compounds, and each kind displays mostly the same set of reactions.
6. (a) isomers, (b) isomers, (c) identical
7. The second compound of part (b).
8. (a) pentane, (b) 2-methylhexane, (c) 2,2-dimethylpentane, (d) 3-methylhexane, (e) 5,7-diethyl-3,5-dimethyldecane

9.

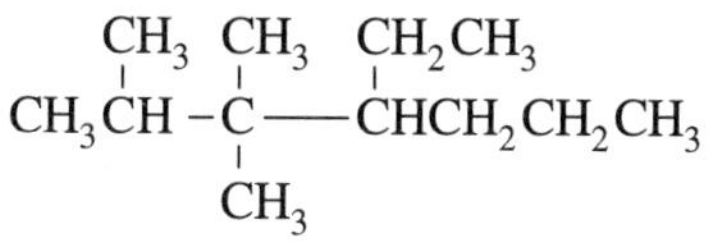

10. $CH_3CH_2CH_2CH_2CH_3$ pentane

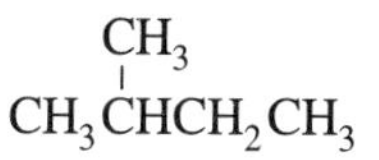

$$CH_3CCH_3 \quad \text{2,2-dimethylpropane}$$

with CH₃ groups above and below the central carbon.

11.   2-ethyl-3-methyl-1-pentene

12.

H₃C and CH–CH₃ (with CH₃) C=C with H, H — **cis-isomer**

H₃C and H, C=C with H and CH–CH₃ (with CH₃) — **trans-isomer**

13.   (a) CH₃CH₂CH₃

(b)   CH₃CH₂CH₂CHCH₂CH₃
                      |
                     OH

(c)   ClCH₂CHCH₂CH₂CH₃
              |
              Cl

(d)   ⬡—Br (+ HBr)

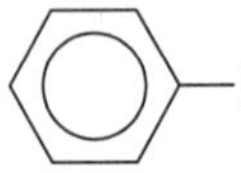

14.   (a) No reaction

(b)   ⬡—OSO₃H forms (the product of an addition reaction

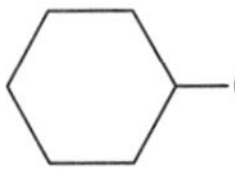

(c)   ⬡—SO₃H forms (the product of a substitution reaction

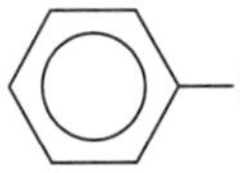

15.   CH₃ CH₃
            | |
      CH₃CH–CHCH₂OH

16.   4-methyl-2-pentanol

17.   (a) CH₃CCH₃ (with O double bond)   (b) CH₃CH₂CHO which is further oxidized to CH₃CH₂CO₂H

(c) no oxidation

18.   (a) B, C, (b) B, (c) A, (d) D, (e) D, (f) A, C, D

19.

(a) ⬡ (cyclohexene)   (b) O (cyclohexanone)

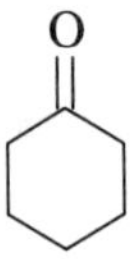

20.   CH₃NH₃⁺Cl⁻

21.   CH₃CH₂NH₂ (+ H₂O + NaCl)

22. (a) 1, 7, 10   (b) 10, 11   (c) 2, 6, 9   (d) 5, 8   (e) 3   (f) 4, 9
23. 3
24. 5, 8
25. 2, 5
26. 6, 8
27. 11
28. 3
29. (a) 1, 2, 5, 6, 8, 9 (Structure 4 is ammonia, not an amine.), (b) 1, (c) 3, 10
30. 5, 8
31. 5 (or 8) and 7
32. 1, 2, 4, 5, 6, 8, 9
33. 7
34. 11
35. The organic compounds present in living cells or that have been made from them.
36. carbohydrates and lipids
37. nucleic acids
38. enzymes
39. b
40. b
41. a
42. c
43. a
44. d
45. c
46. d
47. b
48. a
49. d
50. b
51. a
52. d
53. b
54. c
55. c
56. a
57. d
58. d
59. a
60. codons; bases; ribose
61. DNA; hydrogen bond; A; C
62. DNA double helix
63. exons; introns
64. heterogeneous nuclear RNA (hnRNA)
65. messenger RNA (mRNA)
66. transcription

67. polypeptide
68. transfer RNA (tRNA)
69. virus
70. plasmid; recombinant DNA; genetic engineering

Tools you have learned

Consider removing this chart from the Study Guide so you can have it handy when tackling homework problems.

Tool	How it Works
Functional group concept	If you can recognize a functional group in a molecule, you can place the structure into an organic family and so predict the kinds of reactions the compound can give. You should learn to recognize the functional groups in Table 25.1.
IUPAC rules of nomenclature	As with other rules of nomenclature, you use the rules to write names for compounds based on their compositions and structures. You also use the rules to construct the structure of a compound from its name.